Computational Hydraulics
Elements of the Theory of Free Surface Flows

Computational Hydraulics

Elements of the Theory of Free Surface Flows

M B Abbott

International Institute for Hydraulic and Environmental
Engineering, Delft,
and Danish Hydraulic Institute, Hørsholm

First published 1979 by Pitman Publishing Limited

Reprinted 1980, 1985

Reprinted 1992 with minor changes by

Ashgate

Ashgate Publishing Limited

Gower House

Croft Road

Aldershot

Hants GU11 3HR

England

Ashgate Publishing Company

Old Post Road

Brookfield

Vermont 05036

USA

A CIP catalogue record for this book is available from the British

Library and the US Library of Congress.

ISBN 1 85742 064 0

Printed and bound in Great Britain by

Billing and Sons Ltd, Worcester

Reason only perceives that which it produces after its own design.

Immanuel Kant,
Critique of Pure Reason.

Contents

Preface

Very much as the steam engine was the principal physical instrument of the industrial revolution, so the digital computer is the principal physical instrument of our current 'informational revolution'. The digital computer's capacity to transform semantic information rapidly, reliably and cheaply, changes all technologies, even one with such long historical traditions as hydraulics. One of the first studies of the European enlightenment, investigated by da Vinci, Galileo, Newton and Euler, hydraulics had already been transformed through the industrial revolution into a 'productive benefit'. The current post-industrial revolution, concerned with directing investment into complex, finely balanced, 'high-information' machines and constructions, maintaining 'the highest information level per rate of information destruction', transforms hydraulics again, adapting it to the new demands and the new possibilities of our modern societies.

The new demands placed upon hydraulics are associated, in the first place, with the increasing scale and complexity of hydraulic works. Being so information intensive at the construction stage – so necessitating the rapid concentration and as rapid redispersal of the complex equipment needed for dredging, foundation preparation, heavy construction and control – these works are constructed more rapidly and irreversibly than those of earlier times. By virtue of their increasing scale most modern works influence their immediate environment more strongly while, by virtue of their increased speed of construction and its irreversibility, it is usually too late and too difficult to change the works by the time that their undesirable consequences appear. Accordingly, the future utility of these works and the effects that they have on their environment must be determined in advance with a much greater precision than was previously necessary.

The new demands made upon hydraulics are associated, in the second place, with the elaboration of industrial processes that place new and often extremely severe demands upon machines, structures and containing vessels. One of the best known examples is provided by the design of nuclear reactors, and especially of fast breeder reactors. Problems then occur in computing turbulent flow fields and corresponding temperature fields over a considerable range of operating and reference accident situations, usually under conditions of intense heat exchange, and also of computing the effects of sudden changes of pressure in reactors and other vessels and in their valve and pipeline systems. An apparently very different area of application to industrial facilities is provided by offshore oil exploration and producing operations, with problems ranging from the hydraulics of rapid temperature changes in an oil stream to problems of the mooring of platforms, terminals and ships.

The new demands made upon hydraulics arise, in the third place, from the social and economic transformations of the non-industrial world, where solar energy is plentiful but fossil energy is usually scarce. This world derives its living from agriculture, using solar energy to retain or draw up water against gravity, through the agencies of root tensions and the physics of the non-saturated soil zone, and then using some part of the plants themselves for food, fodder, fuel and cash crops. The intrusion of the philosophy of the enlightenment into this world -- in information theoretic terms 'the use of the cultural channel of information transmission to tap off an increased information flux from the ecological channel' – leads to a further exploitation of the possibility of controlling water movements above, below and in the soil so that the processes of plant physiology can be subjected, in turn, to 'productive controls'. In this case it is important that the controls do not themselves consume too much fossil energy and do not themselves give rise to undesirable side-effects, such as salination or an increase in the incidence of bilharziasis and other sicknesses.

The new possibilities for solving the problems thus posed, and for meeting the still increasing demands of mankind, are centred upon the digital computer. In most cases, the computer is made useful through its ability to obtain quantitative results from various mathematical models of the real world. These models are then run upon a digital computer to provide prognoses of projected real world situations as a means of developing those constructions and operations that provide certain desired real world effects. When a mathematical model is described in terms of the processes of numerical analysis, it is commonly called a 'numerical model'.

The numerical modelling process proceeds, essentially, by describing the physical system with a set of numbers and simulating the laws acting upon it with sets of operations on these numbers. In hydraulics, the numbers usually used are measures of water depths and surface elevations, flow velocities and discharges, bed, surface and perhaps internal stresses, mass exchanges between various components within the model and between the model and its surroundings, salinities, and temperatures. In most cases, the sets of operations transform the description of the system at one time to a description at a later time, so that the numerical model traces the evolution in time of the set of numbers that describes the physical system. The physical evolutions corresponding to various proposed engineering works and operations can be followed accordingly.

By the correct use of numerical modelling in hydraulic, coastal and offshore engineering, dramatic reductions can be made in the initial and running costs of engineering works. There are therefore very great economic and commercial incentives to develop practical numerical modelling methods. This development naturally brings with it a very thorough reformulation of hydraulics to suit the possibilities and requirements of the discrete, sequential and recursive processes of digital computation. The hydraulics that is reformulated to suit digital machine processes in this way is called *computational hydraulics*.

With the new possibilities of computational hydraulics and its modelling realizations, comes the need for more and better data with which to calibrate and verify the models – and the theories and procedures that are behind them. New types of field investigations are economically necessary and economically justified, of a scale quite different from those made only a decade earlier, backed up in turn by a new instrumentarium with its data-interpretation, data-storage and data-retrieval facilities. The numerical models are only as good as the data used to calibrate and verify them, so that the new economic utility of the models translates into a new economic utility for field studies and improved instrumentation.

Similarly, the numerical models open up new possibilities for physical modelling, such as by providing boundary conditions that are more local, and by on-line computer control of these conditions in the laboratory.

As the development of computational hydraulics is so strongly motivated by the needs of practice, so it is natural to start out from that part of practice where computational methods have been most widely applied. This is the area of 'free-surface flows', an area which, moreover, includes many pressurized flows as a simple generalization. Flows of this type occur in systems of channels, canals, rivers, lakes, estuaries, coastal areas and seas. Free-surface flows are associated with such phenomena as short-period waves on beaches and man-made structures, seiching actions in bays and harbours, tides and storm surges, salt-water intrusions, flood waves, flows in sewer systems and much else besides.

This area of free-surface flows is evidently much too wide to be covered in detail in any single book of reasonable proportions, while it involves so many different specializations that, in most cases, only some parts of any such compendium would be of interest to any one reader. Accordingly it seems best to cover this area with a series of books, such that this, the first of the series, outlines the elements of the theory and the successive books cover particular applications of the theory.

This naturally raises the problem of selecting and organizing the material that can be considered as fundamental to the subject. Such material must take the reader from the hydraulics that he already knows into the hydraulics that he needs to know in order to proceed to applications. In such a dynamic, growing subject, this then necessitates some guessing about what the reader will need to know by the time (in three years, or ten years or whatever) that he actually comes to applications.

At this stage of development of the subject, any such selection and organization depends largely upon the judgement of the author. This preface provides an opportunity to explain how this author's judgement has operated in the present case. In this work, the first essential concept to be established is that of conservation, which provides a basis for the formulation of systems of conservation laws. As the processes of digital computation are discrete or countable processes, conservation laws are first introduced in discrete form, in terms of that simplest and most familiar of discontinuities, the standing hydraulic jump. In this way it is possible to introduce the notion of non-equivalence of conservation laws in discrete, countable systems – a notion that is surely fundamental to computational hydraulics – through the notion of energy loss or energy defect across a hydraulic jump when mass and momentum conservation laws are satisfied. That it is the energy that is lost while momentum is conserved, and not vice versa, is shown at this stage to be a consequence of the second law of thermodynamics, which is again surely fundamental to computational hydraulics. And then, by the introduction of Galileo's relativity principle, the essential computational hydraulic dualisms of 'level' and 'flux density' are introduced naturally.

After some examples that should serve to fix ideas and implant some related notions, it is relatively easy to proceed to differential formulations, which are surely familiar, and then to conservation laws, which may not be so familiar. The equivalence of pairs of measures from the infinite vector (mass, momentum, energy-work, Bernoulli, etc.) is then illustrated. This is typical of several demonstrations in this book that would be superfluous in a work on mathematical fluid dynamics, but which still appear desirable in a work on computational hydraulics. For hardly any of the discrete schemes used in computational hydraulics demonstrate this equivalence, and it will later be necessary to see that, while differentiation by parts and other analytical operations are exact (so not 'losing information' or 'producing entropy'), the equivalent discrete operations of differencing by parts and other numerical operations cannot generally be exact (so they 'lose information', or 'produce entropy').

In any such discussion of fundamentals, many aspects can only be sketched in outline, while some must even be neglected altogether. The most notable omissions are probably those of vorticity formulations and the problem of the equivalence of vorticity convection laws and other conservation laws. These aspects must be left to a future work on two- and three-dimensional flows.

The processes of computational hydraulics are not only discrete, but they are also sequential. The notion of order of computation, that is again fundamental to the subject, is introduced in Chapter 3 through an exposition of the method of characteristics. The method is extended to describe kinematic waves and, finally, linearized, superimposable wave motions, or harmonic motions. Several ideas are introduced in this extension, and especially those associated with the processes of Fourier decomposition that are so vital to Chapter 4.

It is really only in Chapter 4 that the reader gets down to the 'serious business' of numerical schemes; and even these are introduced through certain sequencings or codings of the method of characteristics. The main notions of difference schemes, such as consistency, convergence and stability, are only then introduced, initially in terms of the simplest case of kinematic wave propagation. The extension to the full dynamic case follows, leading to the simplest explicit schemes, understood as schemes that can be laid down in any order and, further, to the simplest implicit schemes, understood as schemes that must be laid down in an order dictated by the method of characteristics.

Numerical methods are represented here exclusively by 'difference methods' while 'finite elements methods' are not explicitly described. This is because, on the one hand, there are already two good books on applications of finite elements in fluid mechanics and, on the other hand, finite element methods are very little used in hydraulic engineering practice.

However, most of the material introduced here can be applied equally well to finite element methods, and indeed, from the point of view of the theory of algorithms used in this book, the difference between finite difference and finite element methods is not of a fundamental nature.

Chapter 5 builds upon the earlier chapters to approach the more fundamental aspects of the subject. The case of a discrete discontinuity, the hydraulic jump of Chapter 1, is treated in detail in order to show how a sequence of solutions of a difference equation may converge, as the scale of the discretization tends to zero, to a solution that differs essentially from the solution of the differential equation with which the difference equation is nominally (strongly) consistent. Since the solution of the difference equation is the more realistic physically, this leads to the formulation of a 'resolution paradox' of computational hydraulics. This resolution paradox is then related to the 'stability paradox' of Richtmyer and it is shown that both paradoxes are related to Cantor's famous theorem on the non-equivalence of countable sets and sets of the power of the continuum. It is, in effect, this central theorem of set theory that underlies the descriptive non-equivalences of computational hydraulics, non-equivalences that have, quite naturally, little or no place in the descriptions of classical hydrodynamics. However, some care has been taken to show how this 'new' hydraulics was already implicit in certain formulations of the founders of modern mechanics and mathematical analysis, such as Galileo and Newton. Indeed, at the aesthetic level, the entire book may be seen as an attempt to reclaim a certain heritage of hydraulics – the discrete view of the world – while passing from Galileo's set-theoretic view of motion and Newton's preference for a discrete form for the second law of motion, through the theory of weak solutions of Lax, into the revelation of Cantor's theory of numbers. It traces the thread, running through hydraulics, that connects the second law of thermodynamics to Cantor's theorem.

If the subject of computational hydraulics were a purely academic exercise, then this journey through the aesthetic to the revelation would be sufficient in itself, and the book could as well stop at the end of its fifth chapter. But computational hydraulics is also an instrument of engineering, a guide to action in the real world, a 'productive benefit'. Our increasingly post-industrial societies seek, for better or for worse, to use the aesthetic measure as a productive force. The sixth chapter illustrates this process, by outlining the process of modelling for hydraulic, coastal and offshore engineering practice. The examples have been mostly limited to hydrodynamic aspects, even though many of the models concerned carry elaborate superstructures for tracking the simultaneous processes of transport, dispersion, dilution and biochemical interaction that are often of principle interest in practice. The temptation to introduce material from these applications has been largely resisted, but such applications will naturally form a vital part of specialized works building upon the hydrodynamic basis of this first volume.

M. B. Abbott,
The Hague, The Netherlands

Acknowledgements

The basic ideas for the present work were formulated during my research at Southampton University, U.K., in 1960, arranged and helped by P. B. Morice. A 1963–1964 post-doctoral fellowship at the Mathematical Centre in Amsterdam, under the kind and understanding guidance of H. S. Lauwerier, provided the opportunity to study the underlying set theoretic and functional analytic foundations of numerical analysis in some detail.

In 1966, L. J. Mostertman invited me to teach at the International Institute for Hydraulic and Environmental Engineering in Delft and encouraged me to pursue further my interests in the present subject. Shortly afterwards, G. S. Rodenhuis joined me and entered fully into this work, to excellent effect. The following of my other colleagues at the International Institute participated actively at this formative stage, in chronological order: F. Verhoog, C. Pardo Castro, E. W. Lindeyer, A. N. Jollife, G. Marshall, A. Verwey and I. R. Warren. Research fellows of the International Institute, principally A. G. Barnett, D. O. Hodgins and M. P. Vium, contributed further, while the many participants of the International Institute who have specialized in computational hydraulics have never ceased to be a source of ideas, criticisms and inspirations. On the basis of our combined efforts, it was possible to introduce 'Computational Hydraulics' as a viable specialization in hydraulics, at the 1969 (Kyoto) Congress of the International Association for Hydraulic Research.

In 1970, T. Sørensen and H. Lundgren invited me to set up the Computational Hydraulics Centre, as a part of the Danish Hydraulic Institute, alongside my Delft activities. After initial difficulties, closely shared with Aa. Damsgaard and G. S. Rodenhuis, the Centre developed into a successful and many-sided operation, that contributed greatly to the further development of the subject. Engineers of the Centre, who have provided nearly all the material for Chapter 6, and more besides, were: J. Aa. Bertelsen, O. Brink Kjaer, Aa. Damsgaard, L. Elkin, P. I. Hinstrup, M. Hvidbjerg-Knudsen, A. Kej, U. I. Kroszynski, J. Odgaard, H. M. Petersen, C. H. Rasmussen, G. S. Rodenhuis, O. Skovgaard, M. P. Vium and I. R. Warren.

At the International Institute, the book has benefited particularly from discussions with A. Verwey, while G. Tong checked it all very thoroughly and P. van Daalen organized the text and drawings.

Notation

a acceleration (m s^{-2}); distance in x (m); $Cr \sin \alpha$

A total flow area (m^2); derivative with respect to f of $g = g(f)$ in conservation forms

b breadth of uniform channel (m)

B total width of wetted area of channel (m)

c propagation velocity, or celerity (m s^{-1}), concentration (e.g. ppm)

C alternative for propagation velocity or celerity (m s^{-1}); the coefficient of resistance in the Chézy representation ($m^{\frac{1}{2}}$ s^{-1})

Cr Courant number, either dynamic, viz. $\dot{x}\, \Delta t/\Delta x = \lambda\, \Delta t/\Delta x$, or kinematic, such as $c\, \Delta t/\Delta x$ or $u\, \Delta t/\Delta x$

D total derivative in Eulerian form; amplification factor; diffusion coefficient (m^2 s^{-1})

e direction or unit vector, written as $\mathbf{e}$; energy flux density (kg m s^{-3})

E energy level for short-wave motions (kg s^{-2})

f force (either N or kg m s^{-2}); generalized level vector; friction factor; 'a junction of'

F alternative for force (either N or kg m s^{-2})

Fr Froude number, $= |u/(gh)^{\frac{1}{2}}|$

g acceleration due to gravity (set as 9.8 m s^{-2}); generalized flux density; eigenvalue of amplification matrix

G amplification matrix

h water depth (m)

H bed level relative to some horizontal datum (m); control function; Heaviside step function

i gradient of bed

$\mathrm{Im}(x)$ imaginary part of x

I instantaneous unit hydrograph, $I = I(t)$ (m^3 s^{-1}); unit matrix

j address in x

jj maximum value of j

J Riemann invariant for nearly horizontal flows (m s^{-1}). Alternative to jj

k address in y; volume discharge per unit width, here called 'volume flux density' (m^2 s^{-1}); roughness factor (m); wave number; general constant

kk	maximum value of addresses k
K	alternative to k for wave number
l	length, or extent, of a fluid system (m); range in j or k of a numerical operator
L	mean rate of information loss
m	momentum flux density (kg s^{-2}); occasion of jth increment in inverse schemes
n	address in time
p	mass discharge, per unit width, in x-direction, here called the mass flux density in x-direction, $= \rho uh$ (kg m^{-1} s^{-1}); the x-momentum level (kg m^{-1} s^{-1})
P	complex propagation factor
q	mass discharge per unit width, or mass flux density in y-direction (kg m^{-1} s^{-1}); the y-momentum level (kg m^{-1} s^{-1}); weighting factor
Q	total-section flow discharge volume (m^3 s^{-1}); celerity ratio
r	alternative to Cr for kinematic Courant number $u\ \Delta t/\Delta x$
R	hydraulic radius (m)
$\mathrm{Re}(x)$	real part of x
s	used for x and y indifferently
t	time (s)
T	period (s); fixed time increment (s)
u	velocity in x-direction (m s^{-1})
U	alternative for velocity in x-direction (m s^{-1}), especially for column vector with two u-components
v	velocity in y-direction (m s^{-1})
V	alternative to velocity in y-direction (m s^{-1})
w	velocity in z-direction (m s^{-1}); work (kg m^2 s^{-2}); a test function
x	coordinate in the horizontal plane (m)
$\dot{x}$	characteristic direction in the $x-t$ plane (m s^{-1})
X	eigenvector of amplification matrix
y	coordinate in the horizontal plane orthogonal to x (m)
z	coordinate in the vertical, so orthogonal to x and y, sometimes read as 'height above a given datum' (m)
α	velocity distribution coefficient; degree of dissipation in α-algorithm
β	c/u in kinematic wave theory
γ	Coriolis coefficient (s^{-1}); argument $i2\pi k\ \Delta x/21$ or $i2\pi k\ \Delta x/1$
δ, Δ	increments in a variable; computational celerity (complex); Dirac δ-distribution
ϵ	a distribution, defined on support in x in Equation (1.5.1)
θ	angle of expansion; weighting factor in t
λ	wave length (m); characteristic celerity, $= \dot{x}$ for nearly horizontal flow theory (m s^{-1})
ξ	Fourier coefficient, vector of dependant variables
ρ	density (kg m^{-3}); amplification factor, $= e^{\omega \Delta t}$
σ	ratio of densities; $\sigma_1 = 2\pi K_j \Delta x/2l$, $\sigma_2 = 2\pi K k \Delta y/2l$

τ	shear stress (N m^{-2} or kg m^{-1} s^{-2}); alternative to t(s)
ψ	width of linearized catchment; weighting factor in x
ω	frequency (s^{-1})

1 Discrete forms of conservation laws

1.1 Fundamental laws

This work will be developed initially from the following laws of motion:

(1) A body either remains at rest or continues in a state of uniform rectilinear motion unless acted upon by another body or force.
(2) The instantaneous force applied to a body is equal to the product of the instantaneous mass and the instantaneous acceleration of the body.
(3) Action and reaction are equal and opposite.

These laws are usually called *Newton's laws of motion*, even though, as discussed at length in Chapter 5, Newton did not give the second of them in this form. They will be regarded as fundamental postulates, derived from experience, although they may be regarded from other standpoints, e.g. law (2) may be regarded as a relative definition of force and mass. These laws are by no means the only fundamental set of postulates, of course, and indeed, in classical mechanics, a more satisfactory set may be formed from Galileo's principle of relativity and Hamilton's principle of least action, as demonstrated by Landau and Lifshitz (1960). However, Newton's laws, as expressed above, are by far the most familiar, and this justifies the present choice.

In addition to the above laws, two others will be used. These are the *first and second laws of thermodynamics* (Callen, 1960). The first law postulates the existence of walls that prevent all interactions (and so are, in particular, 'impermeable to the flow of heat') and states that:

(4) In a system isolated by such walls, the work done in taking one state A of a system into another state B is entirely determined by the terminal states A and B. The *internal energy difference* between state A and state B is defined as the mechanical work done in taking either state A into state B or state B into state A, whichever is possible.

The second law of thermodynamics may be conveniently stated, following Zemansky (1957, p. 189), as follows:

(5) There is a tendency on the part of nature to proceed towards a state of greater disorder.

Parallel flows (comprising horizontal flows as a special case and nearly horizontal flows as a generalization) are clearly the most ordered of all flows. The law says, in effect, that when energy of parallel or nearly parallel motion is converted into energy of less parallel motion, essentially short-wave motion and turbulent motion, the energy so converted can never be entirely reconverted into energy of parallel motion. The nearly parallel flows break down 'irreversibly'. The law is often viewed as a consequence of probabilistic reasoning, as for example by Fast (1962).

It is in fact a consequence of the second law that in most processes (and, strictly speaking, in all real processes) only one of the transformations A → B and B → A mentioned in connection with the first law of thermodynamics is in fact possible. The process is then said to be *irreversible*. In the event that a process allows both transformations A → B and B → A (or allows them to a sufficient approximation) the process is said to be *reversible*. More precisely, 'a reversible process is one that is performed in such a way that, at the conclusion of the process, both the system and the local surroundings may be restored to their initial states, without producing any changes in the rest of the universe' (Zemansky, 1957, p. 151).

1.2 Conservation laws and equations: mass, momentum, energy

The object of the present study is to establish relations between hydraulic variables: depths, velocities, pressures, etc. These relations are derived from physical laws used to express the conservation of certain quantities – or combinations of variables. The simplest and the most common of these quantities – or variable combinations – are *mass, momentum* and *energy*.

It is advantageous first to derive the conservation laws for the very special case of a standing (stationary relative to the bed) hydraulic jump in a steady state, over a horizontal bed, as illustrated in Fig. 1.1. As indicated, it is supposed that the flow is nearly horizontal everywhere except in the immediate vicinity of the jump, where large vertical accelerations do occur. From this supposition it follows that *the pressure distribution away from the jump is hydrostatic*. All shear stresses are correspondingly neglected, at least for the moment, so that, away from the jump, the velocity is uniform over the entire channel depth. This corresponds to the assumption that the fluid may be treated as an 'ideal' or 'perfect' fluid (Lamb, 1932, p. 17). Later, this assumption will be modified to account for more realistic velocity distributions. The flow is seen to be completely described by a velocity u_1 and depth h_1 taken at a control section on the left, and a velocity u_2 and depth h_2 taken at a control section on the right. Left to right is taken as the positive direction, in the usual way. The control sections are, of course, taken in the regions of nearly horizontal flow, as schematized in Fig. 1.1.

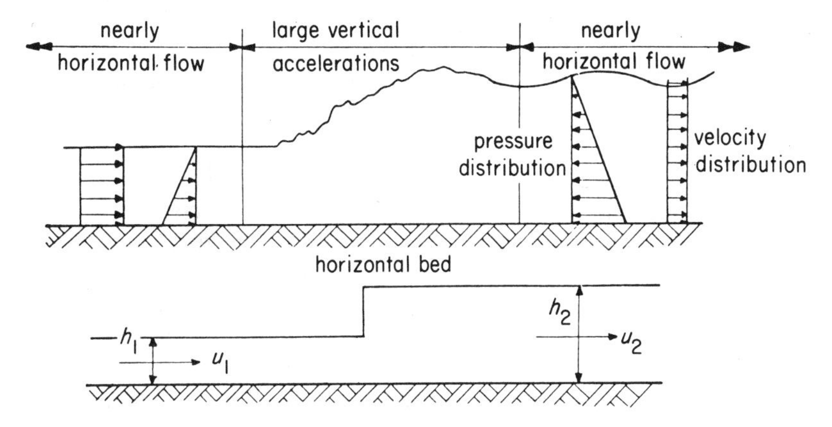

Fig. 1.1. Schematization of the hydraulic jump

The law of conservation of mass may be stated very simply, and quite generally, as follows:

Consider a system of control surfaces enveloping a control volume such that an inside and an outside of the control volume are uniquely defined. Then the net mass of fluid passing from outside to inside through the control surfaces equals the net increase of mass of the control volume.

If this law is applied to the standing hydraulic jump (Fig. 1.1), now supposing that the side walls of the channel containing the jump are parallel and at a distance b apart, the following equality follows:

$$\rho u_1 h_1 b = \rho u_2 h_2 b, \text{ where } \rho \text{ is the fluid density}$$

This equality expresses the fact that the rate at which mass is entering through the left-hand control section is identical to the rate at which mass is leaving through the right-hand control section, nothing being stored in between. Throughout this first section, which is confined to rectilinear flows, all computations will be carried out for a unit width of channel, so that the *equation of mass continuity* becomes simply

$$\rho u_1 h_1 = \rho u_2 h_2 \stackrel{\text{def}}{=} p \qquad (1.2.1)$$

The quantity pb (mass/time) is called the mass flux, while the quantity p itself mass/(time x length) is called the *mass flux density*. In fluid dynamics *per se*, all flux densities are taken per unit *area*, but here they will always be taken per unit *width*, as this corresponds better to the present applications.

The next simplest such law is the *law of conservation of linear momentum*. It is useful first to recall the notion and definition of momentum. Momentum is one description (out of an infinity of possible descriptions) of a state of motion of a system of masses. Such a description can, of course, only be given relative to some initial, or reference, or datum state of motion of the system. It is then supposed that there exists some purely mechanical, perfectly reversible, process whereby the initial (reference, datum) state of motion of the mass system is transformed into the given state. In the case of a 'momentum description' it is supposed that the transformation can be effected by a system of forces acting over certain *times*. Each component product (force x time that force is acting) is called an *impulse*. Since each component force is a vector while time is only a scalar multiplier, each component impulse is also a vector. Thus it is supposed that there exists a system of impulse vectors that effect the transformation from the initial state to the given state. Now, since vectors can be resolved into any direction **e**, it is possible to take, as one description of the given state of motion of the system of masses, the resultant vector in the direction **e**. This resultant vector is clearly a measure of the quantity of motion 'embodied' in the given state, over and above that of the initial state, and this vector measure is called the *momentum* in the direction **e** of the given state. The momentum in the direction **e** is thus also a vector, equal in magnitude and direction to the vector sum in the direction **e** of the impulses effecting the transformation INITIAL → GIVEN. However, whereas 'impulse' refers to the external influences effecting the transformation, 'momentum' refers to the state of motion embodied in the system as a result of the transformation. It will be clear that in general, for the complete momentum description of a system's state of motion in three dimensions, three momentum vectors will be required. It follows from Newton's first law and the laws of linear algebra that by resolving in directions $\mathbf{e}_1$, $\mathbf{e}_2$, $\mathbf{e}_3$ that are *mutually orthogonal*, the process of resolution of any other vectors can be very much simplified.

This can be represented schematically as follows:

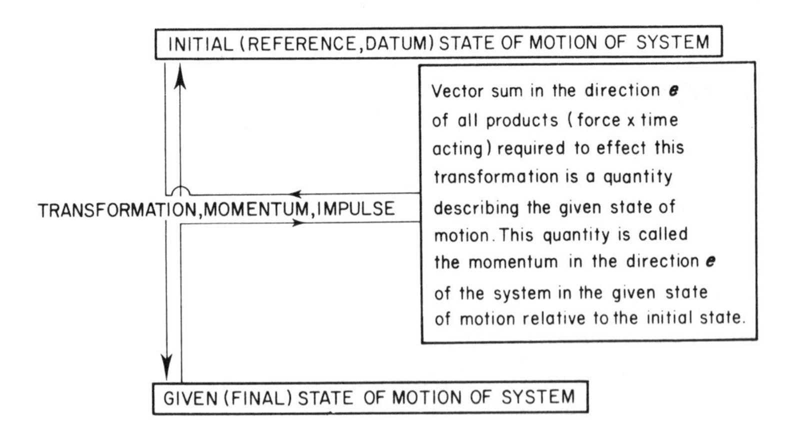

The choice of an initial state is, in principle, quite arbitrary: any convenient initial state can be chosen. Indeed, the momentum being defined only for one state relative to another, the roles of 'initial' and 'final' systems can as well be reversed. In this reversal, it follows from Newton's second and third laws that there will exist impulses providing the transformations FINAL → INITIAL, which are simply equal and opposite to those effecting the transformation INITIAL → FINAL, since the transformation that is used to define momentum is a purely mechanical transformation and is always reversible.

Now if one considers a constant force **f** acting on a constant mass m for time t, it follows from Newton's second law that

$$\mathbf{f}t = (m\mathbf{a})t$$

where **a** is then a constant acceleration. But this may be written as

$$\mathbf{f}t = m(\mathbf{a}t) = m\mathbf{u}$$

where **u** is the increase in velocity after time t, i.e. the increase in velocity over the initial velocity. The quantity which is called momentum is thus also defined by the vector sum of all (mass x velocity) products, and indeed this quantity, which more obviously refers to the amount of motion 'embodied' in a state, is often referred to as the momentum of the state. Since the second law has been enunciated for instantaneous quantities, it can as well be written as

$$\mathbf{f}\,\mathrm{d}t = \mathrm{d}(m\mathbf{u})$$

$$\mathbf{f} = \frac{\mathrm{d}(m\mathbf{u})}{\mathrm{d}t}$$

which states that the applied force vector equals the rate of change of momentum vector – a more usual and generally more convenient formulation in books on classical mechanics (e.g. Goldstein, 1957).

The above definition of momentum can be applied to a control section through the analysis given in Fig. 1.2, taking the velocity of the solid bed as a datum (zero) velocity. Thus, in terms of impulses:

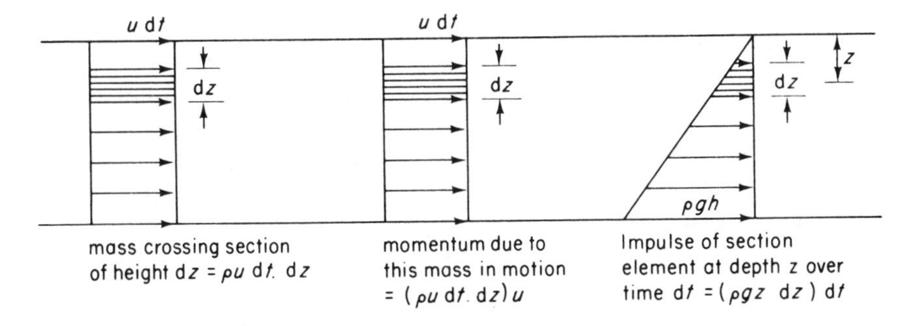

Fig. 1.2. Construction of the momentum flux density for nearly horizontal flow

impulse required to bring this mass from datum (bed) velocity + this impulse or, in terms of momenta:

this momentum + momentum induced by this impulse

Thus, the total *rate* at which momentum (or, putting it the other way, impulse) is transmitted is

$$\int_0^h (\rho gz + \rho u^2)\, dz = \tfrac{1}{2}\rho gh^2 + \rho u^2 h$$

This quantity momentum/(time x length), will be called the *momentum flux density*. Since all shear stresses are ignored, the above quantity comprehends all momenta-impulses, so that, from law (3), the following form for the *law of conservation of linear momentum* between any Sections 1 and 2 of a uniform channel follows.

$$\tfrac{1}{2}\rho g{h_1}^2 + \rho {u_1}^2 h_1 = \tfrac{1}{2}\rho g{h_2}^2 + \rho {u_2}^2 h_2 \tag{1.2.2}$$

It is seen that time cancels out in the discrete formulation (1.2.2), so that this equation can be regarded as a statement that the net force acting over an element, $\tfrac{1}{2}\rho g{h_1}^2 - \tfrac{1}{2}\rho g{h_2}^2$, provides the change in velocity from u_1 to u_2 of the mass passing through the element, $\rho u_1 h_1 = \rho u_2 h_2$. Equation (1.2.2) thus corresponds to *the discrete form of the second law of motion*, i.e. it corresponds to the form of the second law of motion enunciated by Newton in his *Principia*, and discussed in detail in Chapter 5.

From Equations (1.2.1) and (1.2.2) the following explicit equations for u_1 and u_2 in terms of h_1 and h_2 can then be obtained:

$$\begin{aligned} {u_1}^2 &= gh_2 \frac{(h_1 + h_2)}{2h_1} \\ {u_2}^2 &= gh_1 \frac{(h_1 + h_2)}{2h_2} \end{aligned} \tag{1.2.3}$$

The next simplest conservation law is the *law of conservation of energy*. Now energy is another of the possible descriptions of a state of motion, usually generalized, in the presence of a gravitational field, to include the state of vertical displacement – which can then be regarded as the state of capacity for motion of the system.* Again some initial reference or datum state of the system is defined and it is supposed that some purely mechanical process exists whereby this state can be transformed into the given state. In the case of the 'energy description', however, it is supposed that the transformation can be effected by a system of

* The primacy of the notion of kinetic energy and the dependent nature of the potential energy is remarked in several books on mechanics, especially statistical mechanics and histories of mechanics (*see*, for example, Wilson, 1957). The present section owes much to Maxwell's (1877) *Matter and Motion* and the reader is referred to that remarkable work for a more complete discussion of the present concepts. The first three chapters of Landau and Lifshitz's *Mechanics* (1960) gives a more mathematical formulation of the same ideas.

forces acting over certain *distances*. The sum of all *scalar* products (force x distance through which that force acts) is then called the *work done* on the system in effecting the transformation from the initial state to the given state. This scalar quantity is obviously a measure of the state of motion and displacement of a given system, over and above that of the initial state, and this quantity is called the *energy* of the given state. This quantity, being a scalar, provides a description quite different from the momentum: the differences will recur often enough in the sequel! Schematically:

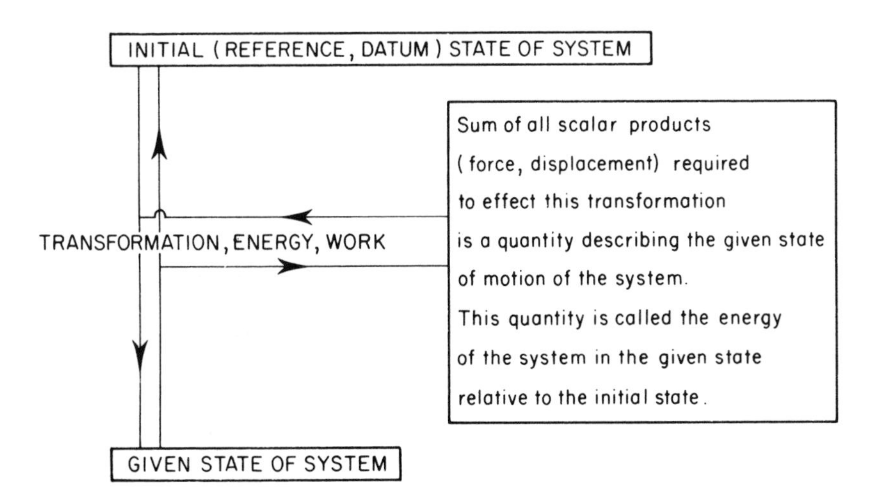

As with momentum, the choice of an initial state is, in principle, quite arbitrary, and again the roles of 'final' and 'initial' can be interchanged. The sum of all (force x displacement) scalar products, or works, providing the transformation FINAL → INITIAL may also be defined to be identical to that providing the transformation INITIAL → FINAL, since transformations are purely mechanical and must again be reversible.

Consider now a constant force **f** acting over time t on a constant mass m situated initially at the origin of displacements with zero velocity. Then, from Newton's second law:

$$\begin{aligned}\text{Energy} &= \mathbf{fs}\\ &= (m\mathbf{a})\,\frac{\mathbf{a}t^2}{2}\\ &= \frac{m(\mathbf{a}t)^2}{2}\\ &= \frac{m\mathbf{u}\cdot\mathbf{u}}{2}, \text{ where } \mathbf{u} \text{ is the velocity after time } t \qquad (1.2.4)\end{aligned}$$

A quantity which is called an energy may be defined by the sum of all (obviously scalar) products ($\frac{1}{2}$ mass x (velocity)^2). This quantity, which more obviously refers to the energy embodied in the motion (as opposed to displacement) part of the state, is called the *kinetic energy*. The energy embodied in displacement is then called the *potential energy*.

The above definitions of energy may be illustrated by computing the energy convected through a control section, taking the bed velocity and level as datum velocity and datum level (Fig. 1.3).

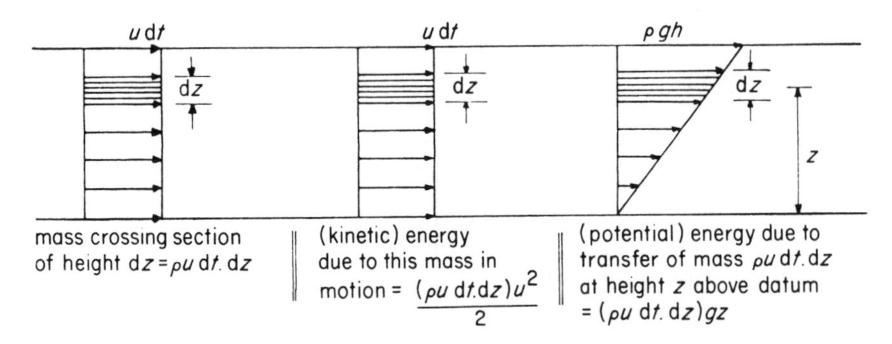

Fig. 1.3. Construction of the convected components of the energy flux density for nearly horizontal flow

The above is entirely in terms of energies. In alternative terms, those of work, it becomes:

work required to bring this mass from datum (bed) velocity + work required to lift this mass into position

Integrating over the depth provides the energy convected across the section:

$$\rho uh(u^2/2)\ \mathrm{d}t + \rho uh(gh/2)\ \mathrm{d}t$$

The above terms account for the nearly horizontal flow energy carried over a section with the fluid. But now consider any process where fluid enters and leaves a system. Any such process is called an *open-flow process* and it is schematized in Fig. 1.4a. In such a process work is done *on* the system in introducing the fluid, while work is done *by* the system in rejecting the fluid.

In the case of nearly horizontal flow in a channel the work done in order to

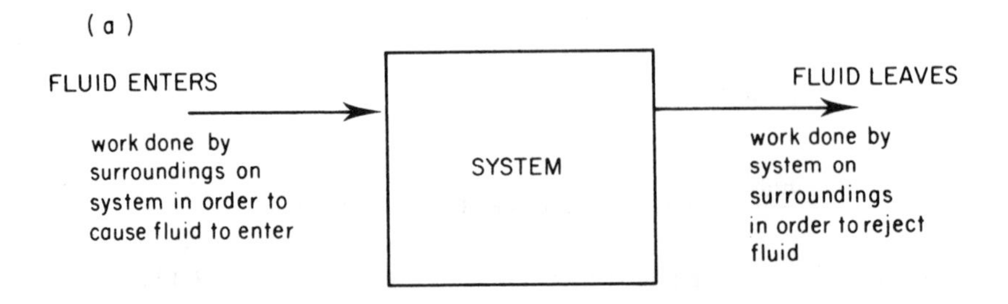

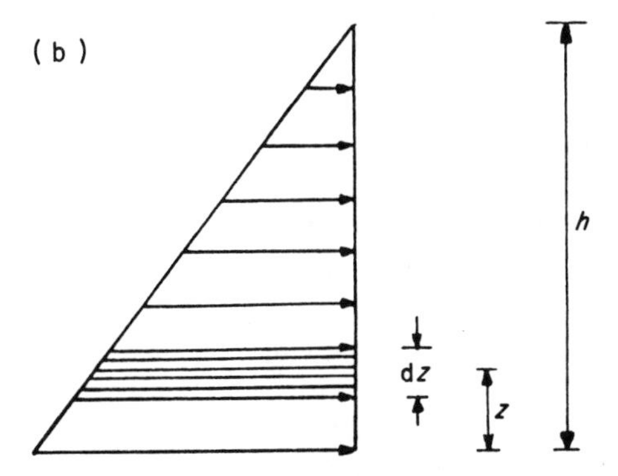

Fig. 1.4. (a) Schematization of an open flow system. (b) Construction of the work component of the energy flux density for nearly horizontal flow

introduce the fluid is given (Fig. 1.4b) by the product $(\rho g(h-z)\,\mathrm{d}z)\cdot(u\,\mathrm{d}t)$. Integrating over depth provides the work done on the system in time $\mathrm{d}t$:

$$\rho u h(gh/2)\,\mathrm{d}t$$

A similar quantity of work is done by the system on its surroundings when fluid leaves the nearly horizontal flow system.

The energy associated with any section then appears as the sum of three quantities:

kinetic energy of fluid passing across the section		potential energy of fluid passing across the section		work done in pushing this fluid across the section
$\rho u h(u^2/2)\,\mathrm{d}t$	+	$\rho u h(gh/2)\,\mathrm{d}t$	+	$\rho u h(gh/2)\,\mathrm{d}t$ (1.2.5)

This sum, energy-work/(time x length), is called the *energy flux density*. In steady accelerated nearly horizontal flows (and in fact under the same circumstances, in the more general 'nearly parallel' flows that are mentioned later), both mass flux and energy flux density are constant, so that the quantity $(u^2/2 + gh)$ is constant over every section of the flow. This quantity is then often called the 'energy head' or 'total head' of the fluid. This situation will be discussed briefly in an example, where it will be related to the weak and strong forms of Bernoulli's law.

Now in the above elementary discussion of mass and momentum conservation laws, it was argued that both mass and linear momentum will be conserved in the ideal flows that are considered. The momentum of the fluid, in particular, will be conserved because all of its component momenta are resolved along the line of direction of the flow. However, this argument provided Equations (1.2.3) and substituting these into the energy flux density difference $(e_1 - e_2)$ between a section on one side of the standing jump and a section on the other side provides

$$\begin{aligned} e_1 - e_2 &= [\rho u h(u^2/2 + gh)]_1^2 = [\rho u h(u^2/2 + gh)]_1 - [\rho u h(u^2/2 + gh)]_2 \\ &= g\rho \cdot \frac{(h_2 - h_1)^3}{4h_1 h_2} \end{aligned} \tag{1.2.6}$$

This 'energy defect' is seen to be non-zero for all finite hydraulic jumps, so that, according to the first law of thermodynamics, energy would have to be lost to the nearly horizontal flow when the flow passed from state 1 to state 2 with $h_2 > h_1$, and it would have to be gained by the nearly horizontal flow when the flow passed from state 1 to state 2 with $h_2 < h_1$. The second law of thermodynamics, however, insists that 'there is a tendency on the part of nature to a state of greater disorder', so that energy can only pass from a more well-ordered to a less well-ordered flow. Newton's laws and the two laws of thermodynamics then insist that when flow passes from state 1 to state 2 with $h_2 > h_1$ then that part of the incident energy flux given by Equation (1.2.6) has to be transformed from nearly horizontal flow energy to energy of a flow that is less well-ordered than a nearly horizontal flow. That is to say, it must be transformed, in the first instance, to the simplest vertically accelerated flows of short-periodic wave trains, and possibly thence, through these short waves and their breaking, to fully turbulent flows. Neither of these energy forms is recognized or *resolved* by the nearly horizontal flow variables so that the energy is effectively lost to the nearly horizontal flow descriptive system. The process itself is, of course, what is observed in nature (e.g. Favre, 1935; Lemoine, 1948; Benjamin and Lighthill, 1954).

When the flow passes from state 1 to state 2 with $h_1 > h_2$, energy has to be gained by the nearly horizontal flow and the second law of thermodynamics then insists that this should come from a flow that is more ordered than the nearly horizontal flow. However, no flow is more ordered than a nearly horizontal flow so that there is no source for this energy. Thus the case where flow passes from state 1 to state 2 with $h_1 > h_2$ is a physical impossibility: the so-called 'negative hydraulic jump' can have no permanent existence. In fact, as described in Chapter 3, such a negative hydraulic jump can be brought into instantaneous existence, as by suddenly reducing the flow into a canal, but the subsequent flow degenerates into a nearly horizontal flow, with only a little short-wave 'noise', through a flow structure that will be identified in Chapter 3 as a 'centred simple wave'.

It should be emphasized that it is only through the introduction of the second law of thermodynamics that the decision can be taken to accept the conservation of linear momentum across the jump and thereby to accept the energy loss of Equations (1.2.6) from the nearly horizontal flow. Without the introduction of the second law of thermodynamics, one could, in principle, just as well have set $e_1 = e_2$ and calculated gains or losses in momentum. Newton's laws of motion, as given in section 1.1, do not, in themselves, suffice to describe the behaviour of the hydraulic jump.

In any description of a flow system in terms of mass and momentum in which processes develop that cannot be resolved by that description, then energy can appear to be lost from the flow system. The simplest and the most common of such energy sinks is in fact associated with the formation of a 'discontinuity', the physical prototype of which is the hydraulic jump in an otherwise smoothly varying, nearly horizontal flow system. As discussed in detail in Chapters 2 and 5, wherever the flow is varying smoothly and so remains nearly horizontal, then momentum and energy are equivalent concepts in any continuum description, in that a momentum formulation in terms of the continuum produces the same answers as an energy formulation in terms of the continuum. But as a discontinuity appears, so momentum and energy are no longer equivalent concepts – even, in a certain limit, in the continuum description – and their equations produce different answers. The principle of non-equivalence of momentum and energy in discrete or discontinuous representations thus appears in its simplest, physical form in the case of the standing hydraulic jump, but its field of application is much wider than this simple example may suggest. Let it suffice for the moment to remark that all of the numerical methods used on discrete, sequential machines correspond in one or another way to discrete, discontinuous representations, so that the principle of non-equivalence of momentum and energy descriptions applies almost universally in computational hydraulics.

1.3 Galilean frames

Throughout the previous section there was a certain arbitrariness in the definition of an initial, reference or datum state. For the formulation of the simplest conservation equations, this state was defined using the bed velocity, and indeed, in its place, this seemed entirely natural, the bed being intuitively 'fixed'. This choice of reference velocity is still, however, no more than a convention, and the conserved quantities could just as well have been defined using some other reference velocity. To each particular choice of reference velocity, the fluid element would then itself appear to have a particular velocity. More precisely, if *frame of reference* B has a uniform rectilinear velocity or 'celerity' c m s^{-1} relative to frame of reference A, then a point seen from frame B appears to be travelling at a velocity c m s^{-1} less than when seen from frame A. This is illustrated diagrammatically in Fig. 1.5.

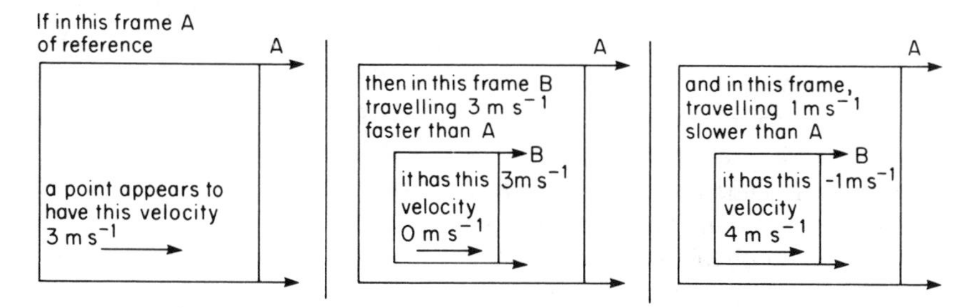

Fig. 1.5. Schematization of a Galilean transformation

More precisely still, if the coordinates of the given point in frame A and frame B are x_A and x_B respectively, and frame B moves with a uniform rectilinear velocity c relative to frame A, then

$$x_A = x_B + ct, \; c = \text{const.} \tag{1.3.1}$$

Equation (1.3.1) defines a transformation from one frame to another. This transformation is called a *Galilean transformation*, while the frames so related are called *Galilean frames*. It has been stated that any one of these frames of reference can be used in the definition of a reference state for mass, momentum and energy. How can one be sure, however, that the conservation equations remain unchanged in form when passing from one such frame to another?

The laws of conservation of momentum and mechanical energy depend upon Newton's laws, so that to answer this question in the case of momentum and energy, it is necessary to determine whether Newton's laws are changed under the transformation (1.3.1). Evidently the first law remains unchanged: a point in uniform rectilinear motion in frame A, which frame is itself in uniform rectilinear motion relative to frame B, will appear to be in uniform rectilinear motion in frame B. The invariance of the second law under the transformation (1.3.1) may be shown by differentiating (1.3.1) twice with respect to time, whence

$$\dot{x}_{\mathrm{A}} = \dot{x}_{\mathrm{B}} + c$$

$$\ddot{x}_{\mathrm{A}} = \ddot{x}_{\mathrm{B}}$$

i.e. the accelerations appear the same in both frames. Since masses and forces are not influenced by Equation (1.3.1), the invariance is demonstrated. Similarly, the third law, involving only forces, or at most 'effect forces', in the sense of the second law is also invariant under the transformation (1.3.1). Thus it follows that *Newton's laws and thereby also the laws of conservation of momentum and mechanical energy are invariant under all Galilean transformations*.

This principle, called *Galileo's relativity principle*, in fact holds quite generally (i.e. wherever Newton's laws are applicable). In particular it holds for reversal of frames, in accordance with the reversible nature of the transformation used above in the momentum and energy definitions.

Consider now the travelling hydraulic jump and take a frame moving with the velocity of the jump (supposed constant for the time of study). Then the absence of storage within the control sections defined by the frames – its steady state in fact – ensures the continued validity of Equations (1.2.1), (1.2.2) and (1.2.6) so long as velocities are corrected to their values within this Galilean frame. Any other Galilean frame would introduce an unsteady state – a control volume with changing storage – and the simple forms of Equations (1.2.1), (1.2.2) and (1.2.6) would no longer obtain (*see* Problem 7). For the travelling hydraulic jump, the following principle can then be introduced:

> *The equations of conservation of mass and momentum and the equation of mechanical energy defect of the hydraulic jump and thereby also all ideal fluid analyses of the jump, are identical for all frames moving with the local velocity of the jump so long as, for each and every such frame, the velocities entering the equations are corrected to those observed within the frame.*

Most studies of the hydraulic jump have to do with travelling jumps, for reasons that will appear much later.

It also follows that, within a frame travelling with the jump just as for the standing jump, the fluid must flow from the smaller depth to the greater depth, $h_2 > h_1$, in order to satisfy the mechanical energy inequality $e_1 > e_2$. From this follows a further general principle. For setting the velocity, usually called the *celerity*, of the jump equal to c and $h_2 - h_1$ equal to Δh, so that Δh is then necessarily positive, gives from Equation (1.2.3):

$$(u_1 - c)^2 = g(h_1 + \Delta h)\frac{(h_1 + (h_1 + \Delta h))}{2h_1} > gh_1$$

$$(u_2 - c)^2 = g(h_2 - \Delta h)\frac{((h_2 - \Delta h) + h_2)}{2h_2} < gh_2$$

Thus, when seen within a frame travelling with the jump (Fig. 1.6), so that the

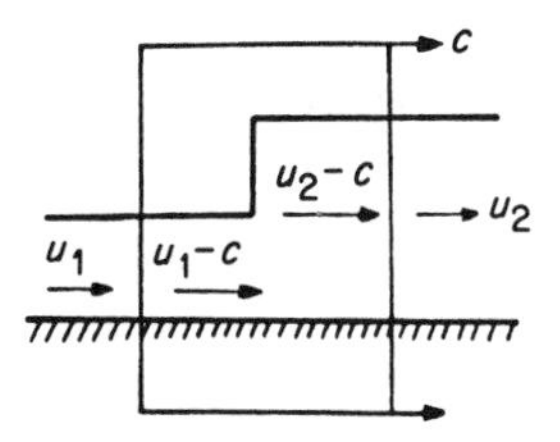

Fig. 1.6. Velocities as seen relative to the bed and within a frame travelling with a hydraulic jump

velocity is the velocity relative to datum corrected by the jump celerity, the value of the *Froude number* $Fr \overset{\text{def}}{=} |\, \text{velocity} . (gh)^{-\frac{1}{2}} \,|$ always appears to pass from $Fr > 1$ to $Fr < 1$ in the direction of flow. This property in itself gives a flow with $Fr = 1$ a special significance. In fact for other reasons besides, a flow with $Fr = 1$ is referred to as a *critical flow,* so that with $Fr > 1$ the flow is said to be *supercritical* and with $Fr < 1$ it is said to be *subcritical.* This principal of the hydraulic jump can then be stated as follows:

> *Within a Galilean frame moving with a hydraulic jump the flow always appears to pass from a supercritical flow to a subcritical flow.* (1.3.2)

It should be emphasized that this principle holds only for $e_1 > e_2$. In the case of a 'negative jump' – in which case the jump is rapidly dispersed – this principle will not hold. In fact it is possible to define a *positive* or *permanent* jump as one for which $e_1 > e_2$, so that it satisfies the principle (1.3.2), while a negative, or 'impermanent', jump can be defined as one for which $e_1 < e_2$, so that it cannot be both steady and satisfy the first and second laws of thermodynamics. By virtue of mass conservation, principle (1.3.2) can also be expressed in the following form:

In a Galilean frame travelling with a permanent hydraulic jump, the flow always appears to pass from the smaller depth to the greater depth. (1.3.3)

1.4 Mass, momentum and energy levels in relation to mass, momentum and energy flux densities

The laws of conservation of mass and momentum of the travelling hydraulic jump may be written in the following form (Fig. 1.6):

$$\rho(u_1 - c)h_1 = \rho(u_2 - c)h_2 \qquad \textbf{mass}$$

$$\rho[(u_1 - c)^2 h_1 + g{h_1}^2/2] = \rho[(u_2 - c)^2 h_2 + g{h_2}^2/2] \qquad \textbf{momentum}$$

However, these may be rearranged and simplified to read

$$\rho c(h_2 - h_1) = \rho(u_2 h_2 - u_1 h_1)$$

$$\rho c(u_2 h_2 - u_1 h_1) = \rho[({u_2}^2 h_2 + g{h_2}^2/2) - ({u_1}^2 h_1 + g{h_1}^2/2)]$$

These equations may then be written, with ρ varying also, simply as

$$c[\rho h]_1^2 = [\rho u h]_1^2$$

$$c[\rho u h]_1^2 = [(\rho u^2 h + g h^2/2)]_1^2$$

or, in vector form,

$$c\begin{bmatrix}\rho h \\ \rho u h\end{bmatrix}_1^2 = \begin{bmatrix}\rho u h \\ \rho(u^2 h + g h^2/2)\end{bmatrix}_1^2 \qquad (1.4.1)$$

Reading down on the right, one observes the mass flux density and momentum flux density relative to the bed, respectively. On the left appear terms ρh and $\rho u h$. The term ρh is then called the *mass level*, it being the mass per unit area of channel. The term $\rho u h$ then corresponds to a *momentum level*. Equation (1.4.1) states that the product of the jump celerity and the change in level across the jump (a sort of 'rate of production or consumption of level') equals the change in corresponding flux across the jump. It then appears as the *discontinuous form*

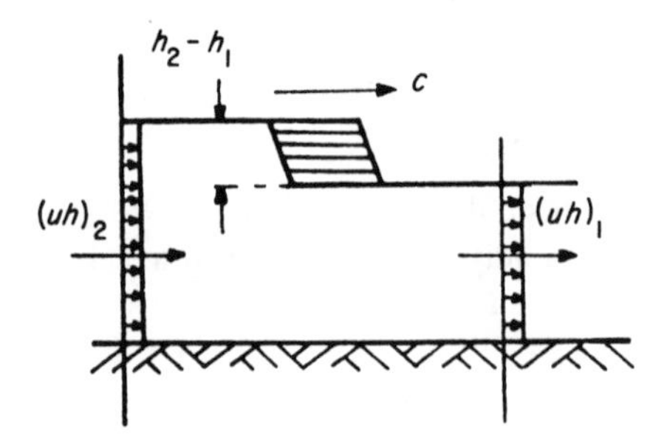

Fig. 1.7. Schematization of the flux-level relations for the volume equation

of the mass and momentum conservation laws. The first law of Equations (1.4.1) is schematized in Fig. 1.7. The corresponding energy inequality becomes

$$c[\rho h(u^2/2 + gh/2)]_1^2 < [\rho uh(u^2/2 + gh)]_1^2 \qquad (1.4.2)$$

where the term on the left evidently corresponds to the *energy level*. It is observed that the energy level lacks the work term $gh/2$ of the energy flux density.

For an infinitesimal jump, Equation (1.4.1) takes the form

$$\begin{aligned} c\,\mathrm{d}h &= \mathrm{d}(uh) \\ c\,\mathrm{d}(uh) &= \mathrm{d}(u^2h + gh^2/2) \end{aligned} \qquad (1.4.3)$$

the first of which is often used to define celerity, as discussed in Section 3.7. From the above derivation it is seen that Equation (1.4.3) can be expressed more precisely through

$$c = \left.\frac{\mathrm{d}(uh)}{\mathrm{d}h}\right|_{x = \text{const.}}, \quad c = \left.\frac{\mathrm{d}(u^2h + gh^2/2)}{\mathrm{d}(uh)}\right|_{x = \text{const.}} \qquad (1.4.4)$$

If the level is written generally as f and the flux density generally as $g, = g(f)$, Equation (1.4.4) generalizes to

$$c = \left(\frac{\partial g(f)}{\partial f}\right)_{x = \text{const.}}$$

1.5 Some worked examples

Problem 1: The following example of the application of Galileo's relativity principle was given by Galileo himself. A body A is travelling with velocity c on a frictionless horizontal surface. It strikes a second body B of the same mass (Fig. 1.8a). What is the motion of A and B subsequent to the collision supposing that the bodies themselves are not deformed in any way by the collision?

Solution: Applying a frame travelling with velocity $c/2$ to the original situation creates a completely symmetric situation (Fig. 1.8b). But then the situation after collision must also be symmetric and, since nothing is changed about the bodies themselves, the motions must be simply reversed (Fig. 1.8c). Removing the frame (Fig. 1.8d) provides the answer to the problem: after collision, body A stops and body B takes the velocity c. In this case nothing need be said about momentum and energy: the supposition that the bodies themselves remain unchanged implies the conservation of both quantities.

Problem 2: An ideal fluid is contained between two parallel vertical walls, originally distance a apart. Forces of magnitude F act on the walls and each causes an infinitesimal displacement $\mathrm{d}a$. Show that, if accelerations are negligible,

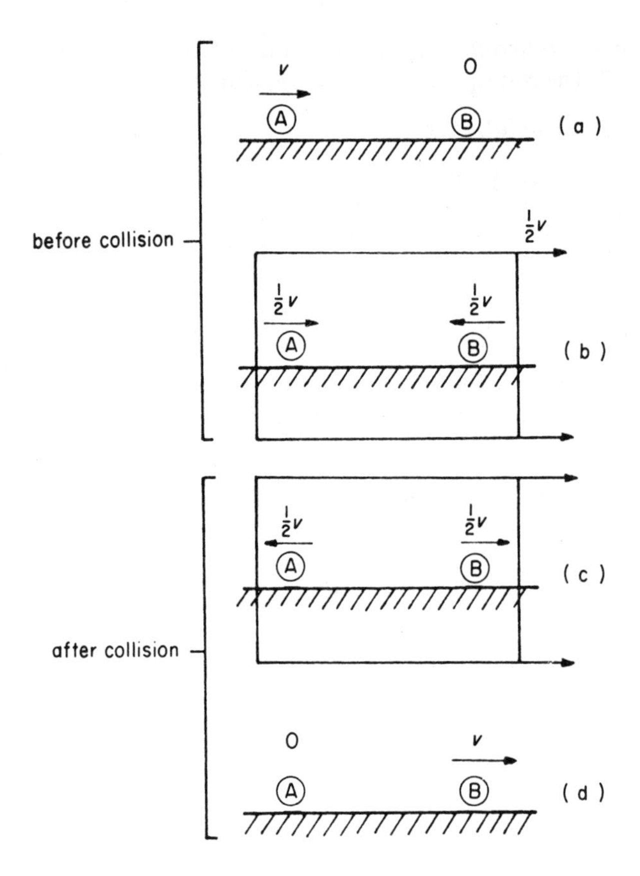

Fig. 1.8. Use of a Galilean frame to create a symmetrical mechanical system

so that the process is *quasi-static*, all work is transformed to potential energy.

Solution: Work done on the fluid $= 2F\,\mathrm{d}a = \dfrac{2\rho g h^2}{2} \cdot \mathrm{d}a$

Increase in the potential energy $= \dfrac{\rho g}{2}\left(\dfrac{ha}{a-2\mathrm{d}a}\right)^2 (a-2\mathrm{d}a) - \dfrac{\rho g h^2}{2}\,a = \dfrac{\rho g h^2}{2}\,2\mathrm{d}a$

This problem demonstrates the essential condition for the reversible transformations used in the definitions of momentum and energy: these transformations must be realized through quasi-static processes. Finite accelerations and decelerations would introduce waves and turbulence, degenerating to heat, and the transformations would not then be purely mechanical.

Problem 3: The potential energy, relative to the bed, of a fluid of depth h is determined, per unit volume, by the following construction (Fig. 1.9). The fluid concerned is first distributed over support $1/a$ so that initially

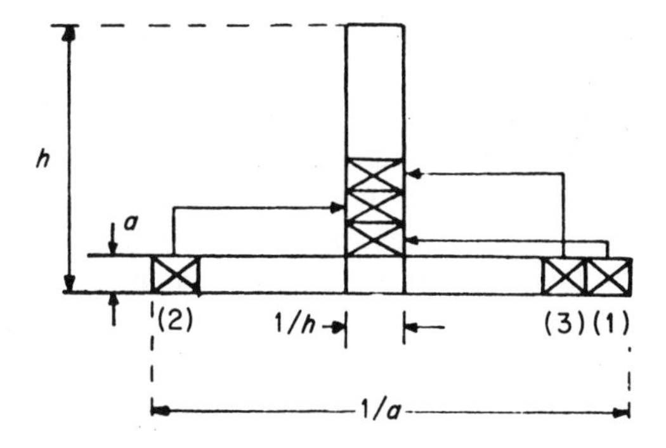

Fig. 1.9. Construction for the potential energy per unit volume, as the limit, as $a \to 0$, of the work done in a countable set of quasistatic lifting operations

$$h_0(x) = a, \ |x| \leqslant 1/2a$$
$$\qquad 0, \ |x| > 1/2a$$

and a is an integral multiple of $1/h$. In Fig. 1.9, the part (1) of support $1/h$ is then lifted and placed without shock on the existing central element. The part (2) is next lifted to rest upon (1) and this process is continued until the element of height h is constructed. The work done e_a in this construction is given by

$$e_a = \rho g\left(\frac{1}{h} \cdot a\right)\left(a + 2a + 3a + \ldots + \left(\frac{h-a}{a}\right)a\right)$$
$$= \frac{\rho g a}{h}\, h\left(\frac{h-a}{2a}\right)$$

How is this related to the true potential energy relative to the bed?

Solution: e_a clearly tends to the true potential energy as $a \to 0$. In this limit the number of elements becomes infinite, while their individual weights tend to 0. In the limit $a \to 0$, the condition that a should be an aliquot part of h becomes redundant, while summing (now synonymous with 'integrating') from either $-\infty$ to 0 or $+\infty$ to 0, one counts up to $\frac{1}{2}$.

The entity $h_0(x)$ obtained as $a \to 0$ is no function, although it does appear as some limit of a sequence of functions. It will be called 'the ϵ-distribution' and defined as that which satisfies

$$\int_{\infty}^{+\infty} \epsilon(x)\, w(x)\, dx \stackrel{\text{def}}{=} \text{average value from } -\infty \text{ to } +\infty \text{ of } w(x) \qquad (1.5.1)$$

where w is any function possessing such an average.

In probability theory, where the integral is called the 'mathematical expectation of w', the ϵ-distribution clearly corresponds to a situation where, speaking very loosely, 'every number is equally probable'. In hydraulics, the ϵ-distribution appears as the unique datum used in the definition, or unique 'fiducial state', for potential energy. It is the only state of mass that has zero potential energy. One can speak intuitively about 'a sheet of fluid of infinitesimal thickness spread out over an infinite length of channel', but this must be defined as a distribution, as in Equation (1.5.1). If the fluid were spread out on one side of the origin only, the ϵ-distribution would correspond to the much more familiar Dirac δ-distribution, defined on h as dependent variable

$$\int_{-\infty}^{+\infty} \delta(h)\, w(h)\, dh \stackrel{\text{def}}{=} w(0) \tag{1.5.2}$$

Problem 4: A drainage canal of rectangular section with length l and breadth b receives a lateral inflow q m^3 s^{-1} per metre run of canal, entering with zero velocity in the direction of the axis of the canal.

The canal is closed at one end, while at the other end it is open with constant depth h_0. Assuming that the system is steady in time, and neglecting all resistances, determine the depth h_e at the closed end.

Solution: Taking the conservation of momentum flux density, i.e. per unit breadth of canal, from one end of the canal to the other, provides

$$\frac{gh_e^{\,2}}{2} = \left(\frac{q \cdot l}{bh_0}\right)^2 \cdot h_0 + \frac{gh_0^{\,2}}{2}$$

or

$$h_e^{\,2} = \left(\frac{q \cdot l}{b}\right)^2 \cdot \frac{2}{gh_0} + h_0^{\,2}$$

It is seen that h_e depends only upon (ql/b) and h_0.

Problem 5: Determine modelling laws for the hydraulic jump, as a relation between depth scales and velocity scales.

Solution: Geometric similarity is assumed so that there exists a unique depth scale $(1:\alpha)$ and a unique velocity scale $(1:\beta)$. Denote model velocities by u_1, u_2 and model depths by h_1, h_2. Then it is seen from Equation (1.2.3) that in the prototype

$$(\beta u_1)^2 = g\alpha h_2 \frac{(h_1 + h_2)}{2h_1}, \qquad (\beta u_2)^2 = g\alpha h_1 \frac{(h_1 + h_2)}{2h_2}$$

so that, if the model is also to obey the relation (1.2.3),

$$\beta = \sqrt{\alpha}$$

Another way of stating this law is to say that the Froude number

$$Fr = |u/(gh)^{\frac{1}{2}}|$$

should be identical at all corresponding points in all similar model systems. For then, in any two systems a and b satisfying these conditions and the conditions of geometric similarity,

$$\frac{u_{1a}^2}{gh_{1a}} = \frac{u_{1b}^2}{gh_{1b}}, \text{ and, since } \frac{h_{1a}}{h_{2a}} = \frac{h_{1b}}{h_{2b}} \text{ by geometric similarity,}$$

$$\frac{u_{2a}^2}{gh_{1a}} = \frac{u_{2b}^2}{gh_{1b}}, \text{ satisfying Equation (1.2.3).}$$

The modelling law then says that *Froude numbers should be identical at corresponding points in model and prototype.* To fix ideas, it is often useful to regard the Froude numbers physically as measures of the ratio of inertial forces to gravitational forces. The magnitudes of the Froude numbers of a flow define the present nearly horizontal flows completely, to within a single scaling factor.

Problem 6: For a given constant discharge, what flow conditions will make:

(a) the momentum flux density a minimum?
(b) the energy flux density a minimum?

Write $uh = k$, a constant. Then the momentum and energy flux densities m and e may be written as

$$m = \rho\left(\frac{k^2}{h} + \frac{gh^2}{2}\right)$$

$$e = \rho k\left(\frac{k^2}{2h^2} + gh\right)$$

so that

$$\frac{dm}{dh} = \rho\left(-\frac{k^2}{h^2} + gh\right) \qquad \frac{d^2m}{dh^2} = \rho\left(\frac{2k^2}{h^3} + g\right)$$

$$\frac{de}{dh} = \rho k\left(-\frac{k^2}{h^3} + g\right) \qquad \frac{d^2e}{dh^2} = \rho k\left(\frac{3k^2}{h^4}\right)$$

The condition that the first derivatives will be zero when the derived quantity is an extremum provides the same relation in each case, $u^2 = gh$. Substituting this relation in the expression for the second derivatives shows that the flow condition providing both minimum momentum flux density and minimum energy flux density, for given discharge, corresponds to a Froude number of unity.

Problem 7: Consider an element of fluid enclosed by walls moving with the stream velocity and containing a standing hydraulic jump, as shown in Fig. 1.10.

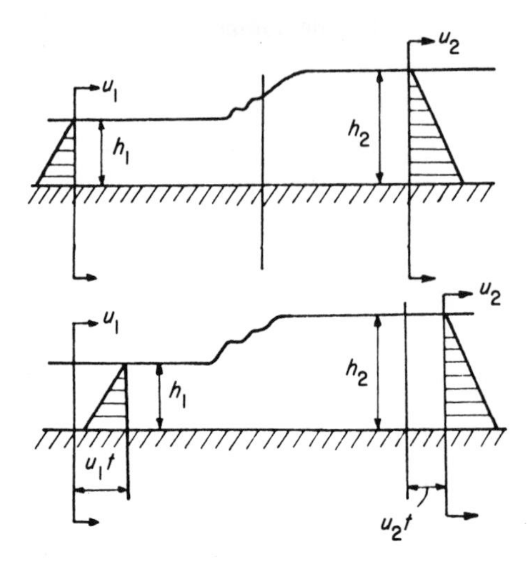

Fig. 1.10. Construction for the energy balance of a hydraulic jump contained within a closed system

Compute the work done, ΔW, *on* the fluid element in time t:

$$\Delta W = \frac{\rho g h_1^2 u_1 t}{2} - \frac{\rho g h_2^2 u_2 t}{2}$$

$$= \frac{gpt}{2}(h_1 - h_2)$$

Since $h_2 > h_1$, it then follows that the system is doing work on its surrounding fluid. How can this be explained?

Solution: Work is indeed being done. This work corresponds to a reduction in the nearly horizontal flow energy of the element chosen. The part of the element with velocity u_1 and depth h_1 loses a length $u_1 t$ over time t, while the other part gains a length $u_2 t$. Thus the *increase* in nearly horizontal flow energy of the element, ΔE, is given by:

$$\Delta E = \rho h_2 u_2 t\left(\frac{u_2^2}{2} + \frac{gh_2}{2}\right) - \rho h_1 u_1 t\left(\frac{u_1^2}{2} + \frac{gh_1}{2}\right)$$

$$= pt\left[\left(\frac{u_2^2}{2} + \frac{gh_2}{2}\right) - \left(\frac{u_1^2}{2} + \frac{gh_1}{2}\right)\right]$$

If Δe is the loss of nearly horizontal flow energy of the element over time t, then it follows from the first law of thermodynamics – the law of conservation of energy – that

$$\Delta W = \Delta E + \Delta e$$

i.e. $\Delta e = \Delta W - \Delta E$

$$= pt\left[(gh_1 - gh_2) - \left(\frac{u_2^2}{2} - \frac{u_1^2}{2}\right)\right]$$

$$= pt\left[\left(\frac{u_1^2}{2} + gh_1\right) - \left(\frac{u_2^2}{2} + gh_2\right)\right]$$

This is the energy defect for an element of fluid containing a standing hydraulic jump. This result is identical to that obtained earlier, the only new aspect being that the jump has now been treated as a *closed-flow system*. In order to assure the continued existence of such a system, it is only necessary to introduce new sections at a suitable distance upstream, and repeat the argument when the section of height h_1 (travelling with the fluid velocity u_1 relative to the jump) arrives at the jump.

Open- and closed-flow systems have other names in fluid mechanics. In general, when attention is fixed upon a definite *element of space*, and the motion of the fluid is considered through that element, then an *Eulerian* scheme is said to be adopted. On the other hand, following a procedure such as that just described, and fixing attention on a definite *element of fluid*, so studying the behaviour of this element in its motion, corresponds to adopting a *Lagrangian* scheme. This terminology, however, is usually applied only to infinitesimal elements of fluid, as discussed extensively for Eulerian schemes in Chapter 2.

Problem 8: An undershot gate passing 6 $m^2 s^{-1}$* provides a steady upstream depth h_1 of 3 m. Calculate the downstream depth h_2 if the flow there is supercritical (Fig. 1.11), and determine also the force F exerted on the gate.

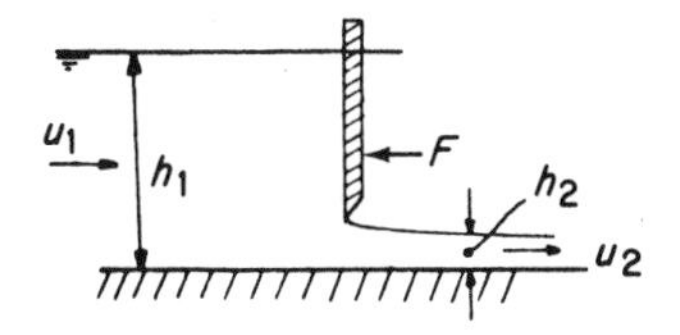

Fig. 1.11. Definition sketch for an undershot gate

Solution: This is a typical problem of steady flow, which will be investigated in considerable detail in order to relate the present approach to the traditional one. For this purpose it is first remarked that if kinetic energy is measured with respect to the gate velocity, so that the gate does no work, then the energy flux density across a suitable region of nearly horizontal flow on the left is equal to that of a similar region on the right. For the existence of such regions, appeal must be made to the theory of vortex squeezing in accelerated flows. The mass flux density is also conserved, while the momentum flux densities differ only by

* i.e., cubic metres per second per metre width of channel.

the force F. Thus the following three relations for the three unknowns (F, h_2, u_2) obtain:

$$[uh]_1^2 = 0 \tag{1.5.3}$$

$$[u^2h + gh^2/2]_1^2 = F/\rho \tag{1.5.4}$$

$$[uh(u^2/2 + gh)]_1^2 = 0 \tag{1.5.5}$$

Equations (1.5.3) and (1.5.5) imply, for any steady flow, that

$$[u^2/2 + gh]_1^2 = 0 \tag{1.5.6}$$

Equation (1.5.6) is an example of a *Bernoulli equation*, in this case expressing the constancy of the sum of kinetic energy, potential energy and work done per unit mass. Dividing through by g provides, in the same order, the familiar 'velocity head', 'position head' and 'pressure head'.

From Equations (1.5.3) and (1.5.6) there follows, taking $g = 9.8\ \mathrm{m\,s^{-2}}$, that

$$h_2 = 0.89\ \mathrm{m} \qquad \text{and} \qquad u_2 = 6.74\ \mathrm{m\,s^{-1}}$$

and thence, from Equation (1.5.4),

$$F = -12.700\ \mathrm{N}.$$

In this simple application, the 'law of Bernoulli' appears as a consequence of overall mass and energy conservation laws. However, the 'law of Bernoulli' can be taken (and usually is taken) to say more than this, namely that the quantity in Equation (1.5.6) is constant *along every streamline* (the 'weak form of Bernoulli's law') and even that the constant concerned is the same for all streamlines (the 'strong form of Bernoulli's law'), so long as no energy goes from rectilinear motion to rotational motion. Again, the existence of streamlines, and even of time-stationary streamlines, is assumed, following known properties of accelerating flows. The present conservation laws will now be used to prove the strong form of Bernoulli's law, under the assumption that time-stationary streamlines exist. (Excellent proofs of more extended 'Bernoulli laws' are given both by Milne-Thomson (1955) and by Landau and Lifshitz (1959).)

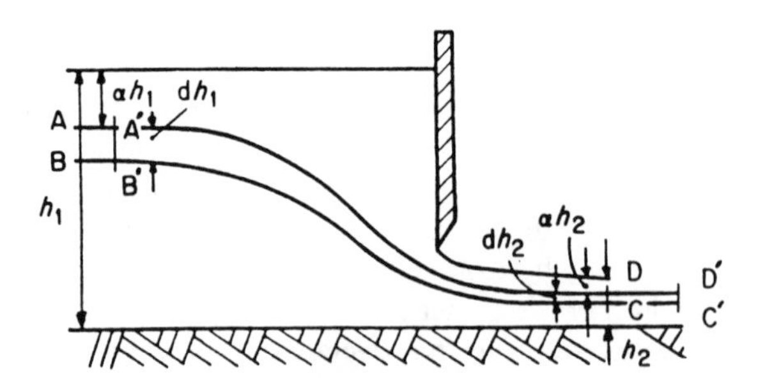

Fig. 1.12. Definition sketch for an open-system and closed-system demonstrations of Bernoulli's law for irrotational flow

Consider the streamtube ABCD shown in Fig. 1.12. This streamtube is first treated as an open system, so that a quantity of fluid $\rho u_1\, dh_1$ enters through AB and quantity $\rho u_2\, dh_2$ leaves through CD in each unit of time. Each unit mass of this fluid carries kinetic energy and potential energy $[u_1^2/2 + g(h_1 - \alpha h_1)]$ into the system and $[u_2^2/2 + g(h_2 - \alpha h_2)]$ out of the system. The work done in unit time *on* the system in order to introduce the volume $u_1\, dh_1$ is $\rho g \cdot \alpha h_1 \cdot u_1\, dh_1$, while the work done *by* the system in rejecting the (same) volume $u_2\, dh_2$ is $\rho g \cdot \alpha h_2 \cdot u_2\, dh_2$. The law of conservation of energy then provides the equality

$$\rho u_1\, dh_1\, [u_1^2/2 \;+ g(h_1 - \alpha h_1) + g\,\alpha h_1] = \rho u_2\, dh_2\, [u_2^2/2 + g(h_2 - \alpha h_2) + g\,\alpha h_2]$$

mass per unit time	kinetic energy per unit mass	potential energy per unit mass	work done per unit mass

upon the assumption that the walls BC and AD are time-stationary, and therefore do no work on the system, and again, that none of the above energy gets transformed into energy of rotational motion. Introducing the law of mass conservation then provides

$$u_1^2/2 + gh_1 = u_2^2/2 + gh_2$$

i.e. the conservation law (1.5.6).

One can as well consider the streamtube ABCD as a closed system, deforming into a system A′B′C′D′ in unit time. This system then loses kinetic energy $\rho u_1^2/2$, and potential energy $\rho g(h_1 - \alpha h_1)$ and gains kinetic energy $\rho u_2^2/2$ and potential energy $\rho g(h_2 - \alpha h_2)$ on every unit mass of the mass $\rho u_1\, dh_1 = \rho u_2\, dh_2$. During the same unit time, work $\rho g\, \alpha h_1 \cdot u_1\, dh_1$ has been done *on* the system in order to push the surface AB to A′B′ while work $\rho g\, \alpha h_2 \cdot u_2\, dh_2$ has been done *by* the system in order to push CD to C′D′. Since no other deformations occur when ABCD goes into A′B′C′D′ and irrotational motion remains irrotational then again energy and work terms can be equated to obtain Equation (1.5.6).

These latter arguments about streamtubes are clearly not restricted to nearly horizontal flows: the quantity (1.5.6) would remain constant at every point along a streamline, and take the same constant on every streamline, so long, of course, as velocity was measured in the stream direction. The restriction to steady streamlines is here essential, however.

Problem 9: Give necessary and sufficient conditions for the direction in which a permanent jump moves in terms of flow conditions relative to the bed on either side of the jump.

Solution: The study of Galilean frames shows that the behaviour of the jump becomes clear and simple in a frame travelling with the jump. Any other frame, such as that fixed by the bed – to which 'upstream' and 'downstream' refer – are generally less simple. Thus, instead of giving the direction in which the jump moves in the usual terms of 'downstream' and 'upstream', conditions must now be given in terms of *b*efore or *b*elow the jump, which will be called side B of the jump, and *a*fter or *a*bove the jump, which will be called side A of the jump. Then taking the direction B → A as the positive direction for celerity and velocity, Fig. 1.13 illustrates how this convention allows a useful symmetry.

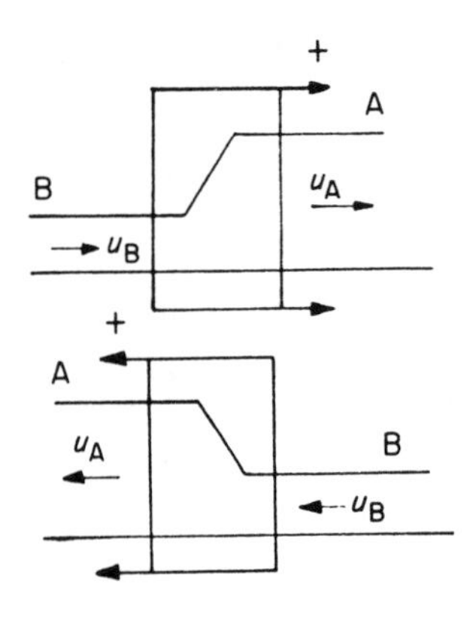

Fig. 1.13. Sign conventions and notation for problem 9

It is observed, from the principle of the hydraulic jump, that $u_B - c > (gh_B)^{\frac{1}{2}}$ and $u_A - c < (gh_A)^{\frac{1}{2}}$. Thus a necessary condition for the jump to travel in the A direction, so that c is positive, is that $u_B > (gh_B)^{\frac{1}{2}}$, while a necessary condition for the jump to travel in the B direction so that c is negative, is that $u_A < (gh_A)^{\frac{1}{2}}$. In other words:

(a) A necessary condition for the permanent jump to travel in the A direction is that the flow on the B side should be supercritical relative to the bed.

(b) A necessary condition for the permanent jump to travel in the B direction is that the flow on the A side should be subcritical relative to the bed.

In order to obtain sufficient conditions it is seen that, if $u_A > (gh_A)^{\frac{1}{2}}$, then c must be positive in order that $u_A - c < (gh_A)^{\frac{1}{2}}$. Similarly, if $u_B < (gh_B)^{\frac{1}{2}}$, c must be negative in order that $u_B - c > (gh_B)^{\frac{1}{2}}$. To summarize:

(c) A sufficient condition for the jump to travel in the A direction is that the flow on the A side should be supercritical relative to the bed.

(d) A sufficient condition for the jump to travel in the B direction is that the flow on the B side should be subcritical relative to the bed.

Note that the necessary conditions may not be sufficient and the sufficient conditions may not be necessary. By taking velocities relative to the bed, the principle of the hydraulic jump is already partly obscured.

Problem 10: The quantity of flow q into one end of a straight uniform canal, length 200 m, is given by:

$$q = \begin{cases} 0 \text{ m}^2 \text{ s}^{-1}* & t < 0 \\ 1 \text{ m}^2 \text{ s}^{-1}* & t \geqslant 0 \end{cases}$$

(The function $q = q(t)$ so defined is called the *unit step function* or *Heaviside step function*.) Given that for $t < 0$ the water in the canal is stationary and of uniform depth 1 m, determine the subsequent flow picture in the canal, assuming that all flows, including inflows, are subcritical.

Solution: Consider, in the first instance, conditions at the end of the canal, where the fluid is entering. Then it can be asserted that there is no hydraulic jump from a smaller to a greater depth. For there exists no sudden reduction in flow, such as could initiate a negative jump, while if any positive jump did occur, it would, by virtue of the subcritical upstream conditions, necessarily propagate upstream. Thus the positive jump, even if it could somehow come into existence, must arrive at the inflow point, 'drowning itself'. The entrance conditions can then at most correspond to a 'drowned hydraulic jump'. In such an inflow (Fig. 1.14), a thorough mixing of the inflowing fluid and the fluid already present

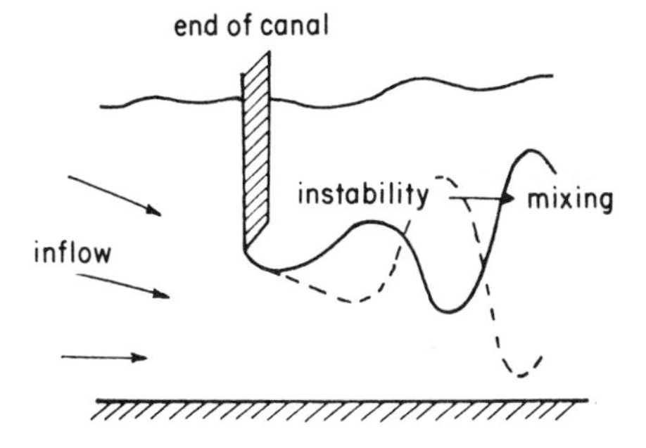

Fig. 1.14. Pictorial representation of the mixing process occurring during flow into a canal. As a result of this mixing the flow behaviour in the canal is determined only by the mass or volume rate of inflow independently of the velocity of this inflow

necessarily occurs. Since the inflow is steady for $t \geqslant 0$ it may now be supposed that conditions adjacent to the inflow are also steady. This steady or 'constant' state must then meet the steady or constant state of the rest of the canal in a sharp transition, which must be accordingly identified as a travelling hydraulic jump with the deeper fluid invading the shallower, as illustrated in Fig. 1.15a. The jump is schematized in Fig. 1.15b whence, by continuity, for $t \geqslant 0$:

$$c(h_2 - h_1), = c(h_2 - 1), = h_2 u_2 = q, = 1 \text{ m}^2 \text{s}^{-1}$$

and, from Equation (1.2.3),

* i.e., cubic metres per second per metre

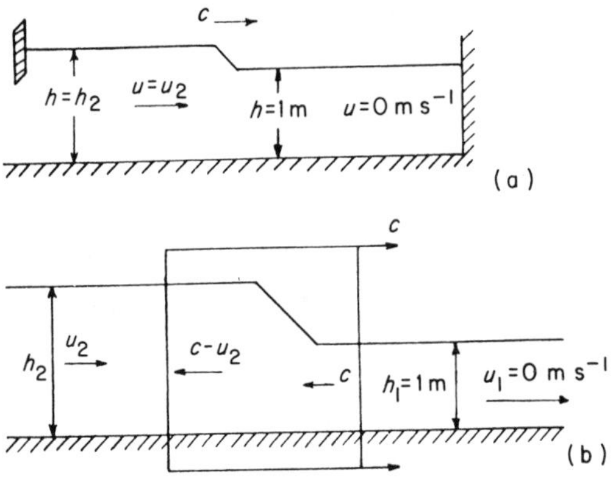

Fig. 1.15. Sketch illustrating the first incident hydraulic jump in problem 10

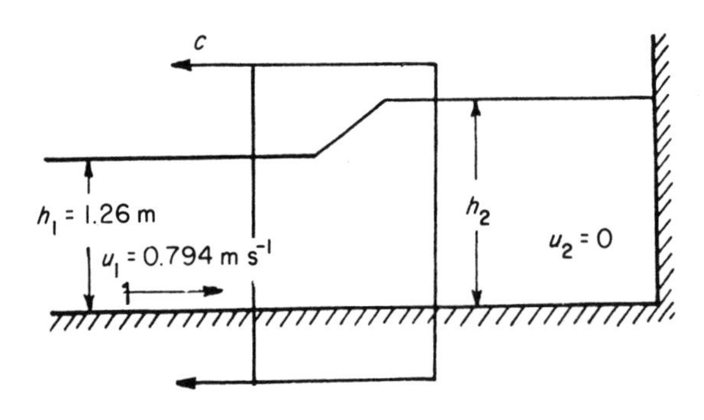

Fig. 1.16. Sketch illustrating the first reflected hydraulic jump in problem 10

$$(c - u_1)^2 = c^2 = \frac{gh_2(h_1 + h_2)}{2h_1}$$

$$= \frac{gh_2(1 + h_2)}{2} \; m^2\,s^{-1}$$

These equations can be solved by iterating on h_2, with or without a graphical solution. Otherwise, the equations can be reduced to a single higher-order system and solved by Newton's method. In the latter case a fourth-order equation is provided:

$$f(h_2) = h_2^4 - h_2^3 - h_2^2 + h_2 - \frac{1}{5} = 0 \quad (\text{using } g = 10\ \mathrm{m\,s^{-2}})$$

giving, for the $(r + 1)$th iteration on a first-order scheme:

$$h_2^{(r+1)} = h_2^{(r)} - \frac{f(h_2^{(r)})}{f'(h_2^{(r)})} = h_2^{(r)} - \frac{(h_2^4 - h_2^3 - h_2^2 + h_2 - \frac{1}{5})^{(r)}}{(4h_2^3 - 3h_2^3 - 2h_2 + 1)^{(r)}}$$

This recurrence is easily set up on a pocket calculator to give $h_2 = 1.26$ m. Thence

$$c = \frac{1}{0.26} = 3.84 \text{ m s}^{-1}, \quad u_2 = \frac{1}{1.26} = 0.794 \text{ m s}^{-1}$$

A positive jump is thus formed, travelling with celerity $c = 3.84$ m s^{-1}. Behind the jump is a *region of constant state* where $h_2 = 1.26$ m and $u_2 = 0.794$ m s^{-1}. This jump will propagate until it arrives at the closed end of the canal, where the closure will impose the condition $u = 0$. From this single condition the reverse flow can be computed (Fig. 1.16). With c taken as positive in the left–right direction:

$$(c - 0.794)^2 = \frac{gh_2(1.26 + h_2)}{2 \times 1.26}$$

and

$$c(h_2 - 1.26) = -1$$

whence $h_2 = 1.56$ and $c = -3.33$ m s^{-1}.

Further reflections will occur later, but since the flow remains subcritical everywhere, these can be calculated exactly as above. The flow picture for the height h can be visualized in a three-dimensional 'block diagram' as shown in Fig. 1.17. The solution can be expressed formally as a series of Heaviside step functions so that the solution reflects the step-like nature of the inflow.

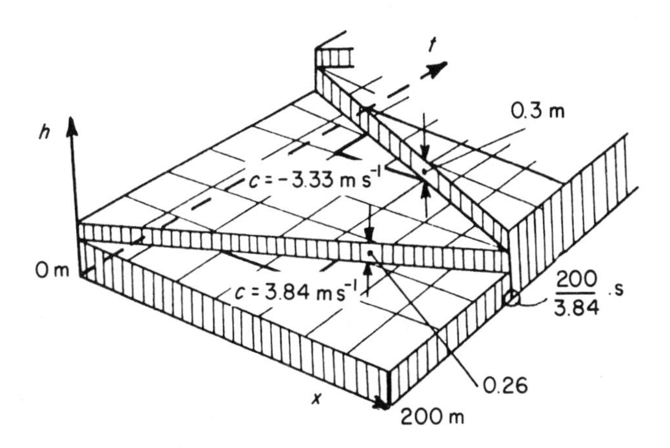

Fig. 1.17. Three-dimensional representation of the results of problem 10, using time as a third dimension, conveyed by perspective drawing

It should be remarked that the above solution implies that fluid is accepted into the system only at one energy flux for a given discharge. Any energy in excess of this 'acceptance energy' must then be dissipated in the submerged jump. This indeed appears to occur in practice – always subject to the condition stated above that the incoming fluid is at a subcritical velocity relative to the control structure. In the event that supercritical flows are introduced, it can happen that a second (positive) hydraulic jump is formed, and propagates (or rather is 'driven') downstream. However, even in this case the calculation remains, in principle, the same. These properties of boundary conditions are taken up again in Chapters 3 and 4.

Problem 11: Compute the heights and lengths of the short waves radiated from the hydraulic jumps of the previous example, on the assumption that the waves are sinusoidal in form.

Solution: The hypotheses described by Lemoine (1948) are adopted, whereby the short-wave system is supposed to be such that its leading element coincides

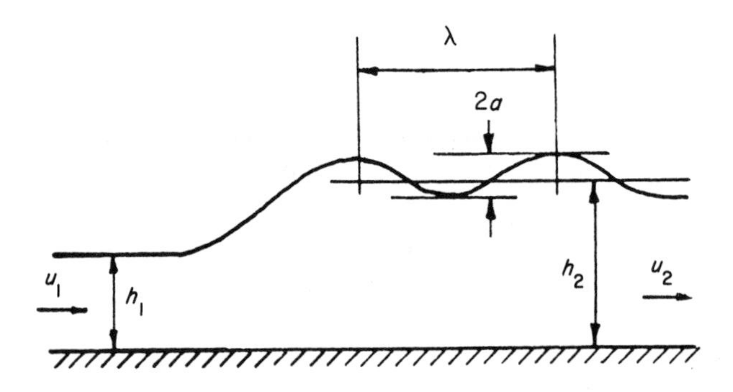

Fig. 1.18. Problem 11 notation

with the leading surface of the hydraulic jump (Fig. 1.18). This is then to say that the wave length λ of the wave is given by (Lamb, 1932, p. 367):

$$c = \left(\frac{g\lambda}{2\pi} \cdot \tanh \frac{2\pi h_2}{\lambda}\right)^{\frac{1}{2}}$$

where c is the celerity of the jump. The energy of the short-wave system per unit length and breadth, which is donoted by E, is given by

$$E = \tfrac{1}{2}\rho g a^2$$

where a is the demi-amplitude of the wave (Lamb, 1932, p. 370). This energy is convected at the group velocity C given by (Lamb, 1932, p. 381):

$$C = \frac{c}{2}\left(1 + \frac{4\pi h_2}{\lambda} \Big/ \sinh \frac{4\pi h_2}{\lambda}\right) \tag{1.5.7}$$

Thus the rate at which energy is radiated away from the jump is

$$E(c - C), = \tfrac{1}{2}\rho g a^2 \cdot \frac{c}{2}\left(1 - \frac{4\pi h_2}{\lambda} \Big/ \sinh \frac{4\pi h_2}{\lambda}\right) \tag{1.5.8}$$

and it is supposed that this may be equated to the energy defect

$$e_1 - e_2, = g\rho \frac{(h_2 - h_1)^3}{4h_1 h_2} \tag{1.5.9}$$

Two independent relations are then available for Equations (1.5.7) and (1.5.8, = 1.2.6) for the two unknowns a and λ. Benjamin and Lighthill (1954) have shown that for $h_2/h_1 < 1.28$ these values are given to two-significant-figure accuracy by

$$\frac{\lambda}{h_2} \approx \frac{2\pi\sqrt{3}}{3}\left(\frac{h_2 - h_1}{h_1}\right)^{-\frac{1}{2}}, \qquad \frac{a}{h_1} \approx \frac{1}{\sqrt{3}}\left(\frac{h_2 - h_1}{h_1}\right)$$

For the previous problem these give

$$\text{(i) } \lambda \approx \frac{2\pi\sqrt{2}}{3} \times \frac{1.26}{\sqrt{0.26}}, \qquad a \approx \frac{1}{\sqrt{3}} \times 1.26 \times 0.26$$

$$= 7.2 \text{ m} \qquad\qquad = 0.19 \text{ m}$$

$$\text{(ii) } \lambda \approx \frac{2\pi\sqrt{2}}{3} \times 1.56 \times \left(\frac{1.26}{0.3}\right)^{\frac{1}{2}}, \qquad a \approx \frac{1}{\sqrt{3}} \times 1.56 \times \frac{0.3}{1.26}$$

$$= 9.5 \text{ m} \qquad\qquad = 0.21 \text{ m}$$

These values may be considerably in error owing to the assumption of a sinusoidal wave train. They do give an indication, however, of the magnitudes of the short waves (*see* the discussion of Benjamin and Lighthill, 1954).

2 Continuous forms of conservation laws

2.1 One-dimensional primitive forms for vertically homogeneous fluids

The simplest situations will first be considered, of mass, momentum and energy conservation laws in primitive form, so stripped of all energy-diffusing terms, such as bed slope, resistance, change of section and wind stress. Correspondingly, a rectangular velocity distribution is employed. An infinitesimal control element

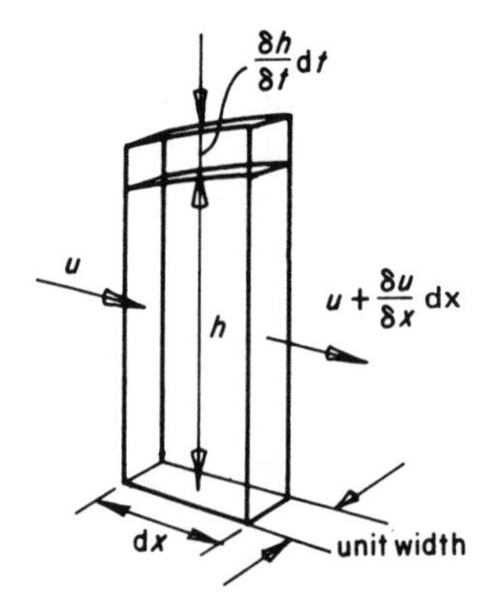

Fig. 2.1. Element used for deriving the primitive one-dimensional conservation equations of nearly horizontal flow

of unit width is considered, as schematized in Fig. 2.1, and the accumulation of mass of the element is computed over time dt as

$$dx \frac{\partial}{\partial t} (\rho h) \, dt$$

The mass entering the element in time dt is

$$\rho u h \, dt$$

while the amount leaving is

$$[\rho uh + \frac{\partial}{\partial x}(\rho uh)\,\mathrm{d}x]\,\mathrm{d}t$$

Thence,

$$\mathrm{d}x\frac{\partial}{\partial t}(\rho h) = (\rho uh\,\mathrm{d}t) - \left[\rho uh + \frac{\partial}{\partial x}(\rho uh)\,\mathrm{d}x\right]\mathrm{d}t$$

or

$$\frac{\partial \rho h}{\partial t} + \frac{\partial}{\partial x}(\rho uh) = 0 \qquad \textbf{mass} \qquad (2.1.1)$$

In the case of a fluid that is also homogeneous in x, ρ is constant in x, and therefore also in t (by the second law of thermodynamics), so that (2.1.1) reduces to the independent volume conservation law of an incompressible fluid

$$\frac{\partial h}{\partial t} + \frac{\partial}{\partial x}(uh) = 0 \qquad \textbf{volume} \qquad (2.1.2)$$

The momentum equation can be treated in much the same way. The accumulation of momentum in the element over time $\mathrm{d}t$ is

$$\mathrm{d}x\frac{\partial}{\partial t}(\rho uh)\,\mathrm{d}t$$

The impulse-momentum applied to and convected into one side of the element of unit width in time $\mathrm{d}t$ is just the momentum flux density multiplied by $\mathrm{d}t$:

$$\rho(u^2h + gh^2/2)\,\mathrm{d}t$$

with, correspondingly,

$$\left[\rho(u^2h + gh^2/2) + \frac{\partial}{\partial x}\rho(u^2h + gh^2/2)\,\mathrm{d}x\right]\mathrm{d}t$$

applied to or convected out of the element. Equating the net impulse-momentum inflow to the momentum accumulation, as corresponds to a locally perfectly reversible process, provides the momentum conservation law

$$\frac{\partial}{\partial t}(\rho uh) + \frac{\partial}{\partial x}(\rho(u^2h + gh^2/2)) = 0 \qquad \textbf{momentum} \qquad (2.1.3)$$

The net impulse in (2.1.3) is in fact composed of a contribution from the surface elevation gradient and a contribution from the density gradient. Referring to Fig. 2.2 it is seen how these can be separated out conceptually, by setting the density gradient to zero and calculating the contribution of the surface gradient and, separately, setting the surface gradient to zero and calculating the contribution of the density gradient. This process gives

$$\rho gh\frac{\partial h}{\partial x}$$

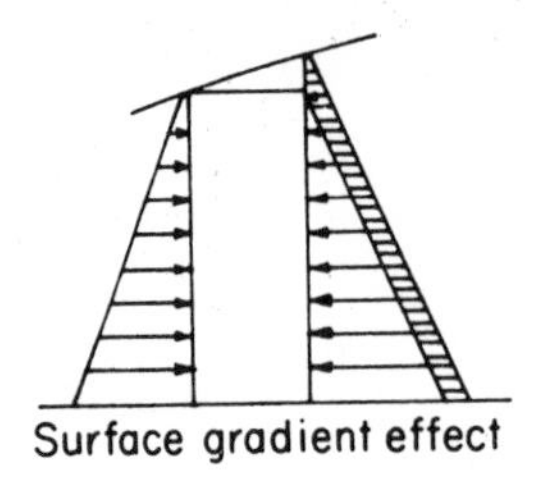

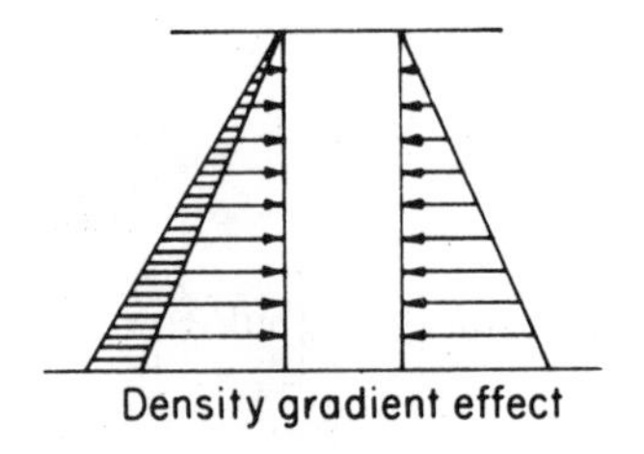

Fig. 2.2. The total impulse from the hydrostatic pressure gradient can be regarded as the sum of two components, the one being due to the surface gradient alone, neglecting the effects of the density gradient, and the other being due to the density gradient alone, neglecting the effects of the surface gradient.

for the first and

$$\frac{gh^2}{2}\frac{\partial\rho}{\partial x}$$

for the second, summing to

$$\rho gh\frac{\partial h}{\partial x}+\frac{gh^2}{2}\frac{\partial\rho}{\partial x}=\frac{\partial}{\partial x}\frac{(\rho gh^2)}{2}$$

For constant ρ, Equation (2.1.3) reduces to

$$\frac{\partial}{\partial t}(uh)+\frac{\partial}{\partial x}(u^2h+gh^2/2)=0 \tag{2.1.4}$$

The derivation of the energy equation is entirely analogous, the accumulation now being taken on energy level

$$\mathrm{d}x\frac{\partial}{\partial t}(\rho h(u^2/2+gh/2))\,\mathrm{d}t$$

with a net influx of energy-work taken on the corresponding energy flux density,

$$\frac{\partial}{\partial x}(\rho uh(u^2/2+gh))\,\mathrm{d}x\cdot\mathrm{d}t$$

to provide, for a locally perfectly reversible process,

$$\frac{\partial}{\partial t}(\rho h(u^2/2+gh/2))+\frac{\partial}{\partial x}(\rho uh(u^2/2+gh)=0 \qquad \textbf{energy} \tag{2.1.5}$$

With ρ constant in x, and therefore also constant in t, this becomes

$$\frac{\partial}{\partial t}(h(u^2/2+gh/2))+\frac{\partial}{\partial x}(uh(u^2/2+gh))=0 \tag{2.1.6}$$

When ρ varies in x a further equation has to be introduced to describe the transport, diffusion and other influences upon this quantity. The simplest such equation is the approximate transport–diffusion equation for fluid density

$$\frac{\partial \rho}{\partial t} + u\frac{\partial \rho}{\partial x} - D\frac{\partial^2 \rho}{\partial x^2} = 0, \quad D \text{ real positive} \tag{2.1.7}$$

which will be discussed further in Chapter 4. The diffusion process is, of course, irreversible.

Putting aside the transport and diffusion processes it is seen that all of the dynamic equations can be written in vector form as

$$\frac{\partial f}{\partial t} + \frac{\partial g(f)}{\partial x} = 0 \tag{2.1.8}$$

where, for the simplest case, of ρ = const.,

$$f = \begin{bmatrix} h \\ uh \\ h(u^2/2 + gh/2) \end{bmatrix} \qquad g = g(f) = \begin{bmatrix} uh \\ u^2h + gh^2/2 \\ uh(u^2/2 + gh) \end{bmatrix} \qquad \begin{matrix} (2.1.2) \\ (2.1.4) \\ (2.1.6) \end{matrix}$$

Individual elements of f and $g(f)$ are denoted by f_i and $g_i(f)$. Laws expressed in the form (2.1.8) are said to be 'expressed in conservation form' and Equations (2.1.2)–(2.1.6) are prototypes of such forms.

Now the above equations are built upon two dependent variables, $h = h(x,t)$ and $u = u(x,t)$, so that it follows from the elementary theory of differential equations that only two independent equations are required to describe the continuous motion. Thus Equation (2.1.8) could be used in mass and momentum form, or in mass and energy form or, for that matter, in momentum and energy form. The question then arises: would one then be saying the same thing with one pair as with another pair? Would these three possible pairs of conservation equations be equivalent: would they all provide the same solution? It is easy to see that in the continuous, reversible case the answers are all in the affirmative, any one of Equations (2.1.2), (2.1.4), (2.1.6) can be derived from the other two: the system (2.1.2), (2.1.4), (2.1.6) is a *non-linearly dependent* system containing two independent equation statements.

For example, Equation (2.1.6) may be opened out to

$$(u^2/2 + gh)\frac{\partial h}{\partial t} - \frac{gh}{2}\frac{\partial h}{\partial t} + h\frac{\partial}{\partial t}(u^2/2 + gh/2) + (u^2/2 + gh)\frac{\partial}{\partial x}(uh) + uh\frac{\partial}{\partial x}(u^2/2 + gh) = 0$$

so that, using Equation (2.1.2) and cancelling terms,

$$h\frac{\partial}{\partial t}(u^2/2) + uh\frac{\partial}{\partial x}(u^2/2 + gh) = 0 \tag{2.1.9}$$

Now

$$\frac{h}{2}\frac{\partial u^2}{\partial t} = uh\frac{\partial u}{\partial t} = u\left(\frac{\partial}{\partial t}(uh) - u\frac{\partial h}{\partial t}\right)$$

giving, from Equation (2.1.2) and (2.1.9),

$$u\frac{\partial}{\partial t}(uh) + u^2\frac{\partial}{\partial x}(uh) + \frac{uh}{2}\frac{\partial u^2}{\partial x} + uh\frac{\partial}{\partial x}(gh) = 0$$

or

$$\frac{\partial}{\partial t}(uh) + u\frac{\partial}{\partial x}(uh) + uh\frac{\partial u}{\partial x} + h\frac{\partial}{\partial x}(gh) = 0$$

Thence the mass conservation equation (2.1.2) and energy conservation equation (2.1.6), for constant density, provide

$$\frac{\partial}{\partial t}(uh) + \frac{\partial}{\partial x}(u^2h + gh^2/2) = 0$$

which is the momentum equation for constant density, Equation (2.1.4). Evidently this argument could just as well be used to derive (2.1.6) from (2.1.2), (2.1.4) or (2.1.2) from (2.1.4), (2.1.6). Moreover it indicates a further conservation form, for Equation (2.1.9) can as well be written as

$$\frac{\partial u}{\partial t} + \frac{\partial}{\partial x}(u^2/2 + gh) = 0 \qquad \textbf{Bernoulli} \qquad (2.1.10)$$

This equation is also often called the 'Euler equation', for reasons outlined in Section 2.5.

It is possible to construct an infinity of further conservation forms through further non-linear transformations. Whitham (1974, p. 459) has given a formal procedure for generating such forms. The first and simplest four elements so generated are Equations (2.1.10), (2.1.2), (2.1.4) and (2.1.6), while the fifth simplest element,

$$\frac{\partial}{\partial t}\left(\frac{1}{3}u^3h + ugh^2\right) + \frac{\partial}{\partial x}\left(\frac{1}{3}u^4h + \frac{3}{2}u^2gh^2 + \frac{1}{3}g^2h^3\right) = 0$$

has not as yet been related to the physical basis.

For the present, immediate purposes, however, the essential feature of continuous flows is that, for the differential (continuum) description of these flows, momentum and energy formulations are equivalent.

2.2 Introduction of lateral inflows, bed slopes, and resistance and driving stresses

Consider first the effect of a lateral inflow of q m^2 s^{-1}* on the one-dimensional primitive equations. This corresponds to lateral flow into a straight horizontal canal of constant rectangular cross section. The canal width is denoted by b and it is supposed that the lateral flow enters the canal with velocity component u_l

*i.e., cubic metres per second per metre length of channel

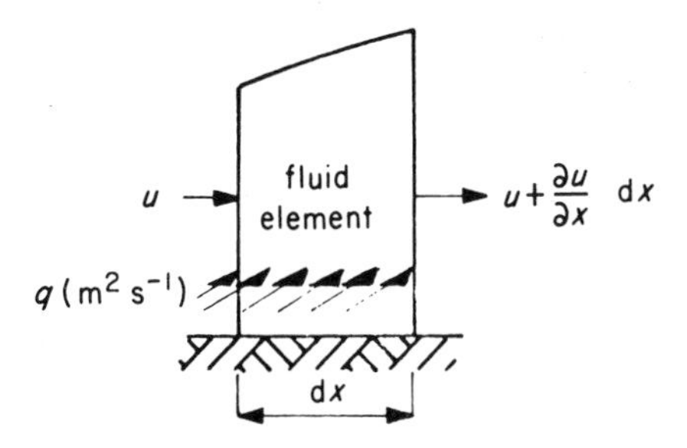

Fig. 2.3. Schematization of lateral inflow

in the positive direction of the canal. Clearly a negative value of q implies outflow.

Referring to Fig. 2.3, it is seen that the accumulation of mass of the fluid element of width b is

$$b \, \mathrm{d}x \frac{\partial(\rho h)}{\partial t} \mathrm{d}t$$

while the net mass entering this element in time $\mathrm{d}t$ is

$$\left\{ b \left[\rho u h - \left(\rho u h + \frac{\partial}{\partial x} (\rho u h) \, \mathrm{d}x \right) \right] + \rho q \, \mathrm{d}x \right\} \mathrm{d}t$$

so that, for constant ρ

$$\frac{\partial h}{\partial t} + \frac{\partial(uh)}{\partial x} - \frac{q}{b} = 0 \tag{2.2.1}$$

Similarly, the accumulation of momentum is

$$\rho b \, \mathrm{d}x \frac{\partial}{\partial t} (uh) \, \mathrm{d}t$$

and the net momentum-impulse input is

$$\left\{ \rho b \left[(u^2 h + g h^2/2) - \left((u^2 h + g h^2/2) + \frac{\partial}{\partial x} (u^2 h + g h^2/2) \, \mathrm{d}x \right) \right] + \rho q u_i \, \mathrm{d}x \right\} \mathrm{d}t$$

so that, for constant ρ

$$\frac{\partial}{\partial t} (uh) + \frac{\partial}{\partial x} (u^2 h + g h^2/2) - \frac{q u_i}{b} = 0 \tag{2.2.2}$$

In several instances, as for example when a river overflows into or is replenished from its flood plains, the lateral flow enters with a velocity component in the x-direction that is almost zero but may leave with practically the full stream velocity u. In this case it is convenient to write

$$\frac{\partial}{\partial t} (uh) + \frac{\partial}{\partial x} (u^2 h + g h^2/2) - \frac{u}{2b} (q - |q|) = 0$$

The energy equation derivation is rather more complicated.

The effect of variations in the bed elevation can be treated in two ways. For a very small bed gradient i (shown for a positive sign of i in Fig. 2.4a) of the type encountered in rivers and canals, something less than 1:1000, it is convenient to take an x axis along the sloping bed (Fig. 2.4a) and to measure water depth

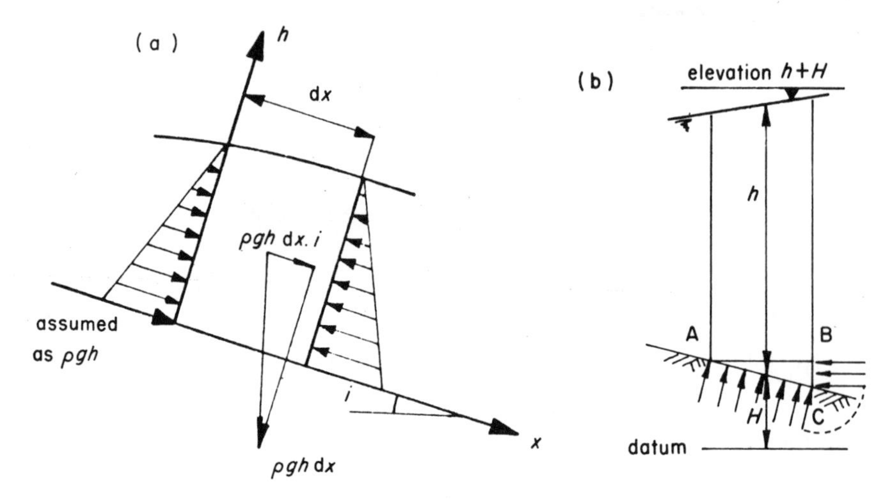

Fig. 2.4. (a) A formulation of the influence of bed slope on the dynamic equations that is often adopted in river engineering. (b) The corresponding view commonly used in coastal engineering

orthogonal to the x axis. By virtue of the small slope, the pressure exerted on the control element can again be assumed to attain a maximum of ρgh at the bed. Clearly, then, the mass equation remains unchanged while, from Fig. 2.4, the momentum equation becomes, for constant ρ,

$$\frac{\partial}{\partial t}(uh) + \frac{\partial}{\partial x}(u^2h + gh^2/2) - ghi = 0 \tag{2.2.3}$$

For the larger slopes it is more usual to use another approach. The bed elevation above the same datum can be represented by H so that, referring to Fig. 2.4b, a balance is still made over a fluid element of depth h but the driving gradient is expressed in terms of $(h + H)$. The mass equation then remains unchanged, but the momentum balance is modified. Since pressures are hydrostatic in nearly horizontal flows, the triangular sub-element ABC is in static equilibrium, with the horizontal component of the force exerted from the bed over AC exactly balancing the horizontal force exerted upon the sub-element over BC. The rate of accumulation of momentum again takes the form

$$dx \cdot \frac{\partial}{\partial t}(\rho uh)\, dt$$

but the impulse term is now seen to take the form (referring to Fig. 1.2):

$$\rho gh \cdot \frac{\partial}{\partial x}(h + H)\,\mathrm{d}x \cdot \mathrm{d}t$$

Thus, what was formerly a conservation law now breaks down, for constant ρ, to

$$\frac{\partial}{\partial t}(uh) + \frac{\partial}{\partial x}(u^2h) + gh\frac{\partial}{\partial x}(h + H) = 0 \tag{2.2.4}$$

Identifying

$$\frac{\partial H}{\partial x} = -i$$

demonstrates the equivalence of Equations (2.2.3) and (2.2.4).

So far, attention has been directed to ideal fluids, incapable of supporting shear stresses. In practice, however, turbulent exchanges of fluid masses within the stream create momentum transfers which in turn appear as shear stresses, and these are transmitted downwards, through a non-turbulent sub-layer, to the bed. In practice, this entire complex is expressed in terms of a 'resistance' to flow, concentrated at the bed. It is usual to represent this resistance as a shear stress τ (Kg $m^{-1}s^{-2}$ or N m^{-2}) at the bed and acting in the direction opposite to the velocity.

Then the momentum equation may be written, for constant ρ and horizontal bed, and retaining a rectangular velocity distribution, as

$$\frac{\partial}{\partial t}(uh) + \frac{\partial}{\partial x}(u^2h + gh^2/2) + \frac{\tau}{\rho}(\text{sign } u) = 0$$

where (sign u) = +1 when u is positive and −1 when u is negative. The hydraulics of the last century has given a number of resistance formulae that express τ in terms of u, h and the physical properties of the bed, at least for steady flows. It is common practice to carry these over to more general nearly horizontal flows without change. Thus the Chézy law gives the form

$$\frac{\partial}{\partial t}(uh) + \frac{\partial}{\partial x}(u^2h + gh^2/2) + \frac{gu|u|}{C^2} = 0 \tag{2.2.5}$$

where C is the well-documented Chézy C.

A driving stress τ, due to wind, can be treated analogously. If this is taken as positive in the positive x-direction, then

$$\frac{\partial}{\partial t}(uh) + \frac{\partial}{\partial x}(u^2h + gh^2/2) - \frac{\tau}{\rho} = 0 \tag{2.2.6}$$

Using the well-known law relating τ to wind speed μ_{10} at 10 m above the water surface, provides

$$\frac{\partial}{\partial t}(uh) + \frac{\partial}{\partial x}(u^2h + gh^2/2) - \frac{k^2\rho_a\mu_{10}^2}{\rho} = 0 \tag{2.2.7}$$

The wind speed μ_{10} can be obtained from the geostrophic wind μ_g through scaling down the magnitude of μ_g and introducing some slight change of the direction vector. In Equation (2.2.7), ρ_a is the density of the air and k may be treated as a constant over a considerable range of wind speeds (Hill, 1962; Hasse, 1974).

2.3 Expansions and contractions in one-dimensional flows

The simplest example of an expanding or a contracting flow is provided by pure radial flow. In itself, of course, pure radial flow is of little practical interest, but it is important because it typifies the problems of representing expansions and contractions in canals, rivers and estuaries in a continuous representation. An idealized situation is shown in Fig. 2.5a. The volume equation follows from the notation of Fig. 2.5b as

$$\frac{\partial}{\partial t}(r\theta h) + \frac{\partial}{\partial r}(r\theta uh) = 0$$

or

$$\frac{\partial h}{\partial t} + \frac{\partial}{\partial r}(uh) + \frac{uh}{r} = 0 \tag{2.3.1}$$

The momentum equation reduces, for constant ρ, to

$$\frac{\partial}{\partial t}(r\theta uh) + \frac{\partial}{\partial r}(r\theta u^2 h) + r\theta\frac{\partial}{\partial r}(gh^2/2) = 0$$

or

$$\frac{\partial}{\partial t}(uh) + \frac{\partial}{\partial r}(u^2 h + gh^2/2) + \frac{u^2 h}{r} = 0 \tag{2.3.2}$$

again owing to the equilibrium of impulses over the sub-element triangles schematized in Fig. 2.5c. To refer again to Fig. 2.5c, the energy equation is seen to take the form

$$\frac{\partial}{\partial t}[\rho r\theta h(u^2/2 + gh/2)] + \frac{\partial}{\partial r}[\rho r\theta uh(u^2/2 + gh)] = 0$$

since the side pressures do no work. For constant ρ, the energy equation can be written in the form

$$\frac{\partial}{\partial t}[h(u^2/2 + gh/2)] + \frac{\partial}{\partial r}[uh(u^2/2 + gh)] + \frac{uh(u^2/2 + gh)}{r} = 0 \tag{2.3.3}$$

and this can be opened up as follows:

$$(u^2/2 + gh)\frac{\partial h}{\partial t} - \frac{gh}{2}\frac{\partial h}{\partial t} + h\frac{\partial}{\partial t}(u^2/2 + gh/2) + (u^2/2 + gh)\frac{\partial}{\partial r}(uh) +$$

$$+ uh\frac{\partial}{\partial r}(u^2/2 + gh) + \frac{uh}{r}(u^2/2 + gh) = 0$$

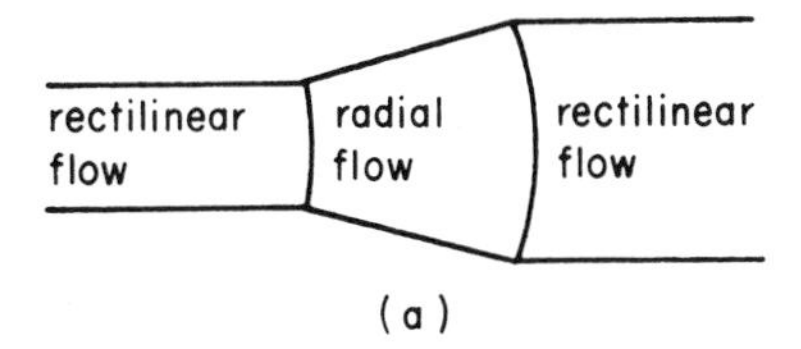

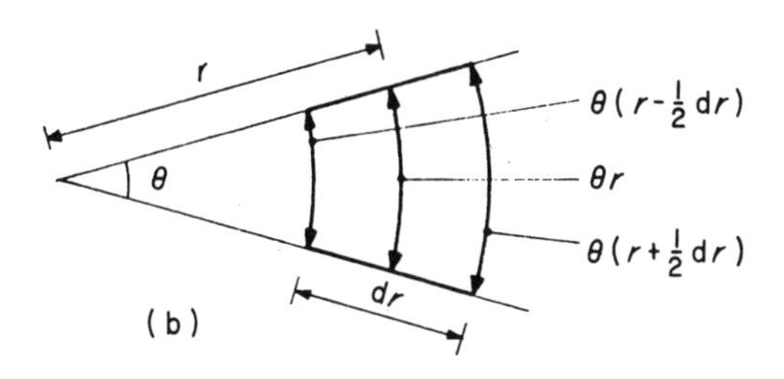

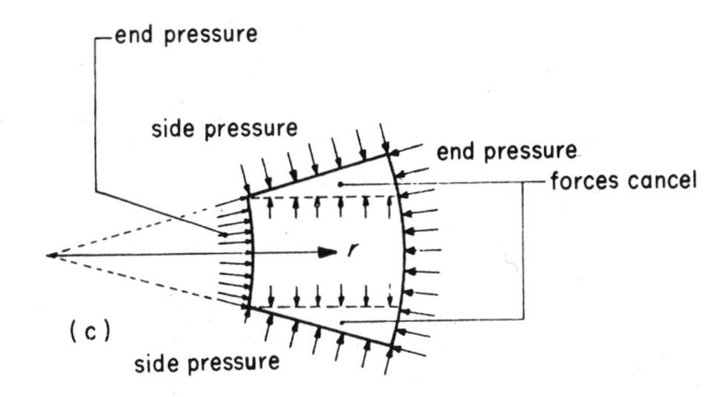

Fig. 2.5. A radial flow can be used to describe a transition from one section to another in one-dimensional flows as schematized in (a). The definition sketch in the radial flow appears as in (b) while (c) schematizes a cancellation of impulse terms that occurs in such a flow

Then, by virtue of Equation (2.3.1),

$$h\frac{\partial}{\partial t}(u^2/2) + uh\frac{\partial}{\partial r}(u^2/2 + gh) = 0$$

which provides

$$\frac{\partial u}{\partial t} + \frac{\partial}{\partial r}(u^2/2 + gh) = 0 \tag{2.3.4}$$

i.e. a form identical to Equation (2.1.10). This particular conservation law thus has the same form for both rectilinear and radial flow. In fact, the steady-state reduction of Equations (2.1.10) and (2.3.4) provides in each case

$$(u^2/2 + gh) = \text{const. for all } x \text{ or } r,$$

which is again simply the strong form of Bernoulli's law.

These notions can be extended to the more general problem of describing flows in a natural watercourse of variable section as follows. A small element can be taken from one such river section, as shown shaded in Fig. 2.6. A velocity dis-

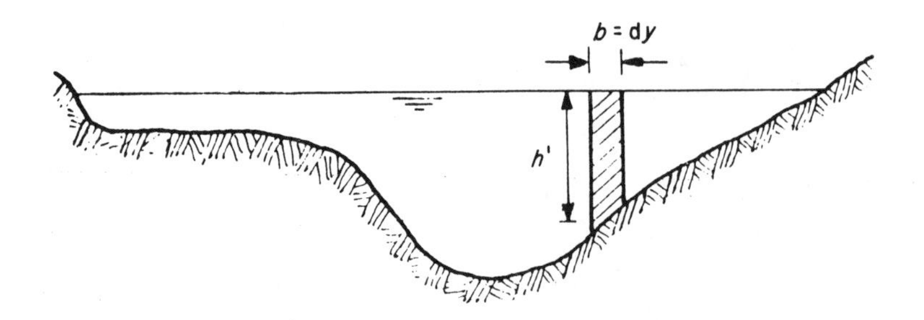

Fig. 2.6. Element used for deriving river section orthogonal to the direction of mean flow

tribution coefficient α may be applied to the depth-averaged velocity $\bar{u}$ to provide a correction to the convected momentum. Mass and momentum conservation laws for this shaded element are then (Verwey, 1972)

$$\frac{\partial}{\partial t}(\rho h'b) + \frac{\partial}{\partial x}(\rho h'b\bar{u}) = 0 \tag{2.3.5}$$

$$\frac{\partial}{\partial t}(\rho h'b\bar{u}) + \frac{\partial}{\partial x}(\alpha'\rho h'b\bar{u}^2 + \tfrac{1}{2}\rho gb(h')^2) - \frac{\partial b}{\partial x}\cdot\tfrac{1}{2}\rho g(h')^2 - gh'bi_b = 0 \tag{2.3.6}$$

where h' is now the depth with i_b the local bed slope. Differentiating out in Equation (2.3.6) and treating ρ as a constant, it is found that impulse terms with $\partial b/\partial x$ cancel out just as for the pure radial flows and Equations (2.3.5), (2.3.6) reduce to

$$\frac{\partial}{\partial t}(h'b) + \frac{\partial}{\partial x}(h'b\bar{u}) = 0 \tag{2.3.7}$$

$$\frac{\partial}{\partial t}(h'b\bar{u}) + \frac{\partial}{\partial x}(\alpha'h'b\bar{u}^2) + gh'b\frac{\partial h}{\partial x} = 0 \tag{2.3.8}$$

where h is then the surface level above some arbitrary horizontal reference level.

An integration can now be carried out across the section under the assumptions that:

(a) $\partial h/\partial x$ is constant across the width of the canal
(b) there is no net loss or gain of mass or momentum of one element from another.

Area A and discharge Q for a total cross section can be defined through

$$A = \int h' \, \mathrm{d}y$$

$$Q = \int h'\bar{u} \, \mathrm{d}y = \bar{\bar{u}}A$$

where the integration is carried across the entire width of the flow section.

Introducing these relations in Equations (2.3.7), (2.3.8) under the given assumptions provides

$$\frac{\partial A}{\partial t} + \frac{\partial Q}{\partial x} = 0 \tag{2.3.9}$$

$$\frac{\partial Q}{\partial t} + \frac{\partial}{\partial x}\left(\alpha \frac{Q^2}{A}\right) + gA\frac{\partial h}{\partial x} = 0 \tag{2.3.10}$$

where α is the (Boussinesq) velocity distribution coefficient for the total cross section, defined as

$$\alpha = \frac{A}{Q^2} \int_A u^2 \, \mathrm{d}A \tag{2.3.11}$$

In order to complete the equations for a natural river, the following terms have to be introduced:

(a) a lateral mass inflow q under an angle ϕ with the streamlines of the river and with mean velocity v

(b) a resistance term, as for example using the hydraulic radius (R) concept of Engelund (1965, 1)

$$R = \frac{1}{A}\int_0^B ff(h')^{\frac{3}{2}} \, \mathrm{d}y \tag{2.3.12}$$

where ff is the ratio of the Chézy resistance in the cross-section point considered to that obtaining for the main bed and B is the total width of the wetted area.

(c) a storage width b_s defined by

$$\frac{\partial A}{\partial t} = b_s \frac{\mathrm{d}h}{\mathrm{d}t}$$

Then Equations (2.3.9) and (2.3.10) become

$$\frac{\partial A}{\partial t} + \frac{\partial Q}{\partial x} - q = 0 \tag{2.3.13}$$

$$\frac{\partial Q}{\partial t} + \frac{\partial}{\partial x}\left(\alpha \frac{Q^2}{A}\right) + gA\frac{\partial h}{\partial x} + \frac{gQ|Q|}{C^2AR} - v\cos\phi q = 0 \tag{2.3.14}$$

2.4 Two-dimensional conservation forms for homogeneous fluids

The fluid element shown in Fig. 2.7 is considered and the mass conservation equation is first considered. The mass accumulation over time dt is

$$dx\, dy \frac{\partial}{\partial t}(\rho h)\, dt$$

while the net mass entering the element in time dt is

$$\left[dy \cdot \frac{\partial}{\partial x}(\rho u h) \cdot dx + dx \cdot \frac{\partial}{\partial y}(\rho v h) \cdot dy\right] dt$$

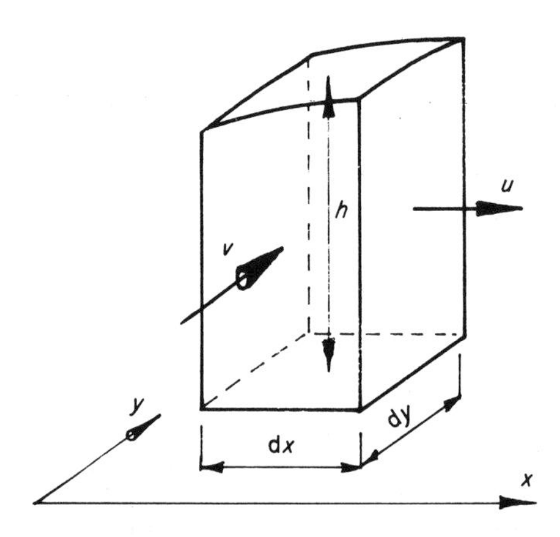

Fig. 2.7. Sketch defining the primitive two-dimensional conservation equations of nearly horizontal flow

Thence

$$\frac{\partial}{\partial t}(\rho h) + \frac{\partial}{\partial x}(\rho u h) + \frac{\partial}{\partial y}(\rho v h) = 0 \qquad \textbf{mass} \qquad (2.4.1)$$

When ρ is constant this reduces to

$$\frac{\partial h}{\partial t} + \frac{\partial}{\partial x}(uh) + \frac{\partial}{\partial y}(vh) = 0 \qquad \textbf{volume} \qquad (2.4.2)$$

For the x-momentum, there will occur a net gain of momentum level over time dt of

$$dx\, dy \cdot \frac{\partial}{\partial t}(\rho u h)\, dt$$

while the net x-momentum flux over the element of length dy during time dt will be, for a horizontal bed

$$-dy \frac{\partial}{\partial x}(\rho u^2 h + \rho g h^2/2)\, dx \cdot dt$$

The net x-momentum transferred over the face of the element of length dx, with velocity v, during time dt can then be computed, using the methods of Chapter 1, as

$$-\,dx\,\frac{\partial}{\partial y}\,(vh\cdot\rho u)\,dy\,dt$$

Equating x-momentum level rates to the sum of the x-impulses and the x-momentum flux density changes in x- and y-directions gives

$$\frac{\partial}{\partial t}(\rho uh)+\frac{\partial}{\partial x}(\rho u^2h+\rho gh^2/2)+\frac{\partial}{\partial y}(\rho uvh)=0\qquad \textbf{\textit{x}-momentum}\qquad (2.4.3)$$

By symmetry alone, the conservation law for y-momenta must then be

$$\frac{\partial}{\partial t}(\rho vh)+\frac{\partial}{\partial y}(\rho v^2h+\rho gh^2/2)+\frac{\partial}{\partial x}(\rho uvh)=0\qquad \textbf{\textit{y}-momentum}\qquad (2.4.4)$$

In the presence of a variable bed elevation H, Equations (2.4.3) and (2.4.4) may be rewritten, following the argument of Section 2.2, as

$$\frac{\partial}{\partial t}(\rho uh)+\frac{\partial}{\partial x}(\rho u^2h)+\rho gh\,\frac{\partial}{\partial x}(h+H)+\frac{gh^2}{2}\,\frac{\partial\rho}{\partial x}+\frac{\partial}{\partial y}(\rho uhv)=0\qquad (2.4.5)$$

$$\frac{\partial}{\partial t}(\rho vh)+\frac{\partial}{\partial y}(\rho v^2h)+\rho gh\,\frac{\partial}{\partial y}(h+H)+\frac{gh^2}{2}\,\frac{\partial\rho}{\partial y}+\frac{\partial}{\partial x}(\rho uvh)=0\qquad (2.4.6)$$

The corresponding simplifications for constant ρ follow directly. The question of the consequences of variation of ρ with x and y can be more complicated than may appear here, as will appear in Section 2.5. It is seen that, for a horizontal bed, Equations (2.4.1), (2.4.5), (2.4.6) can be written analogously to (2.1.8) in vector form as

$$\frac{\partial f}{\partial t}+\frac{\partial g_1(f)}{\partial x}+\frac{\partial g_2(f)}{\partial y}=0\qquad (2.4.7)$$

In the presence of resistance and Coriolis effects Equations (2.4.5) and (2.4.6) may be expanded to the form

$$\frac{\partial}{\partial t}(\rho uh)+\frac{\partial}{\partial x}(\rho u^2h)+\rho gh\,\frac{\partial}{\partial x}(h+H)+\frac{gh^2}{2}\,\frac{\partial\rho}{\partial x}+\frac{\partial}{\partial y}(\rho uvh)+$$
$$+\frac{\rho gu(u^2+v^2)^{\frac{1}{2}}}{C^2}-\gamma\rho hv=0\qquad (2.4.8)$$

$$\frac{\partial}{\partial t}(\rho vh)+\frac{\partial}{\partial y}(\rho v^2h)+\rho gh\,\frac{\partial}{\partial y}(h+H)+\frac{gh^2}{2}\,\frac{\partial\rho}{\partial y}+\frac{\partial}{\partial x}(\rho uvh)+$$
$$+\frac{\rho gv(u^2+v^2)^{\frac{1}{2}}}{C^2}+\gamma\rho hu=0\qquad (2.4.9)$$

where a Chézy form of resistance law is used and the Coriolis coefficient γ is given by $\gamma = 2\omega \sin \phi$, with ω as the angular velocity of the earth and ϕ the north latitude. The argumentation is entirely analogous to that of Section 2.2.

2.5 One-dimensional flows in a stratified fluid

So far, flows have been considered in which the fluid was homogeneous in density in the vertical. In practice, however, variations of density in the vertical may have a decisive influence upon the properties of flow considered. This is especially the case when transport processes are of primary interest. The simplest case of such density variations is that of a fluid stratified into two layers, as schematized in Fig. 2.8.

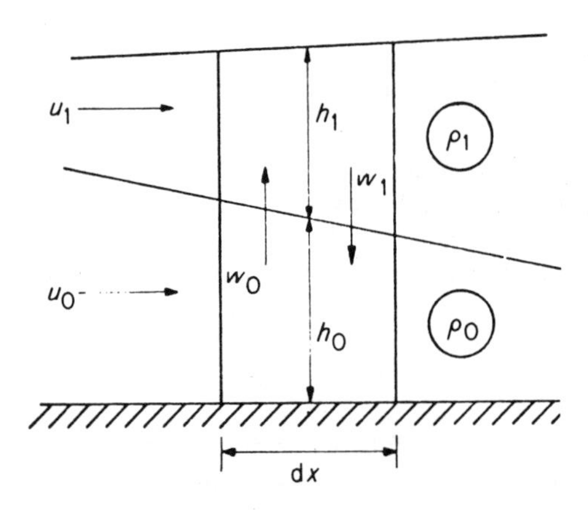

Fig. 2.8. Definition sketch for a stratified fluid

In this schematization the layer depths h_0 and h_1 are treated as variables so that these correspond, in a sense, to the Lagrangian coordinates of the surface and a line in the fluid called the 'interface'. This interface may, in principle, be situated anywhere in the fluid, but it is convenient to place it such that it divides the fluid into regions of distinct behaviour. Thus, the interface may be situated at a rapid change in vertical density when density effects are of central importance or where density (salinity, temperature) is the most easily measured quantity in the field. Similarly, when transport processes are of central importance, the interface may be more conveniently placed at a rapid change in velocities. The two-layer model provides the simplest resolution of internal flows, as compared with the homogeneous, single-layer model, and it is accordingly often used whenever this extra resolution is required.

In the two-layer case, it is convenient to introduce an exchange of fluid across the interface, with w_0(m s^{-1})* crossing from the lower to the upper layer and w_1(m s^{-1})* crossing from the upper to the lower layer, as schematized again in

* i.e., cubic metres per second per square metre.

Fig. 2.8. These exchanges may take the full density of their initiating layer across to their receiving layer, or they may work with only a reduced part of the density difference. In addition to the usual surface (wind) stress τ_w and bed stress τ_b, an interfacial stress τ_i is also introduced. There exists an extensive literature on these exchanges and shear stresses (e.g. Ottesen-Hansen, 1975).

With the schematization of Fig. 2.8, the mass equations for the upper and lower layers can be written simply as

$$\frac{\partial}{\partial t}(\rho_1 h_1) + \frac{\partial}{\partial x}(\rho_1 u_1 h_1) - (\rho_0 w_0 - \rho_1 w_1) = 0 \tag{2.5.1}$$

$$\frac{\partial}{\partial t}(\rho_0 h_0) + \frac{\partial}{\partial x}(\rho_0 u_0 h_0) - (\rho_1 w_1 - \rho_0 w_0) = 0 \tag{2.5.2}$$

when, for the sake of simplicity of exposition the full density defect is supposed transferred across the interface.

It is seen that, in the absence of interfacial mass transfers, Equations (2.5.1) and (2.5.2) both retain the conservation form, individually. Their sum, corresponding to a mass balance over the total fluid, provides also, of course, a conservation form.

Special attention must be given in the momentum equations to the hydrostatic pressure-induced impulses. These impulses can be computed in the lower layer by using the same technique as used in the case of the homogeneous fluid, of adding the variations in impulses due to the surface and interfacial gradients, with densities in both layers fixed, to the variations in impulses due to variations in densities, with the gradients at surface and interface fixed.* This gives a resultant impulse over time dt, for an element of length dx, of

$$\left[h_0\left(g\rho_1 \frac{\partial h_1}{\partial x} + g\rho_0 \frac{\partial h_0}{\partial x}\right) dx + h_0\left(gh_1 \frac{\partial \rho_1}{\partial x} + \frac{gh_0}{2}\frac{\partial \rho_0}{\partial x}\right) dx\right] dt$$

Thence, the momentum equation for the lower layer reads

* In this work, so-called 'engineering derivations' are used for all differential equations, these corresponding to differential descriptions of situations in which certain basic assumptions or schematizations are introduced *a priori* and where the use of mathematical devices such as auxiliary variables is thereby reduced to a minimum. Even in some engineering situations, however, it is preferable to proceed along the traditional hydrodynamic line of setting up a general differential equation, as free as possible from *a priori* assumptions, separating out the mean values and the variations from the means in the manner of Reynolds and then integrating the resulting expansions with auxiliary variables, using Leibnitz' relations between integrals of derivatives and derivatives of integrals (e.g. Nihoul, 1975). This approach often has the advantage of indicating the level of approximation inherent in the equations used. However, such derivations necessitate the introduction of assumptions in the form of one or more 'closure hypotheses', to complete the consequently expanded equation system, and these naturally raise the whole question of where differential descriptions are viable at all. It seems best, for the purpose of this introductory computational hydraulics text, to avoid as far as possible the whole question of the rigour of differential formulations. Only certain aspects, that unavoidably present themselves, are considered in Chapter 5.

$$\frac{\partial}{\partial t}(\rho_0 u_0 h_0) + \frac{\partial}{\partial x}(\rho_0 u_0^2 h_0) + g h_0 \left(\frac{\partial}{\partial x}(\rho_1 h_1) + \rho_0 \frac{\partial}{\partial x}(h_0 + H) + \frac{h_0}{2}\frac{\partial \rho_0}{\partial x}\right) -$$

$$-(\rho_1 u_1 w_1 - \rho_0 u_0 w_0) + (\tau_b - \tau_i) = 0 \qquad (2.5.3)$$

For the upper fluid, the interface can simply be treated as a bed, with shear stress τ_i, so that the momentum equation can be written as

$$\frac{\partial}{\partial t}(\rho_1 u_1 h_1) + \frac{\partial}{\partial x}(\rho_1 u_1^2 h_1) + g h_1 \rho_1 \frac{\partial}{\partial x}(h_1 + h_0 + H) + \frac{g h_1^2}{2}\frac{\partial \rho_1}{\partial x} +$$

$$+ (\rho_1 u_1 w_1 - \rho_0 u_0 w_0) + (\tau_i - \tau_w) = 0 \qquad (2.5.4)$$

It is seen that when ρ_0 and ρ_1 are constant in x, Equations (2.5.3) and (2.5.4) reduce to

$$\frac{\partial}{\partial t}(u_0 h_0) + \frac{\partial}{\partial x}(u_0^2 h_0) + g h_0 \frac{\partial}{\partial x}(h_0 + \sigma h_1 + H) - (\sigma u_1 w_1 - u_0 w_0) +$$

$$+ (\tau_b - \tau_i)/\rho_0 = 0 \qquad (2.5.5)$$

$$\frac{\partial}{\partial t}(u_1 h_1) + \frac{\partial}{\partial x}(u_1^2 h_1) + g h_1 \frac{\partial}{\partial x}(h_0 + h_1 + H) + (\sigma u_1 w_1 - u_0 w_0) +$$

$$+ (\tau_i - \tau_w)/\rho_1 = 0 \qquad (2.5.6)$$

with $\sigma = \rho_1/\rho_0$.

Even when all exchanges, stress and bed variations are excluded, these reduce only to

$$\frac{\partial}{\partial t}(u_0 h_0) + \frac{\partial}{\partial x}\left(u_0^2 h_0 + \frac{g h_0^2}{2}\right) + g\sigma h_0 \frac{\partial h_1}{\partial x} = 0 \qquad (2.5.7)$$

$$\frac{\partial}{\partial t}(u_1 h_1) + \frac{\partial}{\partial x}\left(u_1^2 h_1 + \frac{g h_1^2}{2}\right) + g h_1 \frac{\partial h_0}{\partial x} = 0 \qquad (2.5.8)$$

so that they cannot, individually, reduce to conservation forms. This is, indeed, to be expected from physical considerations, since the one layer is physically coupled to the other through the pressure terms so that momentum and energy will be exchanged between the layers in all dynamic situations. Of course, adding Equation (2.5.7) to $\sigma \times$ (2.5.8) provides

$$\frac{\partial}{\partial t}(u_0 h_0 + \sigma u_1 h_1) + \frac{\partial}{\partial x}\left(u_0^2 h_0 + \sigma u_1^2 h_1 + \frac{g(h_0^2 + 2\sigma h_1 h_0 + \sigma h_1^2)}{2}\right) = 0 \qquad (2.5.9)$$

so providing the conservation form corresponding to the conservation properties of the total fluid.

From this development, it will be seen that the four equations of the individual layers imply the two equations of the total fluid, at least so long as the effects of variations in velocity profiles on the convective momenta are neglected. This leads to the notion that the problem of solving the four equations of a stratified fluid can be reduced to a problem of first solving the two equations of the total fluid and thence solving the two equations of one of the component layers. The total fluid model then corresponds to what oceanographers call a *barotropic* model while the internal model then corresponds to a *baroclinic* model (Hill, 1962).

2.6 Eulerian forms and algorithmic forms

The conservation forms are the most used differential forms of computational hydraulics but, of course, they are hardly if ever treated in works on classical hydraulics. Classical hydraulics uses Eulerian forms, almost exclusively, so that the relation between the differential formulations of computational hydraulics and the corresponding formulations of classical hydraulics proceeds through the relations between conservation forms and Eulerian forms.

The basis of the Eulerian point of view was outlined in Section 1.5 as one in which attention was fixed upon a definite element of space and the behaviour of the fluid was considered as it moved through this space. Consider any quantity ϕ that is convected with the fluid through such a space-element as schematized in the x–t plane of Fig. 2.9. Then the fluid within the reference element at $x = \xi$ and $t = \tau$ originated at $x = \xi - u\,\Delta t$ at time $\tau - \Delta\tau$. Correspondingly, the value of ϕ at the reference element, at $x = \xi$ and $t = \tau$ is determined by the variation of ϕ in t and the variation of ϕ in x such that

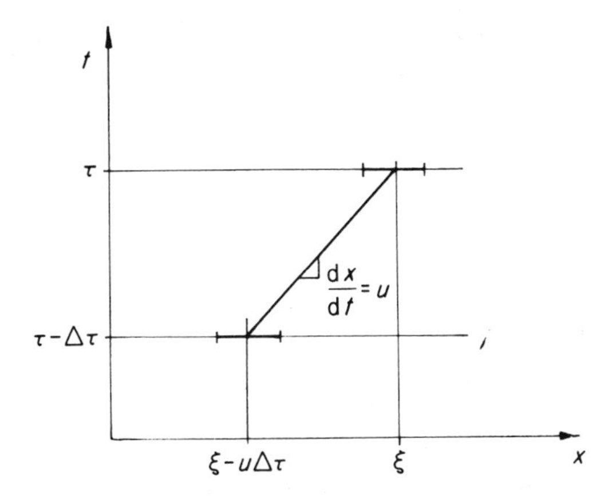

Fig. 2.9. Schematization for the convective terms of Eulerian forms

$$d\phi = \frac{\partial \phi}{\partial x}\Delta x + \frac{\partial \phi}{\partial t}\Delta t + \text{higher-order terms}$$

Since $\Delta x/\Delta t$ in this situation converges simply to the convective velocity u, then

$$\frac{d\phi}{dt} = \frac{\partial \phi}{\partial t} + u\frac{\partial \phi}{\partial x}$$

The derivative on the left is sometimes called the 'total time derivative'.

An acceleration equation can then be constructed by expressing the total change of velocity over time dt as the sum of the change with fixed x and the change due to translation through $u\ dt$ in a region where $\partial u/\partial x$ remains constant over time dt (Fig. 2.9). Then the total change of velocity over time dt is

$$\frac{\partial u}{\partial t}\,dt + \frac{\partial u}{\partial x}\cdot u\,dt$$

so that the fluid particle acceleration is

$$\frac{\partial u}{\partial t} + u\frac{\partial u}{\partial x}$$

The mass of the element is, on average, $\rho h\ dx$, so that the product of mass and acceleration, or the effect force, is

$$\rho h\,dx\left(\frac{\partial u}{\partial t} + u\frac{\partial u}{\partial x}\right)$$

The driving force on the element is provided by the hydrostatic pressure difference

$$\rho g h\frac{\partial h}{\partial x}\,dx$$

Since no momentum is convected into the element travelling as shown in Fig. 2.9 then the local driving force can be equated to the product of local mass and acceleration to give

$$\frac{\partial u}{\partial t} + u\frac{\partial u}{\partial x} + g\frac{\partial h}{\partial x} = 0 \tag{2.6.1}$$

This type of acceleration equation is often called an 'Euler equation'.

A corresponding volume equation can be constructed, albeit rather awkwardly, along the same lines. The fluid within the control element of Fig. 2.9 then increases in mass, to a first approximation, by $dx\ (\partial(\rho h)/\partial t)\,\Delta t$, by virtue of the increase of h with t at fixed x, and by $dx\ (\partial(\rho h)/\partial x)u\ \Delta t$, by virtue of the increase in h caused by translation though distance $u\ \Delta t$ in a region with surface gradient $\partial h/\partial x$. Thus the increase in volume of the element tends to

$$\left(\frac{\partial h}{\partial t} + u\frac{\partial h}{\partial x}\right)dx\cdot dt \qquad \text{as } \Delta t \to dt$$

The net volume inflow into the element is given to a first approximation by

$$-\left[u\frac{\partial h}{\partial x} + h\frac{\partial u}{\partial x}\right]dx\,\Delta t$$

where the brackets [ ] indicate that the enclosed quantities are measured within the frame moving with velocity u shown in Fig. 2.9. But then, for limiting $\Delta t \to dt$,

$$[u] = 0, \quad \left[\frac{\partial h}{\partial x}\right] = \frac{\partial h}{\partial x}, \quad \left[h\frac{\partial u}{\partial x}\right] = h\frac{\partial u}{\partial x}$$

Equating this to the volume increase gives the equation

$$\frac{\partial h}{\partial t} + u\frac{\partial h}{\partial x} + h\frac{\partial u}{\partial x} = 0 \tag{2.6.2}$$

Both this and the Euler equation itself

$$\frac{\partial u}{\partial t} + u\frac{\partial u}{\partial x} + g\frac{\partial h}{\partial x} = 0 \qquad (2.6.3, = 2.6.1)$$

are examples of *Eulerian forms*. It is intriguing, as Lamb pointed out, that (2.6.1) has always been derived through the use of what is really a Lagrangian point of view!

Much the same argument can be used in two dimensions to obtain a total derivative of any quantity ϕ convected with a velocity (u,v) given by

$$\frac{d\phi}{dt} = \frac{\partial \phi}{\partial t} + u\frac{\partial \phi}{\partial x} + v\frac{\partial \phi}{\partial y}$$

The primitive volume equation then becomes

$$\left(\frac{\partial h}{\partial t} + u\frac{\partial h}{\partial x} + v\frac{\partial h}{\partial y}\right) + h\left(\frac{\partial u}{\partial x} + \frac{\partial v}{\partial y}\right) = 0$$

while the corresponding primitive acceleration equations are

$$\left(\frac{\partial u}{\partial t} + u\frac{\partial u}{\partial x} + v\frac{\partial u}{\partial y}\right) + g\frac{\partial h}{\partial x} = 0 \tag{2.6.4}$$

$$\left(\frac{\partial v}{\partial t} + u\frac{\partial v}{\partial x} + v\frac{\partial v}{\partial y}\right) + g\frac{\partial h}{\partial y} = 0 \tag{2.6.5}$$

The Eulerian forms are obtained from the conservation forms by differentiating these out and, when necessary, simplifying. Thus Equations (2.6.2) and (2.6.3) are just (2.1.2) and (2.1.10) differentiated out.

The general form of the one-dimensional equations of nearly horizontal flow in a Eulerian frame of reference is seen to be

$$\frac{\partial f}{\partial t} + A\frac{\partial f}{\partial x} = 0 \tag{2.6.6}$$

where it is seen, referring to Equation (2.1.8), that A has elements a_{ij} given by $a_{ij} = \partial g_i(f)/\partial f_j$. For example, for vertically homogeneous flows,

$$\text{with } f = \begin{bmatrix} h \\ u \end{bmatrix}, \qquad A = \begin{bmatrix} u & h \\ g & u \end{bmatrix} \tag{2.6.7}$$

$$f = \begin{bmatrix} h \\ uh \end{bmatrix}, \qquad A = \begin{bmatrix} 0 & 1 \\ (gh - u^2) & 2u \end{bmatrix} \tag{2.6.8}$$

All such forms are, of course, equivalent, in that the one pair of equations is simply a non-linear transformation of any other pair, just as for the conservation forms.

In the same way, for a primitive one-dimensional two-layered flow, Equations (2.5.1), (2.5.2), (2.5.7) and (2.5.8) provide (not necessarily from a conservation form):

$$\text{with } f = \begin{bmatrix} h_0 \\ h_1 \\ u_0 \\ u_1 \end{bmatrix}, \qquad A = \begin{bmatrix} u_0 & 0 & h_0 & 0 \\ 0 & u_1 & 0 & h_1 \\ g & \sigma g & u_0 & 0 \\ g & g & 0 & u_1 \end{bmatrix} \tag{2.6.9}$$

and

$$\text{with } f = \begin{bmatrix} h_0 \\ h_1 \\ u_0 h_0 \\ u_1 h_1 \end{bmatrix}, \qquad A = \begin{bmatrix} 0 & 0 & 1 & 0 \\ 0 & 0 & 0 & 1 \\ (gh_0 - u_0^2) & \sigma g h_0 & 2u_0 & 0 \\ gh_1 & (gh_1^2 - u_1^2) & 0 & 2u_1 \end{bmatrix} \tag{2.6.10}$$

The primitive equations of two-dimensional, vertically homogeneous, nearly horizontal flow can be expressed in the form

$$\frac{\partial f}{\partial t} + A_1 \frac{\partial f}{\partial x} + A_2 \frac{\partial f}{\partial y} = 0 \tag{2.6.11}$$

with, for example,

$$f = \begin{bmatrix} u \\ h \\ v \end{bmatrix}, \qquad A_1 = \begin{bmatrix} u & g & 0 \\ h & u & 0 \\ 0 & 0 & u \end{bmatrix}, \qquad A_2 = \begin{bmatrix} v & 0 & 0 \\ 0 & v & h \\ 0 & g & v \end{bmatrix} \tag{2.6.12}$$

corresponding to Equation (2.4.2) written in Eulerian form and the corresponding Euler equations (2.6.4), (2.6.5).

Now for the purpose of reducing the coefficients of the unknowns of the difference schemes corresponding to these equations to a diagonal matrix form (as discussed in detail in Section 4.7), an *algorithmic* form is introduced:

$$A^{-1} \frac{\partial f}{\partial t} + \frac{\partial f}{\partial x} = 0, \qquad \text{Det } A \neq 0 \tag{2.6.13}$$

for one-dimensional forms, or

$$A_1^{-1}\frac{\partial f}{\partial t}+\frac{\partial f}{\partial x}+A_1^{-1}A_2\frac{\partial f}{\partial y}=0, \qquad \text{Det } A_1 \neq 0$$

$$A_2^{-1}\frac{\partial f}{\partial t}+A_2^{-1}A_1\frac{\partial f}{\partial x}+\frac{\partial f}{\partial y}=0, \qquad \text{Det } A_2 \neq 0 \tag{2.6.14}$$

for the alternating direction approach to two-dimensional flows, that will be presented in Chapter 4. Evidently there is a large set of possible algorithmic forms and any subset of equations sufficient to define a solution is equivalent to any other such subset, in the differential formulation.

The Eulerian forms are often used in a *linearized form* for situations where $h = h_0 + h'$ with h_0 constant and $h' \ll h_0$ or, correspondingly, $u \ll gh$. The primitive equations (2.6.2) and (2.6.3) then reduce to

$$\frac{\partial h'}{\partial t}+h_0\frac{\partial u}{\partial x}=0 \tag{2.6.15}$$

$$\frac{\partial u}{\partial t}+g\frac{\partial h'}{\partial x}=0 \tag{2.6.16}$$

which in turn imply that

$$\frac{\partial^2 h'}{\partial t^2}-c^2\frac{\partial^2 h'}{\partial x^2}=0 \tag{2.6.17}$$

$$\frac{\partial^2 u}{\partial t^2}-c^2\frac{\partial^2 u}{\partial x^2}=0 \tag{2.6.18}$$

with $c^2 = gh_0$.

In two-dimensional flow, the volume equation reduces to

$$\frac{\partial h'}{\partial t}+h_0\left(\frac{\partial u}{\partial x}+\frac{\partial v}{\partial y}\right)=0 \tag{2.6.19}$$

while the Euler equations (2.6.4) and (2.6.5) become

$$\frac{\partial u}{\partial t}+g\frac{\partial h'}{\partial x}=0 \tag{2.6.20}$$

$$\frac{\partial v}{\partial t}+g\frac{\partial h'}{\partial y}=0 \tag{2.6.21}$$

These in turn provide

$$\frac{\partial^2 h'}{\partial t^2}-c^2\left(\frac{\partial^2 h'}{\partial x^2}+\frac{\partial^2 h'}{\partial y^2}\right)=0 \tag{2.6.22}$$

2.7 Eulerian forms of Boussinesq equations

The assumption that flows are nearly horizontal, which is used in most of this book, corresponds to the assumption that vertical accelerations are negligible or the pressure distribution is hydrostatic. On the basis of this assumption, vertically integrated equations are obtained (e.g. Equations (2.6.1) and (2.6.2)) that are particularly amenable to computation, as outlined in the succeeding chapters. However, the process of vertical integration of the more general, three-dimensional equations of hydrodynamics is by no means limited to nearly horizontal flow processes, but can as well be extended to flows where some vertical acceleration is allowed. One of the simplest and most widely applicable of such extensions is to the case where the magnitude of the vertical velocity is supposed to increase linearly from zero at the bed to a maximum at the free surface. In the case of such a flow, it is easily seen that the mass and volume conservation laws (Equations (2.1.1) and (2.1.2)) and hence the Eulerian form (2.6.2) remain unchanged. The momentum-impulse equation, however, is changed by virtue of the superposition of the effect forces induced by the vertical accelerations now introduced upon the hydrostatic pressure forces. Any extended momentum-impulse equation that accounts for the vertical accelerations in this simplest way is called a *Boussinesq equation* (Boussinesq, 1872).

There are many procedures for deriving the Boussinesq equations and almost as many resultant forms of these equations in consequence. The main reasons for this variety have been given by Peregrine (1974). In the simplest and therefore first place, there is a variety created by the choice of variables that may be used in the equations. For example, the velocity used may be a depth-averaged value, the value at still-water level or the value at the surface. In the second place there is the quite fundamental observation of Long (1964) that *one form of the Boussinesq equation can be transformed into another form simply by use of the nearly horizontal flow equations*. This property is discussed further in Section 4.12.

In the third place come various possible restrictive conditions upon the general Boussinesq equation. Of these restrictions, the most notable is that propagation is restricted to one direction only, leading in one case to the Korteweg and de Vries (1895) type of equation. Some consequences of this last restriction are investigated in Section 3.7.

For the present introductory purposes, an 'engineering derivation' of the Boussinesq equation, along the lines used in the rest of this work, seems most appropriate. This will be restricted to the simplest case of one-dimensional irrotational flow of a homogeneous fluid over a horizontal bed. As stated above, it is assumed that the magnitude of the vertical velocity w increases linearly from zero at the bed to a maximum w_s at the surface, where then

$$w_s = \frac{dh}{dt} = \frac{\partial h}{\partial t} + u\frac{\partial h}{\partial x}$$

Neglecting the convective term $u\,\partial h/\partial x$ gives

$$w = w(z) = \frac{\partial h}{\partial t}\frac{z}{h} \tag{2.7.1}$$

The vertical component of the Euler equation reads

$$\frac{dw}{dt} = -\frac{1}{\rho}\frac{\partial p}{\partial z} - g$$

which again reduces, through neglect of the convective terms, to

$$\frac{\partial w}{\partial t} = -\frac{1}{\rho}\frac{\partial p}{\partial z} - g \tag{2.7.2}$$

Introducing Equation (2.7.1) into (2.7.2) provides

$$\frac{z}{h}\frac{\partial^2 h}{\partial t^2} - \frac{z}{h^2}\left(\frac{\partial h}{\partial t}\right)^2 = -\frac{1}{\rho}\frac{\partial p}{\partial z} - g \tag{2.7.3}$$

Following a methodology of Boussinesq (1872), the powers of derivatives are neglected as compared with derivatives themselves, so that Equation (2.7.3) reduces to

$$\frac{z}{h}\frac{\partial^2 h}{\partial t^2} = -\frac{1}{\rho}\frac{\partial p}{\partial z} - g \tag{2.7.4}$$

Integrating Equation (2.7.4) from the free surface, $z = h$, where $p = 0$, to an elevation z, provides the pressure $p = p(z)$:

$$\frac{p(z)}{\rho} = g(h - z) + \frac{\partial^2 h}{\partial t^2}\frac{(h^2 - z^2)}{2h} \tag{2.7.5}$$

The horizontal component of the Euler equation·

$$\frac{du}{dt} = -\frac{1}{\rho}\frac{\partial p}{\partial x}$$

can then be introduced in equation (2.7.5) to give the equivalent depth-local Euler equation

$$\left.\frac{du}{dt}\right|_z = -g\frac{\partial}{\partial x}(h - z) - \left(\frac{\partial^3 h}{\partial x\,\partial t^2}\;\frac{(h^2 - z^2)}{2h}\right) - \left(\frac{\partial^2 h}{\partial t^2}\frac{\partial}{\partial x}\left(\frac{h^2 - z^2}{2h}\right)\right) \tag{2.7.6}$$

This equation must then be integrated over the total depth of the fluid in order to provide an equation in the depth-averaged velocity $\overline{u}$ given by

$$\overline{u} = \frac{1}{h}\int_0^h u \, dz$$

Again neglecting the product of derivatives in Equation (2.7.6), the integration takes the form

$$\frac{d\overline{u}}{dt} = \frac{1}{h}\int_0^h \left(-g\frac{\partial}{\partial x}(h - z) - \frac{\partial^3 h}{\partial x\,\partial t^2}\;\frac{(h^2 - z^2)}{2h}\right) dz$$

$$= -g\frac{\partial h}{\partial x} - \frac{h}{3}\frac{\partial^3 h}{\partial x\,\partial t^2}$$

or

$$\frac{\partial \bar{u}}{\partial t} + \bar{u}\frac{\partial \bar{u}}{\partial x} + g\frac{\partial h}{\partial x} + \frac{h}{3}\frac{\partial^3 h}{\partial x\,\partial t^2} = 0 \tag{2.7.7}$$

It follows, from the linearized nearly horizontal flow equation (2.6.15):

$$\frac{\partial h}{\partial t} + h \cdot \frac{\partial \bar{u}}{\partial x} = 0$$

that

$$\frac{\partial^3 h}{\partial t^2\,\partial x^2} = -h\frac{\partial^3 \bar{u}}{\partial t\,\partial x^2}$$

so that Equation (2.7.7) can as well be written (with the same order of approximation as before) as

$$\frac{\partial \bar{u}}{\partial t} + \bar{u}\frac{\partial \bar{u}}{\partial x} + g\frac{\partial h}{\partial x} - \frac{h^2}{3}\frac{\partial^3 \bar{u}}{\partial x^2\,\partial t} = 0 \tag{2.7.8}$$

It is seen that the Boussinesq equation differs from the Euler equation of nearly horizontal flow by the last third-order term of Equation (2.7.7), or (2.7.8). More rigorous derivations have been made by Abbott and Rodenhuis (1972) and Peregrine (1972), following an approach initiated by Friedrichs (1948).

In two dimensions, the volume and Boussinesq equations generalize, for a variable bed with elevation H, to

$$\frac{\partial h}{\partial t} + \frac{\partial}{\partial x}(uh) + \frac{\partial}{\partial x}(vh) = 0 \tag{2.7.9}$$

$$\frac{\partial u}{\partial t} + u\frac{\partial u}{\partial x} + v\frac{\partial u}{\partial y} + g\frac{\partial(h+H)}{\partial x} = \frac{h}{2}\left(\frac{\partial^3(uh)}{\partial x^2\,\partial t} + \frac{\partial^3(vh)}{\partial x\,\partial y\,\partial t}\right) - \frac{h}{6}\left(\frac{\partial^3 u}{\partial x^2\,\partial t} + \frac{\partial^3 v}{\partial x\,\partial y\,\partial t}\right) \tag{2.7.10}$$

$$\frac{\partial v}{\partial t} + u\frac{\partial v}{\partial x} + v\frac{\partial v}{\partial y} + g\frac{\partial(h+H)}{\partial y} = \frac{h}{2}\left(\frac{\partial^3(vh)}{\partial y^2\,\partial t} + \frac{\partial^3(uh)}{\partial x\,\partial y\,\partial t}\right) - \frac{h}{6}\left(\frac{\partial^3 v}{\partial y^2\,\partial t} + \frac{\partial^3 u}{\partial x\,\partial y\,\partial t}\right) \tag{2.7.11}$$

where the bar over the velocity denoting a depth-average value is suppressed (Peregrine, 1967).

It may be shown (Abbott, Petersen and Skovgaard, 1978) that the Boussinesq equations have a very wide range of application, providing especially good approximations to short waves with water depth to wave length ratios of up to 0.2 (also for waves not of permanent form) and for ratios of wave height to wave length (wave steepness) up to the conditions of the breaking zone. When provided with an appropriate dissipative mechanism (Sections 4.6 and 5.8), they can also be used to approximate the breaker zone as well, providing 'wave thrusts' (Lundgren, 1962) or 'radiation stresses' (Longuet-Higgins and Stewart, 1960, 1962) and thus wave-induced currents (Abbott, Petersen and Skovgaard, 1978).

3 The method of characteristics

3.1 The role of the method of characteristics in computational hydraulics

The method of characteristics is a method of solving the equations that have been developed in the last chapter and, in principle at least, such solutions can all be realized using a digital computer. However, it is usual in practice to use other methods of solution with a digital machine, and primarily the finite difference methods that are the subject of the next chapter.

When using difference methods, the method of characteristics comes to play another role, essentially that of guiding and steering calculations. The present chapter is first intended to present the essentials of the method, as a means of calculation, but in such a way as to lead into the steering and control applications of the method in Chapter 4. At the same time, some fundamental notions of finite computational schemes can be introduced, for further use in Chapter 5.

3.2 Characteristics and invariants

In Chapter 1, the four types of hydraulic jump illustrated in Fig. 3.1 were identified. It was there shown that, among these types there are jumps of a permanent kind and jumps that are essentially impermanent. It was seen that these types can be most conveniently identified within a Galilean frame travelling with the jump and in fact the entirely general condition was formulated that, in a jump of permanent type, the flow must appear to pass from a supercritical to a subcritical flow.

Putting aside this question of permanence of type for the moment, consider simply the limiting cases of all the jumps illustrated in Fig. 3.1, which limiting cases may be schematized in an $(h;x,t)$ space as shown in Fig. 3.2. In the latter figure each jump running from left to right is associated with its 'symmetric' or 'conjugate' jump running from right to left. A similar figure could be drawn for the velocity, i.e. in the $(u;x,t)$ space. Evidently the situations depicted in Fig. 3.2 could be realized by introducing a constant infinitesimal source or sink, initially situated at $x = 0$ and subsequently travelling with the stream velocity. Equations

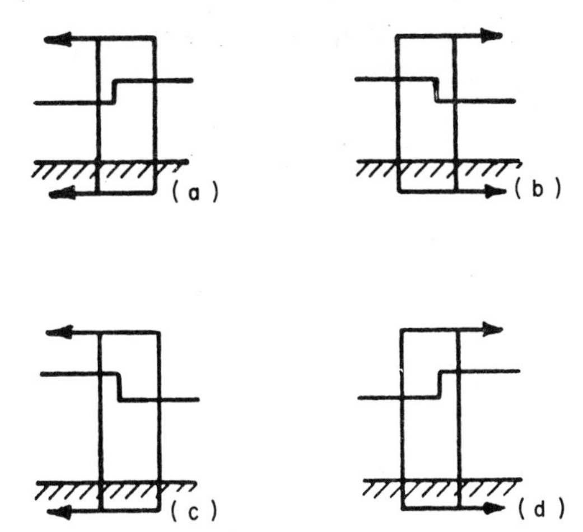

Fig. 3.1. The four configurations of an elementary hydraulic jump

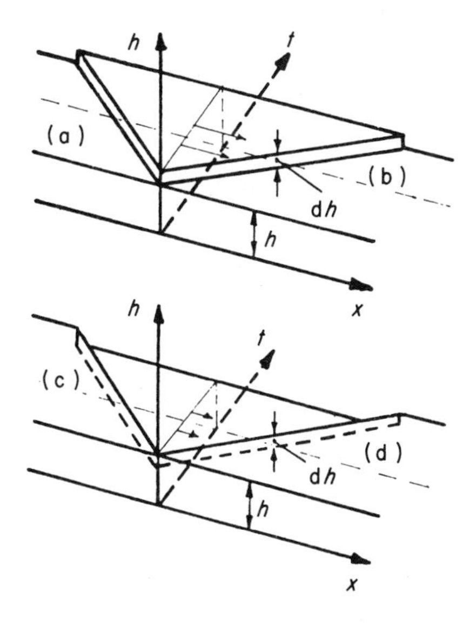

Fig. 3.2. Three-dimensional representation of the four elementary configurations

(1.2.3) provide the following relations between the celerity of the jump c and the dependent variables u and h as $u_1, u_2 \to u, h_1, h_2 \to h$:

$$(c-u_1)^2, \to (c-u)^2, = \frac{gh_2(h_1+h_2)}{2h_1}, \to gh$$

$$(c-u_2)^2, \to (c-u)^2, = \frac{gh_1(h_1+h_2)}{2h_2}, \to gh$$

i.e. the flow converges, within a frame travelling with the jump, from the supercritical side and from the subcritical side upon a critical flow. The infinitesimal jump celerities are thus given by

$$c_{\pm} = u \pm (gh)^{\frac{1}{2}} \tag{3.2.1}$$

This result is schematized in Fig. 3.3, from which the symmetry of the result clearly appears. Across the infinitesimal jump, $\mathrm{d}h$ and $\mathrm{d}u$ may be related as follows:
The first equation of (1.4.4) gives

$$\mathrm{d}(uh) = c\ \mathrm{d}h$$

whence, from Equation (3.2.1)

$$\mathrm{d}(uh) = (u \pm (gh)^{\frac{1}{2}})\ \mathrm{d}h$$

However,

$$\mathrm{d}(uh) = u\ \mathrm{d}h + h\ \mathrm{d}u$$

Comparing these expressions for $\mathrm{d}(uh)$ it is seen that:

(a) across an infinitesimal jump travelling with celerity $c = u + (gh)^{\frac{1}{2}}$,

$$\mathrm{d}u = +\left(\frac{g}{h}\right)^{\frac{1}{2}} \mathrm{d}h$$

or (3.2.2)

$$\mathrm{d}u - \left(\frac{g}{h}\right)^{\frac{1}{2}} \mathrm{d}h = 0, \qquad \text{or} \qquad \mathrm{d}u - \mathrm{d}(2(gh)^{\frac{1}{2}}) = 0$$

(b) across an infinitesimal jump travelling with celerity $c = u - (gh)^{\frac{1}{2}}$,

$$\mathrm{d}u = -\left(\frac{g}{h}\right)^{\frac{1}{2}} \mathrm{d}h$$

or (3.2.3)

$$\mathrm{d}u + \left(\frac{g}{h}\right)^{\frac{1}{2}} \mathrm{d}h = 0, \qquad \text{or} \qquad \mathrm{d}u + \mathrm{d}(2(gh)^{\frac{1}{2}}) = 0$$

The flow situation of Fig. 3.2, with $\mathrm{d}u$ and $\mathrm{d}h$ related as in Equations (3.2.2) and (3.2.3), constitutes a differential form of 'fundamental solution' of the nearly horizontal flow problem, in that all other solutions can be built up from it. The trace of an infinitesimal jump (step, or discontinuity) of the type illustrated in Fig. 3.2 is called a 'characteristic'. The projection of this trace onto the x–t plane is properly called a *base characteristic*, but it is usually again called simply 'a characteristic'. It is seen that two characteristics may be associated with any point in the x–t plane, corresponding to one infinitesimal jump running from left to right and another such jump running from right to left.

The special computational significance of the characteristics appears when

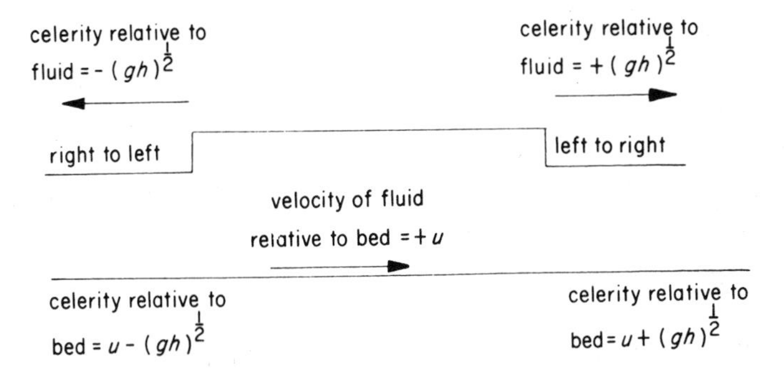

Fig. 3.3. Relativization of the celerities $\pm (gh)^{\frac{1}{2}}$ by the flow velocity u to provide the characteristic directions $u \pm (gh)^{\frac{1}{2}}$

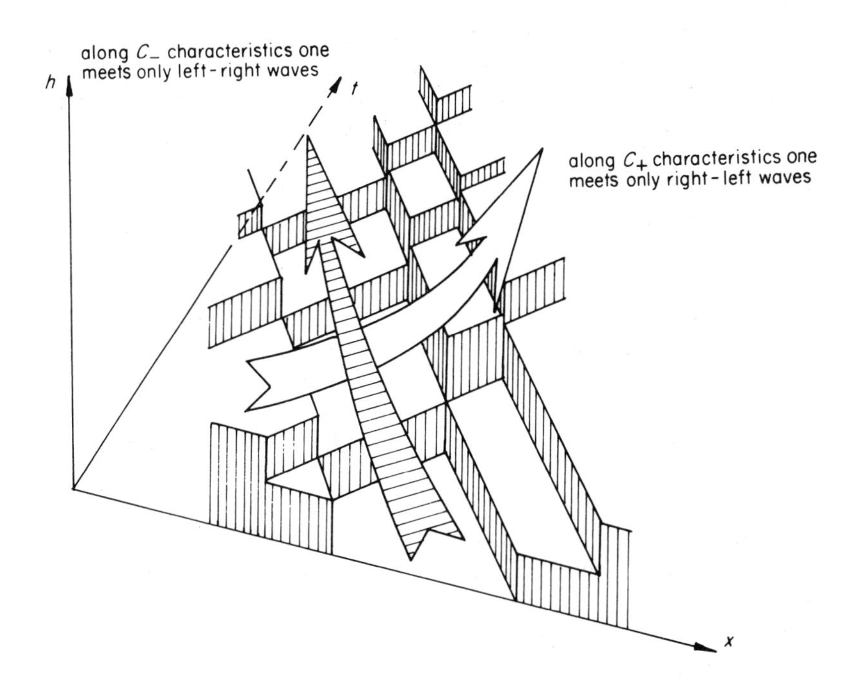

Fig. 3.4. Three-dimensional representation of the physical basis of the method of characteristics: when following the paths of jumps propagating from right to left only jumps propagating from left to right are encountered and when following the paths of jumps propagating from left to right only jumps propagating from right to left are encountered

they are considered together as schematized in Fig. 3.4. It is seen from Fig. 3.4 that when proceeding along a C_+ characteristic only elementary jumps of the C_- type (travelling from right to left relative to the stream) are encountered. It is then possible to integrate all infinitesimal elements in Equation (3.2.3) along this characteristic in order to obtain

$$u + 2(gh)^{\frac{1}{2}} = \text{const.}, \quad = J_+ \tag{3.2.4}$$

along a C_+ characteristic (i.e. one with $c_+ = u + (gh)^{\frac{1}{2}}$). Similarly, proceeding along a C_- characteristic only elementary jumps of the C_+ type (travelling from left to right relative to the stream) are encountered, so that

$$u - 2(gh)^{\frac{1}{2}} = \text{const.}, \quad = J_- \tag{3.2.5}$$

along a C_- characteristic (i.e. characteristic with $c_- = u - (gh)^{\frac{1}{2}}$). The quantities J_+ and J_- that are invariant along C_+ and C_- characteristics respectively, are called the *Riemann invariants* of the fluid motion. As is shown in the next section, the *characteristic directions* $\mathrm{d}x/\mathrm{d}t_\pm = c_\pm$, as given in Equation (3.2.1), and the Riemann invariant relations, as given in Equations (3.2.4) and (3.2.5), provide an exceedingly powerful method for solving problems of nearly horizontal flow. This method is called *the method of characteristics*.

In a work that is so concerned with the relation between the continuous and the discrete views of processes, it is instructive to consider the above process in a little more detail. Fundamental solutions were constructed and it was implicitly supposed that any variations in $h = h(x,t)$ and $u = u(x,t)$ that satisfied the conservation laws could be approximated as closely as required by 'staircases' of fundamental solutions. It was then found that, proceeding along either one of the characteristic directions, only steps in the conjugate characteristic direction were encountered. Thus, by following either one of the characteristic directions, all variations of one type (the conjugate type) were 'filtered out', so to speak, from variations of the other type (the type of the direction of travelling). By virtue of this filtering process, it was possible simply to integrate (in effect, 'add up an infinite number of times') to obtain the Riemann invariants. This is a process of proceeding from the discrete and countable to the continuous and, correspondingly, uncountable, by 'formally' integrating $\mathrm{d}u$ and $(g/h)^{\frac{1}{2}}\,\mathrm{d}h = \mathrm{d}(2(gh)^{\frac{1}{2}})$. Although all this is probably sufficiently rigorous (i.e. invoking sufficient conviction or credulity) for most, it is interesting to follow the discrete view somewhat further.

Consider a system of 'staircase' constructions, the first of which approximates the true solution surface in the $(h,u;x,t)$ space with only a few steps. The next construction uses twice as many steps to approximate the system, the next twice as many again, and so on. If the approximation is chosen in a reasonable way, sections along any jump path (which becomes a characteristic path in the limit) will appear as in Fig. 3.5, such that the approximation improves monotonically as one proceeds from a finer to finer 'net'. By taking a sufficiently fine net, the path can be approximated to any desired accuracy, even when the steps are taken as equal in magnitude.

The jth jump in h is denoted by $\Delta_j h$ and the corresponding jth jump in $2(gh)^{\frac{1}{2}}$ by $\Delta_j 2(gh)^{\frac{1}{2}}$ and in u by $\Delta_j u$ where, of course, $\Delta_j (gh)^{\frac{1}{2}}$ and $\Delta_j u$ may take positive or negative values. Then, for any path, across the transformed space

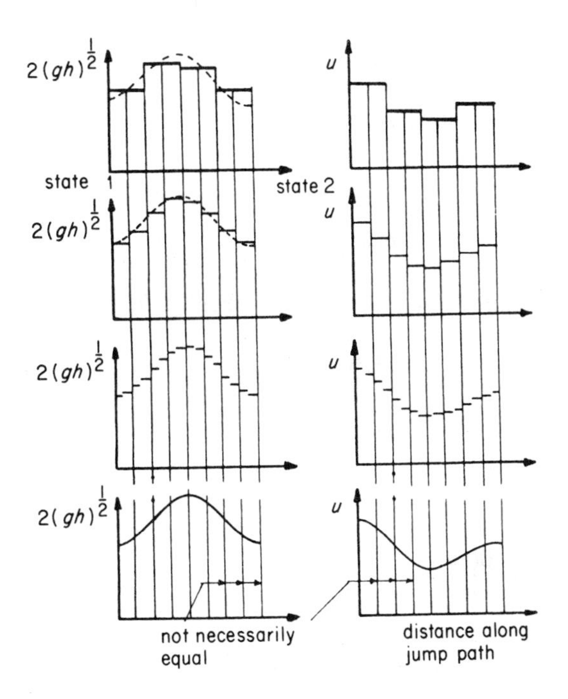

Fig. 3.5. Staircase schematization of the variation of flow properties along a jump path, illustrating the process of refinement of the net

$(2(gh)^{\frac{1}{2}}, u; x, t)$, between net-resolved states $(u_1, 2(gh_1)^{\frac{1}{2}})$ and $(u_2, 2(gh_2)^{\frac{1}{2}})$

$$u_1 + \sum_{j=1}^{N} \Delta_j u = u_2$$

and

$$2(gh_1)^{\frac{1}{2}} + \sum_{j=1}^{N} \Delta_j 2(gh)^{\frac{1}{2}} = 2(gh_2)^{\frac{1}{2}}$$

Each of the $\Delta_j u$ or $\Delta_j 2(gh)^{\frac{1}{2}}$ behaves as a 'weight' or 'measure', which is added in as one proceeds along any path. Thus in Fig. 3.5 a positive $\Delta_j u$ would cause an increase by $\Delta_j u$ in u as one passes from left to right, in that case along the jump path. However, if in fact the jump path is followed, then Equations (3.2.2) and (3.2.3) imply that (*see also* Section 3.4):

$$\Delta_j u = \pm \Delta_j 2(gh)^{\frac{1}{2}} + 0((\Delta_j h)^2, (\Delta_j h\ \Delta_j u)) = \pm \Delta_j 2(gh)^{\frac{1}{2}} + 0((\Delta_j h)^2)$$

Thus, if the limit of the jump path is a C_+ characteristic, so that only steps of the C_- type are encountered

$$\Delta_j u = -\Delta_j 2(gh)^{\frac{1}{2}} + 0(\Delta_j h)^2$$

and, therefore,

$$u_1 - u_2 = \sum_j \Delta_j u = -\sum_j \Delta_j 2(gh)^{\frac{1}{2}} + \sum_j 0((\Delta_j h)^2) = 2(gh_2)^{\frac{1}{2}} -$$

$$- 2(gh_1)^{\frac{1}{2}} + \sum_j 0((\Delta_j h)^2)$$

where the sum is taken over all steps between 1 and 2. Now without loss of generality, all steps can as well be taken as equal in magnitude so that, *so long as they remain distinct*, (i.e. so long as they do not coalesce during net refinement, which is to say that *the approximated flow is continuous*), then $|\Delta_j h| = \Delta h$ for every j and $0((\Delta_j h)^2) = 0(\Delta h)^2$ for every j and $\sum_{j=0}^{J} 0((\Delta_j h)^2) = J \cdot 0(\Delta h)^2$.

But, under this same restriction of flow continuity,

$$J = 0\left[\frac{h_{\text{sup}} - h_{\text{inf}}}{\Delta h}\right] = \frac{(h_{\text{sup}} - h_{\text{inf}})}{0(\Delta h)}$$

so that $\Sigma 0((\Delta_j h)^2) = (h_{\text{sup}} - h_{\text{inf}}) \dfrac{0(\Delta h)^2}{0(\Delta h)}, \rightarrow 0$ as $\Delta h \rightarrow 0$.

Thence

$$u_1 + 2(gh_1)^{\frac{1}{2}} \rightarrow u_2 + 2(gh_2)^{\frac{1}{2}} \quad \text{or} \quad [u + 2\,(gh)^{\frac{1}{2}}\,]_1{}^2 \rightarrow 0$$

The construction along the C_- characteristic proceeds in exactly the same way.

The above construction can be repeated at the differential level: once again, following lines $\dot{x} = u \pm (gh)^{\frac{1}{2}}$, only the variations of the conjugate directions will appear. For example, starting out from the Eulerian forms (2.6.2) and (2.6.3),

$$\frac{\partial h}{\partial t} + u\frac{\partial h}{\partial x} + h\frac{\partial u}{\partial x} = 0$$

$$\frac{\partial u}{\partial t} + u\frac{\partial u}{\partial x} + g\frac{\partial h}{\partial x} = 0$$

a transformation to $\dot{x} = u \pm (gh)^{\frac{1}{2}}$ can be made by multiplying the first equation by g, introducing $c = (gh)^{\frac{1}{2}}$, differentiating out for c and simplifying,

$$\frac{\partial c^2}{\partial t} + u\frac{\partial c^2}{\partial x} + c^2\frac{\partial u}{\partial x} = 2c\frac{\partial c}{\partial t} + 2uc\frac{\partial c}{\partial x} + c^2\frac{\partial u}{\partial x} = 0$$

or

$$2\frac{\partial c}{\partial t} + 2u\frac{\partial c}{\partial x} + c\frac{\partial u}{\partial x} = 0$$

entering $gh = c^2$ and differentiating out in the second equation,

$$\frac{\partial u}{\partial t} + u\frac{\partial u}{\partial x} + 2c\frac{\partial c}{\partial x} = 0$$

and then alternately adding and subtracting these reduced equations, provides:

$$\frac{\partial}{\partial t}(u + 2c) + (u + c)\frac{\partial}{\partial x}(u + 2c) = 0$$
$$\frac{\partial}{\partial t}(u - 2c) + (u - c)\frac{\partial}{\partial x}(u - 2c) = 0 \qquad (3.2.6)$$

These are then indeed seen to have the form of total differentials:

$$\mathrm{d}(u + 2c) = \frac{\partial}{\partial t}(u + 2c)\,\mathrm{d}t + \frac{\partial}{\partial x}(u + 2c)\,\mathrm{d}x$$

$$\mathrm{d}(u - 2c) = \frac{\partial}{\partial t}(u - 2c)\,\mathrm{d}t + \frac{\partial}{\partial x}(u - 2c)\,\mathrm{d}x$$

Thus, along directions $\mathrm{d}x/\mathrm{d}t = u + c, = u + (gh)^{\frac{1}{2}}$,

$$\mathrm{d}(u + 2c) = \mathrm{d}(u + 2(gh)^{\frac{1}{2}}) = 0$$

or

$$u + 2(gh)^{\frac{1}{2}} = \text{const. or } [u + 2\,(gh)^{\frac{1}{2}}]_1^{\,2} = 0$$

while along directions $\mathrm{d}x/\mathrm{d}t = u - c, = u - (gh)^{\frac{1}{2}}$,

$$\mathrm{d}(u - 2c) = \mathrm{d}(u - 2(gh)^{\frac{1}{2}}) = 0$$

or

$$u - 2(gh)^{\frac{1}{2}} = \text{const. or } [u - 2\,(gh)^{\frac{1}{2}}]_1^{\,2} = 0$$

The equivalence of the discrete and continuous representations in the formulation of the method of characteristics can be rounded off by considering what happens to the energy defect in the limiting case discussed above. After all, as the magnitude of each jump tends to zero so the number of jumps tends to infinity. How does the total energy defect then tend to zero, as is required by the equivalent continuum representation?

For the purpose of this investigation it suffices to consider the construction of Fig. 3.4 in the case of a unidirectional wave distributed over any given interval. This wave can be divided into sub-intervals in x such that h varies monotonically across each sub-interval, as shown in Fig. 3.6. If the flow is everywhere a nearly

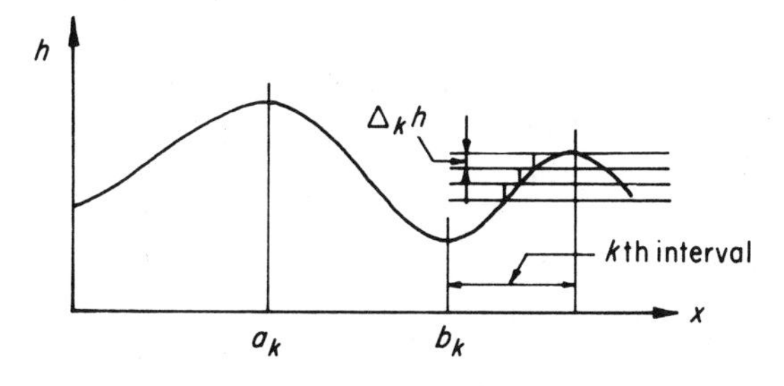

Fig. 3.6. Staircase schematization of the variation of flow properties along a jump path used in the computation of total energy defect

horizontal flow, the number of these sub-intervals is clearly finite over any given finite interval so that they can be identified by the sequence $1, 2, \ldots, k, \ldots, K$. The flow is again approximated over each sub-interval by a sequence of hydraulic jumps and intervening horizontal flows. The approximation can evidently be made in many ways, but it is convenient here to take equal step heights over each of the monotonic sub-intervals, so that each step height $\Delta_k h$ is an aliquot part of the total height variation over the kth sub-interval and in such a way that the bottom-most point of each step coincides with the approximated function $h = h(x, t)$. If J_k is the number of elementary jumps in the kth sub-interval, the total rate of energy loss to the nearly horizontal flow, $\mathrm{d}E/\mathrm{d}t$, is always less than

$$\sum_{k=1}^{K} \sum_{j=1}^{J_k} \left(\frac{\rho g |u_j|}{4h_j}\right)_k |\Delta_k h|^3$$

since in this expression all energy defects are counted as positive, even for 'negative' elementary hydraulic jumps and no jumps coalesce. However, if $[a_k, b_k]$ is the support in x of the kth monotonic sub-interval

$$\frac{\mathrm{d}E}{\mathrm{d}t} < \sum_{k=1}^{K} \sum_{j=1}^{J_k} \left(\frac{\rho g |u_j|}{4h_j}\right)_k |\Delta_k h|^3 \leqslant \sum_{k=1}^{K} \left[\operatorname*{Max}_j \left(\frac{\rho g |u_j|}{4h_j}\right)\right]_k J_k |\Delta_k h|^3$$

$$= \sum_{k=1}^{K} \left[\operatorname*{Max}_j \left(\frac{\rho g |u_j|}{4h_j}\right)\right]_k \left| h\big|_{a_k} - h\big|_{b_k} \right| (\Delta_k h)^2$$

$$\leqslant \operatorname*{Max}_{a_1 < x < b_K} \left(\frac{\rho g |u(x)|}{4h(x)}\right) \sum_{k=1}^{K} \left| h\big|_{a_k} - h\big|_{b_k} \right| (\Delta_k h)^2$$

$$\leqslant K \operatorname*{Max}_{a_1 < x < b_K} \left(\frac{\rho g |u(x)|}{4h(x)}\right) \operatorname*{Max}_k \left[\left| h\big|_{a_k} - h\big|_{b_k} \right| (\Delta_k h)^2\right]$$

Thence for any finite K, i.e. any finite interval, the energy defect tends to zero at the rate $(\Delta h)^2$ as $\Delta h \to 0$. Since K is generally no function of Δh, this property holds for arbitrary K and so holds as well for an infinite interval. That the sequence of step functions shown in Fig. 3.6 is capable of converging to any required function in the limit, $J_k \to \infty$ for every K, corresponds to a central theorem of measure theory, and is, of course, independent of the assumption of continuity of the approximated function. (In measure theory, the step functions that are superimposed to provide the constructions of Figs 3.4 and 3.6 are called characteristic functions (e.g. Weir, 1973). This terminology, of course, is essentially independent of the present use of the term 'characteristics'.) This is indeed as expected, since any such limit must correspond to an ideal fluid in nearly horizontal motion, as described by the primitive differential equations (*see* Chapter 2), and this is energy-conserving.

This latter type of argument, through sets of measures, ordered and added in various ways, makes no demand upon the assumption or even the notion of differentiability. At the same time, it also emphasizes the essential restriction of the method of characteristics to continuous flow processes.

3.3 Regions of state

The method of characteristics will be described in terms of the following example. A uniform canal of length 400 m initially contains static water of depth 5 m. Fluid is discharged from one end of the canal according to the following quantity/time programme:

$$k = \begin{cases} -\dfrac{t}{10}\ \mathrm{m^2\,s^{-1}}* & 0 \leqslant t \leqslant 60 \\ -\left(6 - \dfrac{(t-60)}{10}\right)\ \mathrm{m^2\,s^{-1}}* & 60 \leqslant t \leqslant 80 \\ -4\ \mathrm{m^2\,s^{-1}}* & 80 \leqslant t < \infty \end{cases}$$

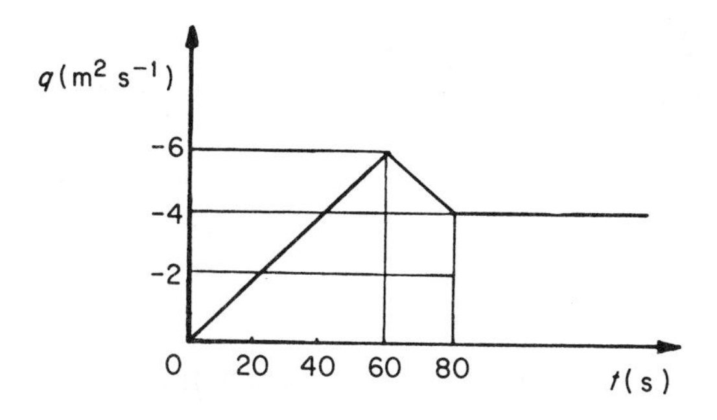

Fig. 3.7. Programme of variation of discharge with time at one end of the canal

This programme is illustrated in Fig. 3.7. The problem is posed of determining flow conditions in the canal for all $t \geqslant 0$ assuming that the other end of the canal is closed. Consider first the $x-t$ plane shown in Fig. 3.8. It is seen that the first intimation of any change at the discharging end travels along the C_+ characteristic AB. The region AOB of the $x-t$ plane then comprises all distance–time pairs not influenced by the disturbance. This property is schematized for one distance along the canal and also for one time in Fig. 3.8. Clearly the region AOB is a region in which h remains constant (at 5 m), and u remains constant (at 0 m s^{-1}), i.e. it is a *region of constant state*.

The celerity of the characteristic AB is that of the region AOB, i.e.

$$\left(\frac{dx}{dt}\right)_{AB} = u + (gh)^{\frac{1}{2}} = 0 + (9.81 \times 5)^{\frac{1}{2}} \approx 7\ \mathrm{m\,s^{-1}}$$

* i.e., cubic metres per second per metre width of channel.

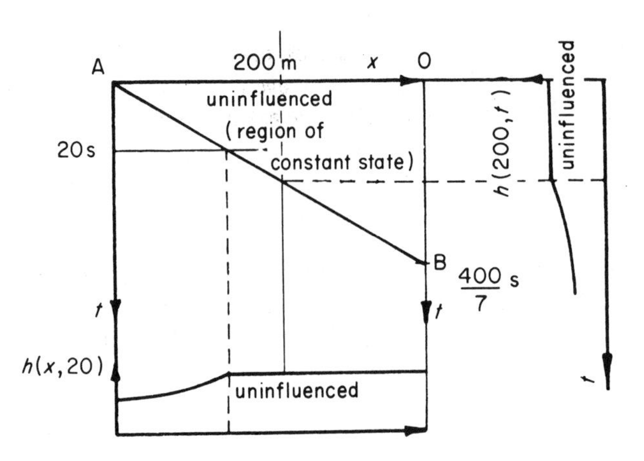

Fig. 3.8. Formation of the region of constant state

so that B is situated at (400/7) s.

Now consider conditions along the discharging end (Fig. 3.9). Following a C_- characteristic from the region of constant state, it is seen from Equation (3.2.5) that

$$u_d - 2(gh_d)^{\frac{1}{2}} = u_0 - 2(gh_0)^{\frac{1}{2}} \tag{3.3.1}$$

or

$$u_d - 2(gh_d)^{\frac{1}{2}} = -14 \text{ m s}^{-1}$$

for all points as far as the point C on the C_- characteristic through B. However, the condition

$$uh = k = k(t)$$

is also provided along the discharging end, at least as far as the point C, so that there are two independent equations for the two unknowns u and h. Computing (by iteration on h or Newton's method again) gives the discharging end conditions as:

t (s)	0	10	20	30	40	50	60
h_d (m)	5.00	4.85	4.70	4.53	4.36	4.18	3.97
u_d (m s^{-1})	0	−0.206	−0.426	−0.662	−0.917	−1.20	−1.51

The flow is seen to be entirely subcritical. The list can obviously be extended as far as point C – although at the moment C has not been placed exactly. It is possible, however, to compute conditions within the region ABC. For, referring to Fig. 3.10 it is seen that, following the C_+ characteristic from the discharging end to point P,

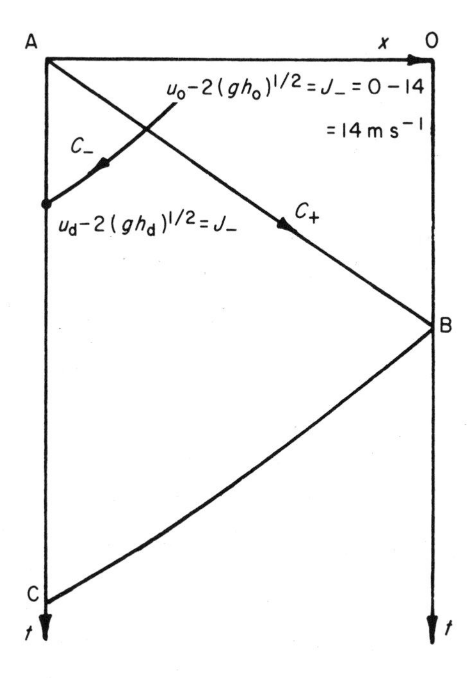

Fig. 3.9. Construction of the inflow end conditions

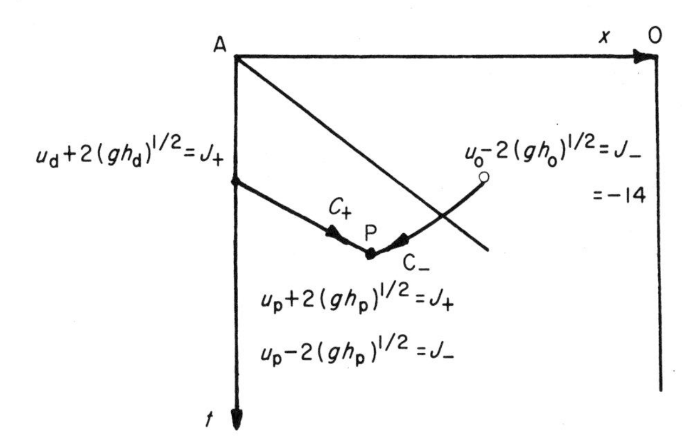

Fig. 3.10. Construction of the simple-wave region

$$u_p + 2(gh_p)^{\frac{1}{2}} = u_d + 2(gh_d)^{\frac{1}{2}} \quad (3.3.2)$$

while following the C_- characteristic from the region of constant state to point P,

$$u_p - 2(gh_p)^{\frac{1}{2}} = u_0 - 2(gh_0)^{\frac{1}{2}} \quad (3.3.3)$$

From the two relations (3.3.2) and (3.3.3) it follows that

$$2u_p = (u_d + 2(gh_d)^{\frac{1}{2}}) + (u_0 - 2(gh_0)^{\frac{1}{2}})$$

or, recalling the formation of u_d and h_d by Equation (3.3.1):

$$u_p = u_d$$

Similarly from Equations (3.3.1), (3.3.2) and (3.3.3) it is seen that

$$h_p = h_d$$

This is then to say that the end conditions are propagated unchanged along their C_+ characteristics. Accordingly, along each C_+ characteristic u and h will be constant, the characteristic directions

$$\left(\frac{dx}{dt}\right)_\pm = u \pm (gh)^{\frac{1}{2}}$$

will also be constant and the C_+ characteristics are straight lines intersected by C_- characteristics at one definite angle for each C_+ characteristic. Summarizing the above as a procedure, it is seen that the region ABC is computed as follows:

(1) Determine u_d, h_d from $u_d h_d = q_d = q(t)$, given, and $u_d - 2(gh_d)^{\frac{1}{2}} = u_0 - 2(gh_0)^{\frac{1}{2}}$.

(2) Thence determine $(dx/dt)_\pm = u_d \pm (gh_d)^{\frac{1}{2}}$.

(3) Draw C_+ characteristics and their corresponding 'field' of intersecting C_- characteristics elements.

(4) Connect C_- characteristics elements, giving (which is all that is really required from (3)), the ultimate characteristic and the point C.

The procedure is schematized in Fig. 3.11.

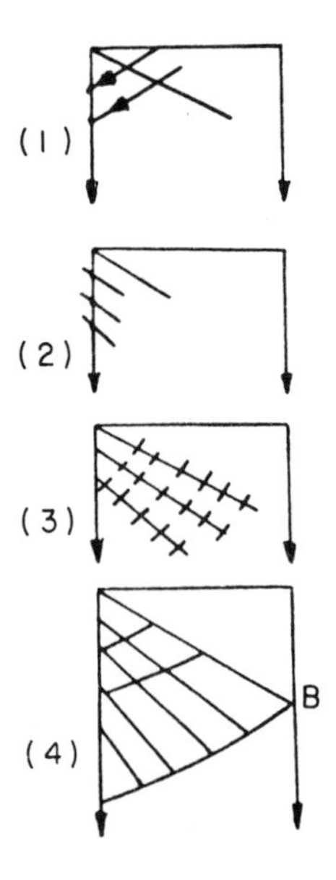

Fig. 3.11. Schematization of the procedure used in the construction of the simple-wave region

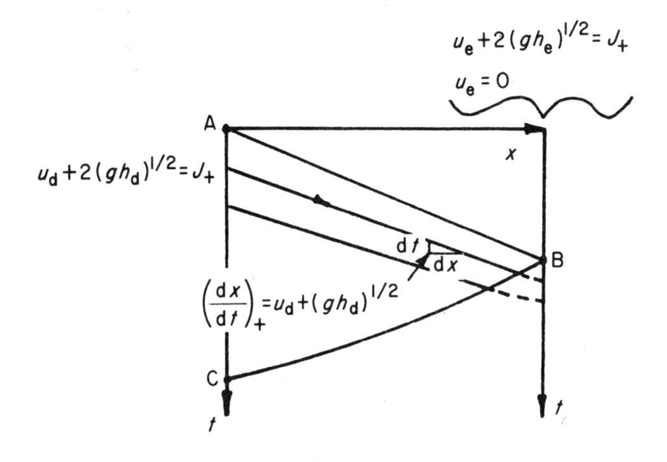

Fig. 3.12. Construction of the closed-end conditions

A region of this last type, in which one family of characteristics consists of straight lines, is called a *simple-wave region*. It is seen from the above construction that simple-wave regions always occur next to a region of constant state. Such regions will be encountered several times in the sequel.

The problem is now posed of computing beyond the line BC. First enter the closed-end boundary conditions, as indicated in Fig. 3.12. Thus

$$u_d + 2(gh_d)^{\frac{1}{2}} = u_e + 2(gh_e)^{\frac{1}{2}}$$

with $u_e = 0$, as given, so that

$$2(gh_e)^{\frac{1}{2}} = u_d + 2(gh_d)^{\frac{1}{2}}$$

It follows that, at the boundary point,

$$\left(\frac{dx}{dt}\right)_{\pm} = \pm(gh_e)^{\frac{1}{2}}$$

If a boundary point is initially chosen that is near to B, then that point can evidently be located to a very good approximation by using the mean characteristic slope:

$$\left(\frac{dx}{dt}\right)_+ = \frac{u_d + (gh_d)^{\frac{1}{2}}}{2} + \frac{(gh_e)^{\frac{1}{2}}}{2} = \frac{u_d + (gh_d)^{\frac{1}{2}}}{2} + \frac{u_d + 2(gh_d)^{\frac{1}{2}}}{4} = \frac{3}{4}u_d + (gh_d)^{\frac{1}{2}}$$

Having thus determined an end point near to B, conditions at a point such as P can be determined, as indicated in Fig. 3.13. For then

$$u_p + 2(gh_p)^{\frac{1}{2}} = u_d + 2(gh_d)^{\frac{1}{2}}$$

$$u_p - 2(gh_p)^{\frac{1}{2}} = u_e - 2(gh_e)^{\frac{1}{2}}$$

giving two independent expressions for the two unknowns u_p, $(gh_p)^{\frac{1}{2}}$.

Having determined u_p, $(gh_p)^{\frac{1}{2}}$ one can obtain

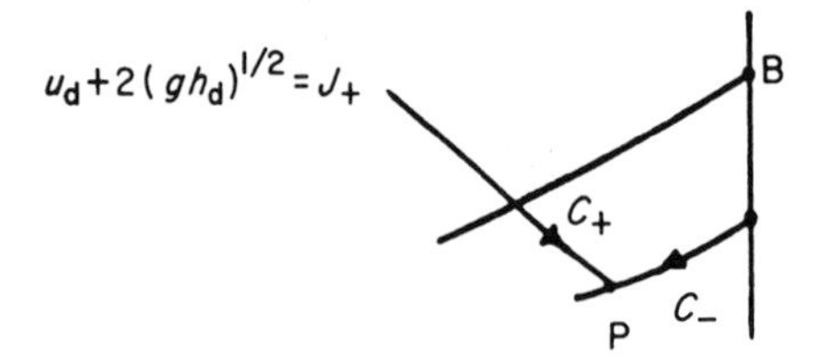

Fig. 3.13. Construction of the complex-state region

$$\left(\frac{dx}{dt}\right)_{\pm} = u_P \pm (gh_P)^{\frac{1}{2}}$$

and enter the position of P, again using mean characteristic directions. Proceeding in this way, the entire nearly horizontal flow region beyond BC can be completed, using, for example, the order schematized in Fig. 3.14.

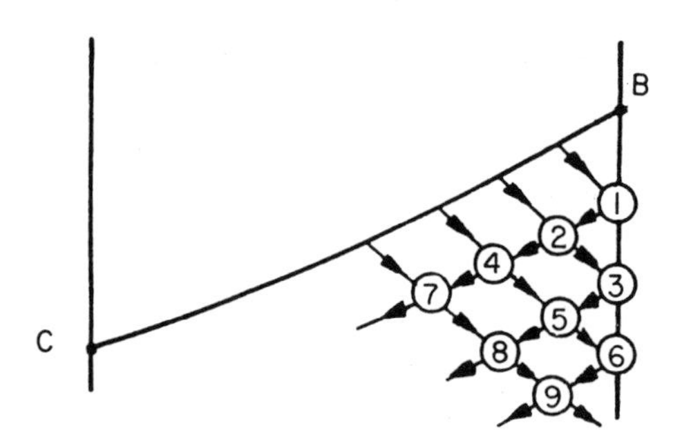

Fig. 3.14. Typical construction sequence in the complex-state region

It will be clear that in this region both families of characteristics will be curved. Such a region, in which neither family of characteristics is composed of straight lines, is called a *region of complex state.*

This complex-state region completes the nearly horizontal flow computation, as all subsequent regions will be of one or the other of the above types. In Fig. 3.18 (*see below*), the *physical plane characteristics* are illustrated for the above problem, together with a computational table.

The above method of solution can be simplified (and subsequent solutions very much simplified) by introducing the *hodograph plane*. This has coordinates u and $2(gh)^{\frac{1}{2}}$ so that Equations (3.2.4) and (3.2.5) correspond to lines with slopes of ±1, i.e.

$$u \pm 2(gh)^{\frac{1}{2}} = J_{\pm} \text{ implies } \left[\frac{d2(gh)^{\frac{1}{2}}}{du}\right]_{\pm} = \mp 1$$

Referring to Fig. 3.15, it is seen that in this plane the entire region of constant state transforms into the single point P. Similarly the simple wave region ABC

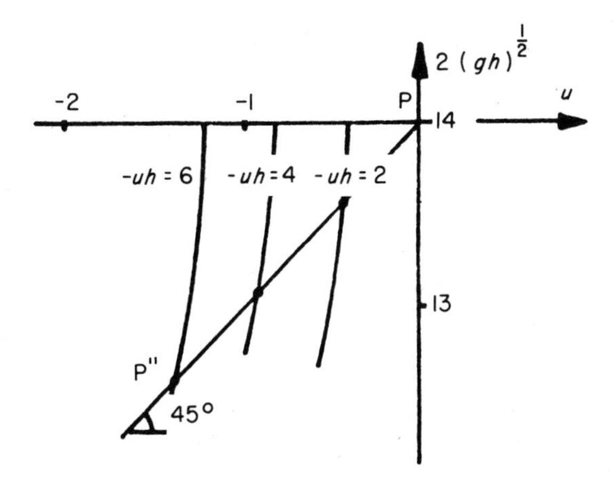

Fig. 3.15. Simple-wave region in the hodograph plane

transforms into the line PP″, as, for any point in this region,

$$u - 2(gh)^{\frac{1}{2}} = J_- = -14$$

while (uh) ranges from 0 $m^2\ s^{-1}$ at (t = 0 s) down to $-6\ m^2\ s^{-1}$ (at t = 60 s). Up to this stage the only advantage of the hodograph plane is that it simplifies the process of entering lines uh = constant, to determine the conditions $(u_d, (gh_d)^{\frac{1}{2}})$.

The hodograph plane becomes more advantageous for computing the complex-wave region (where in fact most of the computational work lies). Thus conditions at the closed end may be determined by drawing lines

$$u + 2(gh)^{\frac{1}{2}} = J_+$$

to the $2(gh)^{\frac{1}{2}}$ axis (where u = 0) as shown in Fig. 3.16. Then conditions at any

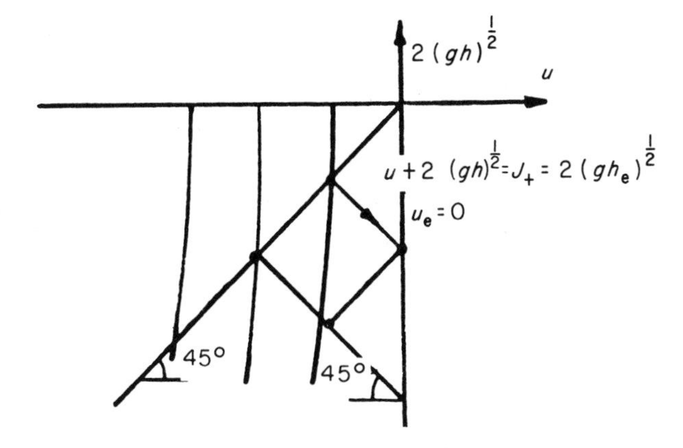

Fig. 3.16. Complex-state region in the hodograph plane

point P can be determined through the intersection of the line

$$u - 2(gh)^{\frac{1}{2}} = J_-$$

through the closed end point, where $u = u_e = 0, (gh)^{\frac{1}{2}} = (gh_e)^{\frac{1}{2}}$ and the line

$$u + 2(gh)^{\frac{1}{2}} = J_+$$

through the open end point, where $u = u_d$, $(gh)^{\frac{1}{2}} = (gh_d)^{\frac{1}{2}}$ (Fig. 3.16). The *hodograph transformation* of Fig. 3.14 has been schematized in Fig. 3.17, while Fig. 3.18 illustrates the complete physical and hodograph planes for the above problem.

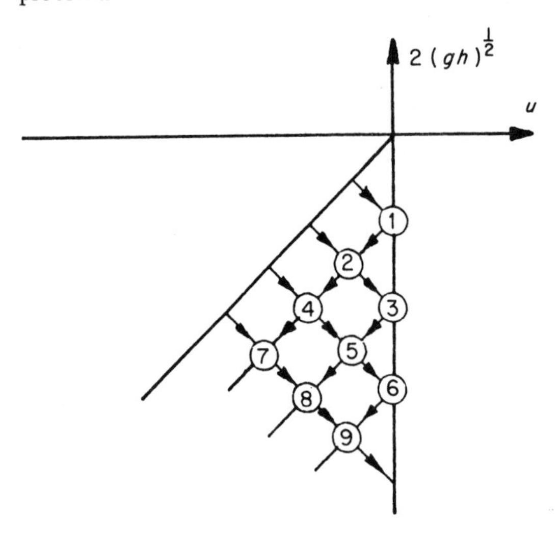

Fig. 3.17. Correspondence of points in the hodograph plane with the sequential points of Fig. 3.14

The preceding problem has been concerned only with subcritical flows, and for these the solution is uniquely determined when two sets of 'initial' data are given at $t = 0$ (for example, depth and velocity) and one set of 'boundary' data (h, u or uh or some relation between these that is independent of the Riemann relations) is given at each of the boundaries $x = 0$ and $x = l$. It is then usual to speak of 'two-point initial conditions' and 'one-point boundary conditions'. Although these data have been posed to the problem as data along lines t = const., x = const., they could clearly have been posed along any lines *other than characteristic lines*. If the data are posed along characteristic line boundaries, the

Fig. 3.18. (*opposite*). The complete physical and hodograph planes with calculation table for the example of the elementary method of characteristics

Physical Plane

Point	$u + 2(gh)^{\frac{1}{2}}$	$u - 2(gh)^{\frac{1}{2}}$	u	$(gh)^{\frac{1}{2}}$	h	$u + (gh)^{\frac{1}{2}}$	$u - (gh)^{\frac{1}{2}}$
7	+13.15	−13.15	0	6.60	4.40	+6.60	−6.60
b 7	+12.17	−13.15	−0.50	6.35	4.10	+5.85	−6.85
c 7	+11.00	−13.15	−1.08	6.05	3.75	+4.97	−7.13
d 7	+12.17	−13.15	−0.50	6.35	4.10	+5.85	−6.85
e 7	+12.17	−13.15	−0.50	6.35	4.10	+5.85	−6.85
8	+12.17	−12.17	0	6.10	3.75	+6.10	−6.10
c 8	+11.00	−12.17	−0.58	5.80	3.45	+5.22	−6.38
d 8	+12.17	−12.17	0	6.10	3.75	+6.10	−6.10
9	+11.00	−11.00	0	5.50	3.05	+5.50	−5.50
d 9	+12.17	−11.00	+0.58	5.80	3.45	+6.38	−5.22
10	+12.17	−12.17	0	6.10	3.75	+6.10	−6.10

Hodograph plane

$h =$	5.0	4.8	4.6	4.4	4.2	4.0	3.8	3.6
$2(gh)^{\frac{1}{2}} =$	14.0	13.7	13.4	13.2	12.5	12.5	12.2	11.9
$uh = -2; u =$	−0.40	−0.42	−0.44	−0.46	−0.48	−0.50	−0.53	−0.56
$= -4; u =$	−0.80	−0.84	−0.87	−0.91	−0.95	−1.00	−1.05	−1.11
$= -6; u =$	−1.20	−1.25	−1.30	−1.36	−1.43	−1.50	−1.58	−1.66

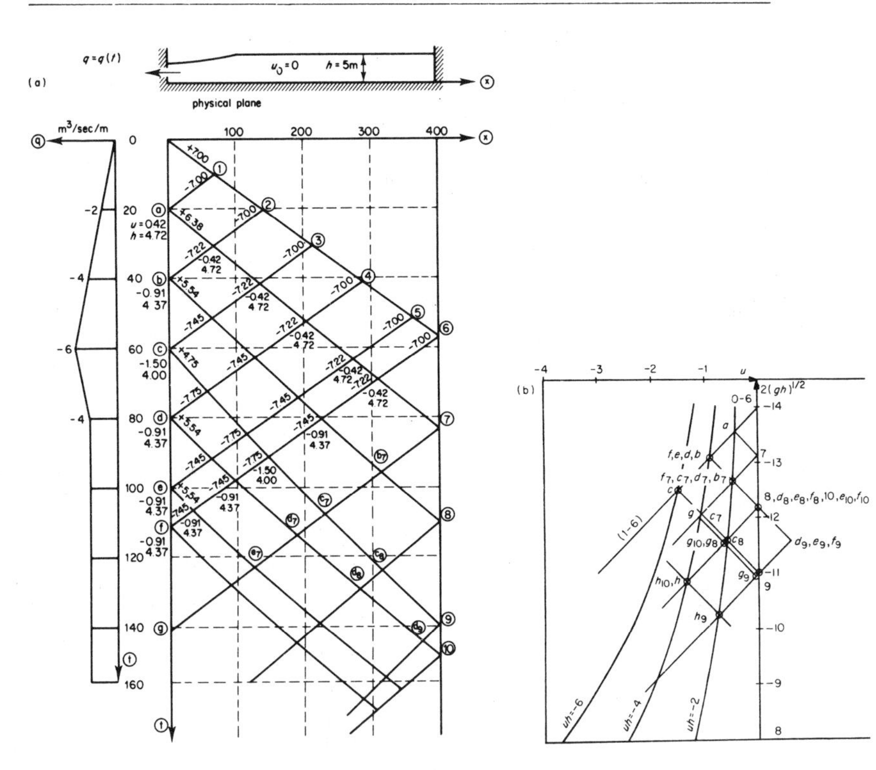

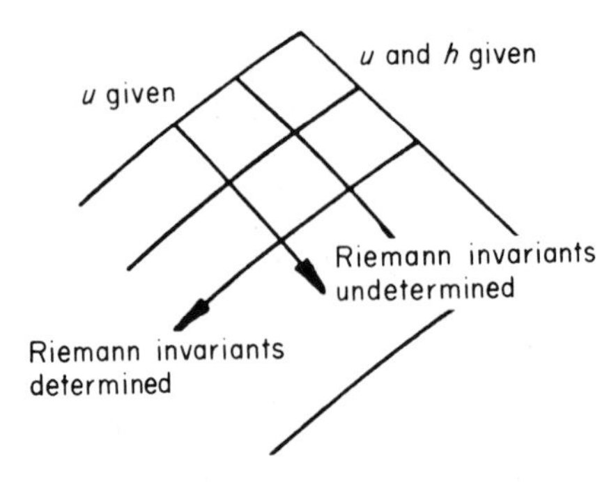

Fig. 3.19. Ill-conditioning of the method of characteristics when one-point data are presented upon a characteristic line

solution is defined only when two-point data are provided along both such boundaries (Fig. 3.19).

The case of characteristically posed data arises in practice when the line x = const. = x_0, say, is characteristic, i.e. $u \pm (gh)^{\frac{1}{2}} = 0$, or $u = \mp(gh)^{\frac{1}{2}}$ and *the flow is critical*. Such a flow that receives one-point information (in the form of Riemann invariants) from one direction (from $x < x_0$) without transmitting any information back from the other direction (from $x > x_0$) occurs whenever the flow is, so to speak, 'free to choose its own level', as at a sudden drop. This situation occurs, in particular, over broad-crested weirs, where it is, however, usually studied through the conditions that the 'head' should be a minimum for given discharge (problem (4) of section 1.5). The reason for this preference is that the latter criterion is more easily extended to account for boundary layer effects which modify the critical flow condition of elementary nearly horizontal flow theory.

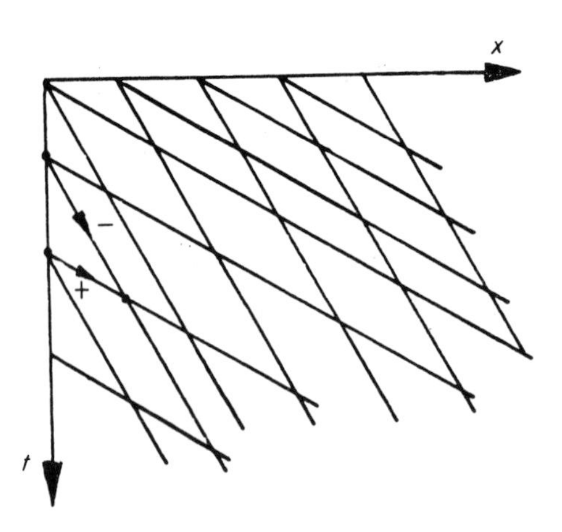

Fig. 3.20. Typical configuration of characteristics for supercritical flow

In the case of *supercritical flows* two-point initial data will again be posed, again usually along t = const., but now the upstream data will have to be of two-point type (Fig. 3.20). Indeed these data are, in a sense, as much 'initial' to the problem as are data along t = const. Downstream, one can impose no condition at all, however, and data will be of 'zero-point' type. Indeed any attempt to impose such downstream data will usually lead to a hydraulic jump advancing upstream, forming a subcritical region behind it. In natural channels, where slopes and resistances are such as to establish subcritical flows, a standing jump may, of course, be formed, and indeed this is often observed at the outlets of control structures when the outlet energy flux is greater than can be accepted by the subcritical flow as a 'drowned jump'.

3.4 Formation and computation of the hydraulic jump

The above problem illustrates the application of the method of characteristics for all possible primitive, uniform channel regions of nearly horizontal flow. That it does not complete even this simplest computation, however, will be clear from Fig. 3.18, where it is seen that the characteristics 'form up' to provide ultimately a multi-valued solution in one part of the reflected simple wave region. Since each characteristic carries a constant h value the behaviour of the wave profile can be visualized during the early part of this process, as indicated in Fig. 3.21.

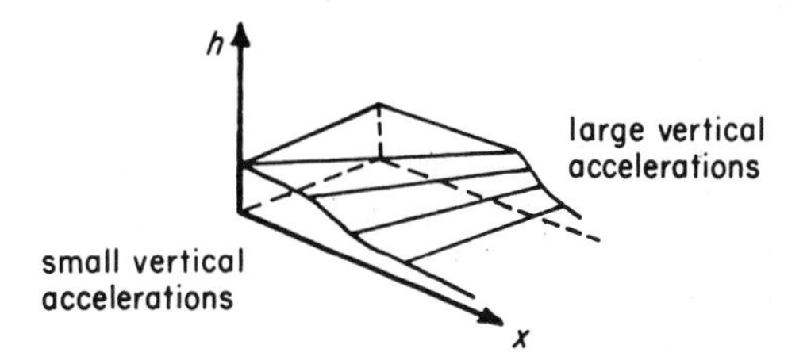

Fig. 3.21. Three-dimensional (h, x, t) representation of the formation of a hydraulic jump

It is seen that this forming up of the characteristics corresponds to a breaking down of the underlying assumptions of negligible vertical accelerations. Clearly these accelerations must now become significant. Then, as was remarked earlier, either these accelerations will tend to balance the tendency towards breaking of the wave so that a 'short-wave' system is formed, or else they will become so pronounced as to introduce a hydraulic jump with intense turbulent energy dissipation. In many cases, however, it is possible to continue the computation by using overall continuity and momentum equations across the undulating or breaking 'jump', while still using the method of characteristics on either side of the jump. For this purpose the following simplifications can be used, using the notation of Fig. 3.22:

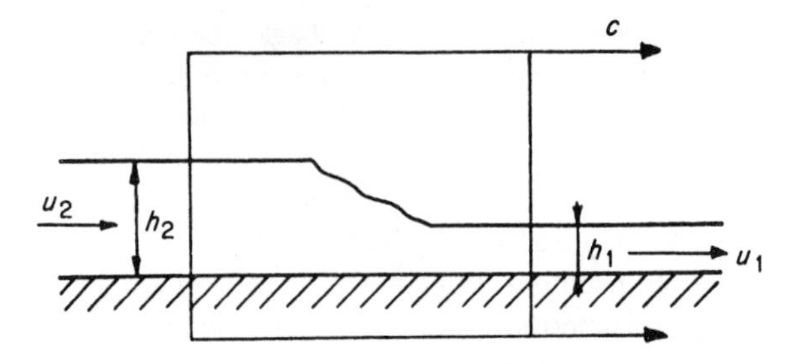

Fig. 3.22. Definition sketch of the hydraulic jump used to derive approximate jump relations

$$(c-u_1)^2 = \frac{gh_2(h_1+h_2)}{2h_1} \text{ implies that } c = u_1 \pm \left(\frac{gh_2(h_1+h_2)}{2h_1}\right)^{\frac{1}{2}}$$

$$(c-u_2)^2 = \frac{gh_1(h_1+h_2)}{2h_2} \text{ implies that } c = u_2 \pm \left(\frac{gh_1(h_1+h_2)}{2h_2}\right)^{\frac{1}{2}}$$

i.e.

$$c = \tfrac{1}{2}\left[u_1 \pm \left(\frac{gh_2(h_1+h_2)}{2h_1}\right)^{\frac{1}{2}} + u_2 \pm \left(\frac{gh_1(h_1+h_2)}{2h_2}\right)^{\frac{1}{2}}\right]$$

Writing $h_2 - h_1 = \Delta h$ this may be expressed as

$$c = \tfrac{1}{2}(u_1 \pm (gh_1)^{\frac{1}{2}} + u_2 \pm (gh^2)^{\frac{1}{2}} \pm 0(\Delta h)^2)$$

Thus for moderate hydraulic jumps, the jump celerity is given, to a good approximation, by the mean of the characteristic directions of the same sense on either side of it (Craya, 1945/1946, originated this approach.)

Similar means can be used to determine the error involved in using the Riemann invariants across the jump instead of the overall mass and momentum equations. A second-order error is again found. Consider, for example, the expansion of the right-hand side of

$$c(h_2 - h_1) = [(uh)_2 - (uh)_1]$$

along a C_+ characteristic, as follows:

$$\frac{u_2h_2 - u_1h_1}{2} + \frac{u_2h_2 - u_1h_1}{2} = \frac{u_2h_2 - (u_2 + 2(gh_2)^{\frac{1}{2}} - 2(gh_1)^{\frac{1}{2}})h_1}{2} +$$
$$+ \frac{(u_1 + 2(gh_1)^{\frac{1}{2}} - 2(gh_2)^{\frac{1}{2}})h_2 - u_1h_1}{2}$$

$$= \frac{u_2h_2 - u_1h_1}{2} + \frac{u_1h_2 - u_2h_1}{2} +$$
$$+ ((gh_1)^{\frac{1}{2}} - (gh_2)^{\frac{1}{2}})(h_2 + h_1)$$

$$= \frac{u_2 h_2 - u_1 h_1}{2} + \frac{u_1 h_2 - u_2 h_1}{2} - \left(\frac{(gh_1)^{\frac{1}{2}} + (gh_2)^{\frac{1}{2}}}{2}\right) \times$$

$$\times (h_2 - h_1) + 0((g/h)^{\frac{1}{2}} (h_2 - h_1)^2)$$

$$= \tfrac{1}{2}(u_1 - (gh_1)^{\frac{1}{2}} + u_2 - (gh_2)^{\frac{1}{2}})(h_2 - h_1) +$$

$$+ 0((g/h)^{\frac{1}{2}}(h_2 - h_1)^2)$$

This, however, is just the left-hand side of the conservation law, again with the jump celerity given as the mean of the appropriate characteristic directions, which in this case are the C_- characteristic directions. This last result shows that when using such approximations for the discontinuity, not only is the energy defect in some error but also some mass may appear to be lost to the flow system. Evidently, better approximations can be obtained by introducing correction terms to the Riemann invariants, along the lines discussed in Section 3.6.

3.5 Indeterminacy conditions for the method of characteristics

In the one-dimensional problems so far discussed in this chapter, the method of characteristics has appeared as a process of filtering the effects of waves propagating in one direction from the effects of waves propagating in the other, or conjugate, direction. In the differential formulation of Section 3.2 the filtering was realized by transforming to a new coordinate system that followed the lines of wave propagation in the x–t space, these lines being then called the characteristics of the differential equation. Now this process of transforming to new coordinates can be greatly facilitated by considering the indeterminacy of derivatives across the characteristics. This indeterminacy can be visualized, at the simplest level, from the observation that, along any characteristic, the solution surface may fold to provide an 'edge' (Fig. 3.23). This is possible along each and

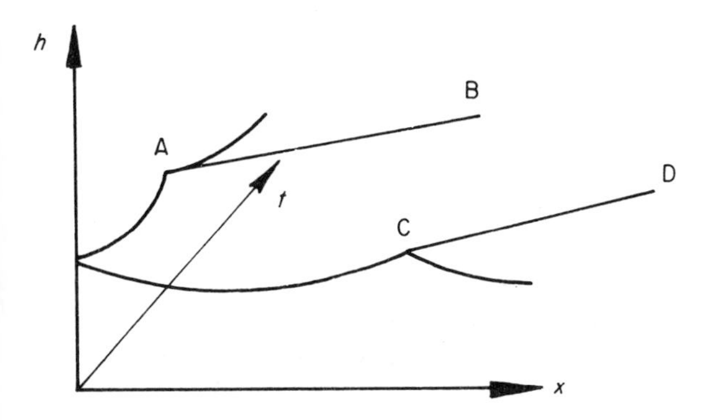

Fig. 3.23. The entire surface in (h, x, t) or, more generally, in (h, u, x, t) is covered by characteristic lines, across all of which surface slopes or derivatives may be discontinuous

every characteristic line, as each such edge may be visualized as being so small as not to be resolved in a calculation. It follows that, across any characteristic, the solution surface must admit indeterminate gradients, or derivatives.

Since derivatives are indeterminate across characteristics, it is impossible to develop a Taylor series across a characteristic. Thus, even when a function and all its derivatives are known at any one point in a propagation problem, the values at surrounding points cannot be determined. Solutions of these propagation problems lack the property of 'analytic continuation' of equilibrium problems (Hadamard, 1923; Abbott, 1966). Equations such as (2.6.1) and (2.6.2) that generate such solutions, and that possess an even number of real and distinct characteristic directions, are said to be of 'hyperbolic' type.

Now the solution surface is generated from the governing equations, which in the simplest nearly horizontal flow case can be taken in Eulerian form

$$\frac{\partial h}{\partial t} + u\frac{\partial h}{\partial x} + h\frac{\partial u}{\partial x} = 0 \qquad (3.5.1, = 2.6.1)$$

$$\frac{\partial u}{\partial t} + u\frac{\partial u}{\partial x} + g\frac{\partial h}{\partial x} = 0 \qquad (3.5.2, = 2.6.2)$$

and the equations of variation that are conditioned by these equations to form the solution surface,

$$\begin{aligned} \frac{\partial h}{\partial t}\,\mathrm{d}t + \frac{\partial h}{\partial x}\,\mathrm{d}x &= \mathrm{d}h \\ \frac{\partial u}{\partial t}\,\mathrm{d}t + \frac{\partial u}{\partial x}\,\mathrm{d}x &= \mathrm{d}u \end{aligned} \qquad (3.5.3)$$

The solution surface at any point is then defined by the equation system

$$\begin{bmatrix} 1 & u & 0 & h \\ 0 & g & 1 & u \\ \mathrm{d}t & \mathrm{d}x & 0 & 0 \\ 0 & 0 & \mathrm{d}t & \mathrm{d}x \end{bmatrix} \begin{bmatrix} \dfrac{\partial h}{\partial t} \\ \dfrac{\partial h}{\partial x} \\ \dfrac{\partial u}{\partial t} \\ \dfrac{\partial u}{\partial x} \end{bmatrix} = \begin{bmatrix} 0 \\ 0 \\ \mathrm{d}h \\ \mathrm{d}u \end{bmatrix} \qquad (3.5.4)$$

The derivatives become indeterminate when the equation system does not admit of solutions other than the indeterminate case of '0/0'. In this situation there are no longer four independent equations in (3.5.4): any one equation is a linear combination of the other three. It then follows from the rules of determinants and matrices that not only is the determinant of the left-hand coefficient matrix zero (to provide the zero in the denominator of the '0/0' relation) but, when any column vector of this matrix is replaced by the right-hand column vector of

Equations (3.5.4), the determinant of the resulting matrix will also be zero (to provide the zero in the numerator of the '0/0' relation). Thus, the derivatives $(\partial h/\partial x, \partial u/\partial t, \partial u/\partial t, \partial u/\partial x)$ become indeterminate when

$$\begin{vmatrix} 1 & u & 0 & h \\ 0 & g & 1 & u \\ \mathrm{d}t & \mathrm{d}x & 0 & 0 \\ 0 & 0 & \mathrm{d}t & \mathrm{d}x \end{vmatrix} = 0 \tag{3.5.5}$$

and when, for example,

$$\begin{vmatrix} 1 & u & 0 & 0 \\ 0 & g & 1 & 0 \\ \mathrm{d}t & \mathrm{d}x & 0 & \mathrm{d}h \\ 0 & 0 & \mathrm{d}t & \mathrm{d}u \end{vmatrix} = 0 \tag{3.5.6}$$

In this case the determinants are easily evaluated, but it is convenient for further applications to remark the following simplifications (condensations) arising from the theory of determinates. The first follows from the rule that 'a determinant is unaltered in value when to any row, or column, is added a constant multiple of any other row or column' (e.g. Aitken, 1939, p. 40). Thus, in Equation (3.5.6) the third and fourth rows can be divided by $\mathrm{d}t$ and the resulting third row subtracted from the first row, and the resulting fourth row subtracted from the second row to provide

$$\begin{vmatrix} 0 & u-\dot{x} & 0 & h \\ 0 & g & 0 & u-\dot{x} \\ 1 & \dot{x} & 0 & 0 \\ 0 & 0 & 1 & \dot{x} \end{vmatrix} = 0$$

The second rule is that interchange of two rows or two columns of a determinate merely alters its sign. Interchanging first columns 1 and 2 and then interchanging columns 2 and 4 provides

$$\begin{vmatrix} u-\dot{x} & h & 0 & 0 \\ g & u-\dot{x} & 0 & 0 \\ \dot{x} & 0 & 0 & 1 \\ 0 & \dot{x} & 1 & 0 \end{vmatrix} = 0$$

Considering then the sub-determinants, it is seen that

$$\begin{vmatrix} u-\dot{x} & h \\ g & u-\dot{x} \end{vmatrix} = 0 \tag{3.5.7}$$

or

$$(u - \dot{x})^2 - gh = 0$$

or

$$\dot{x} = u \pm (gh)^{\frac{1}{2}}$$

It is seen that the $\dot{x}$ appear in Equations (3.5.7) as eigenvalues of the matrix A in the form of

$$\frac{\partial f}{\partial t} + A \frac{\partial f}{\partial x} = 0$$

with

$$f = \begin{bmatrix} h \\ u \end{bmatrix}, \qquad A = \begin{bmatrix} u & h \\ g & u \end{bmatrix}$$

Tracing back through the above construction, it will be clear that this is a general property of one-dimensional flows: *the characteristic directions of any system of one-dimensional equations will be the eigenvalues of the matrix A in the form (2.6.6).* Correspondingly, the equation system will be of hyperbolic type if its matrix A in the form (2.6.6) has real and distinct eigenvalues.

Repeating the above process with the determinant (3.5.6) provides

$$\begin{vmatrix} 0 & u - \dot{x} & 0 & -\dot{h} \\ 0 & g & 0 & -\dot{u} \\ 1 & \dot{x} & 0 & \dot{h} \\ 0 & 0 & 1 & \dot{u} \end{vmatrix} = 0$$

and thence

$$\begin{vmatrix} u - \dot{x} & -\dot{h} \\ g & -\dot{u} \end{vmatrix} = 0$$

giving the *Riemann differential equation*

$$(u - \dot{x})\dot{u} - g\dot{h} = 0 \qquad (3.5.8)$$

When $\dot{x} = u + (gh)^{\frac{1}{2}}$, Equation (3.5.8) reduces, via multiplying through by dt and rearranging, to

$$\mathrm{d}u + \left(\frac{g}{h}\right)^{\frac{1}{2}} \mathrm{d}h = 0$$

that is,

$$u + 2(gh)^{\frac{1}{2}} = \text{const.}$$

and when $\dot{x} = u - (gh)^{\frac{1}{2}}$, Equation (3.5.8) reduces similarly to

$$du - \left(\frac{g}{h}\right)^{\frac{1}{2}} dh = 0$$

that is,

$$u - 2(gh)^{\frac{1}{2}} = \text{const.}$$

The matrix and determinant formulation really comes into its own when dealing with stratified fluids. Equations (2.6.9), for example, provide the matrix equation

$$\begin{bmatrix} 1 & u_0 & 0 & h_0 & 0 & 0 & 0 & 0 \\ 0 & g & 1 & u_0 & 0 & \sigma g & 0 & 0 \\ 0 & 0 & 0 & 0 & 1 & u_1 & 0 & h_1 \\ 0 & g & 0 & 0 & 0 & g & 1 & u_1 \\ dt & dx & 0 & 0 & 0 & 0 & 0 & 0 \\ 0 & 0 & dt & dx & 0 & 0 & 0 & 0 \\ 0 & 0 & 0 & 0 & dt & dx & 0 & 0 \\ 0 & 0 & 0 & 0 & 0 & 0 & dt & dx \end{bmatrix} \begin{bmatrix} \partial h_0/\partial t \\ \partial h_0/\partial x \\ \partial u_0/\partial t \\ \partial u_0/\partial x \\ \partial h_1/\partial t \\ \partial h_1/\partial x \\ \partial u_1/\partial t \\ \partial u_1/\partial x \end{bmatrix} = \begin{bmatrix} 0 \\ 0 \\ 0 \\ 0 \\ dh_0 \\ du_0 \\ dh_1 \\ du_1 \end{bmatrix}$$

The characteristic directions $\dot{x}$ appear in the role of eigenvalues again:

$$\begin{vmatrix} u_0 - \dot{x} & h_0 & 0 & 0 \\ g & u_0 - \dot{x} & \sigma g & 0 \\ 0 & 0 & u_1 - \dot{x} & h_1 \\ g & 0 & g & u_1 - \dot{x} \end{vmatrix} = 0$$

giving the *characteristic quartic* in $\dot{x}$,

$$[(u_0 - \dot{x})^2 - gh_0]\,[(u_1 - \dot{x})^2 - gh_1] - \sigma g^2 h_0 h_1 = 0 \qquad (3.5.9)$$

The four roots of this equation, defining the four characteristic directions, can be calculated graphically as follows. The variables $\theta_0^2 = (u_0 - \dot{x})^2/gh_0$ and $\theta_1^2 = (u_1 - \dot{x})^2/gh_1$ are introduced so that Equation (3.5.9) reduces to

$$(1 - \theta_0^2)(1 - \theta_1^2) - \sigma = 0 \qquad (3.5.10)$$

Then, for given σ, a unique curve is defined in the (θ_0, θ_1) plane (Fig. 3.24). This is seen to have an inner closed branch and an outer open branch. Now θ_0 and θ_1 are also related by definition, whereby

$$(gh_0)^{\frac{1}{2}}\theta_0 - (gh_1)^{\frac{1}{2}}\theta_1 = u_0 - u_1$$

defining a straight line in the (θ_0, θ_1) plane. The intersection of this line with the line of Equations (3.5.10) provides the values of four characteristic directions. It is seen that the two of these generated at the closed inner branch, often called the 'inner characteristics', will generally be considerably smaller in magnitude

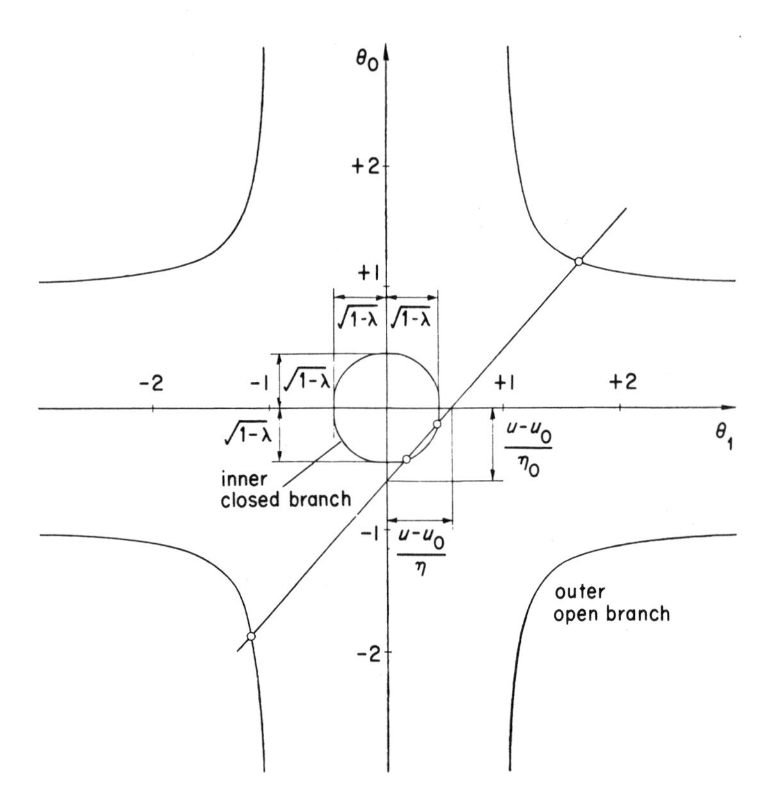

Fig. 3.24. Construction for the characteristic directions for a two-layered stratified fluid from the characteristic quartic

than the other two, often called the 'outer characteristics', that are defined on the open, outer branch.

In the case of the homogeneous fluid, three distinct 'characteristic structures' were identified, corresponding to subcritical flow (one characteristic direction positive and the other negative), supercritical flow in the positive x-direction (both characteristic directions positive) and supercritical flow in the negative x-direction (both characteristic directions negative). Corresponding to these three distinct characteristic structures are three distinct data structures: the subcritical flow calls for one-point data at each boundary, the positive x-direction supercritical flow calls for two-point data at the lower x-boundary and zero-point data at the upper x-boundary while the negative x-direction supercritical flow calls for zero-point data at the lower x-boundary and two-point data at the upper x-boundary. In the two-layer stratified fluid case there are five distinct characteristic structures, as shown in Fig. 3.25. Once again, the number of data-points presented at each boundary, for an initial-value computation working forward in time, equals the number of characteristics initiated from any point on the boundary and heading forward in time. Thus, a flow corresponding to Fig. 3.25a would

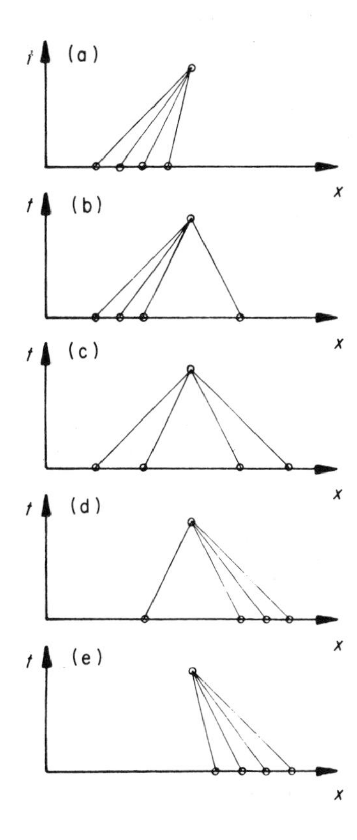

Fig. 3.25. The five basic characteristic structures of a two-layered fluid

require four-point boundary data on the left and zero-point boundary data on the right; a flow corresponding to Fig. 3.25b would require three-point left boundary data and one-point right boundary data, and so on. However, this is not all.

For consider the following limiting conditions on Equation (3.5.9):

(1) $(h_1/h_0) \to 0$ with h_1 finite and thus $h_0 \to \infty$ implies that

$$\dot{x} = u_1 \pm ((1 - \sigma)gh_1)^{\frac{1}{2}}$$

(2) $(h_1/h_0) \to 0$ with h_0 finite and thus $h_1 \to 0$ implies that

$$\dot{x} = u_0 \pm (gh_0)^{\frac{1}{2}}$$

(3) $(h_1/h_0) \to \infty$ with h_1 finite and thus $h_0 \to 0$ implies that

$$\dot{x} = u_1 \pm (gh_1)^{\frac{1}{2}}$$

(4) $(h_1/h_0) \to \infty$ with h_0 finite and thus $h_1 \to \infty$ implies that

$$\dot{x} = u_0 \pm ((1 - \sigma)gh_0)^{\frac{1}{2}}$$

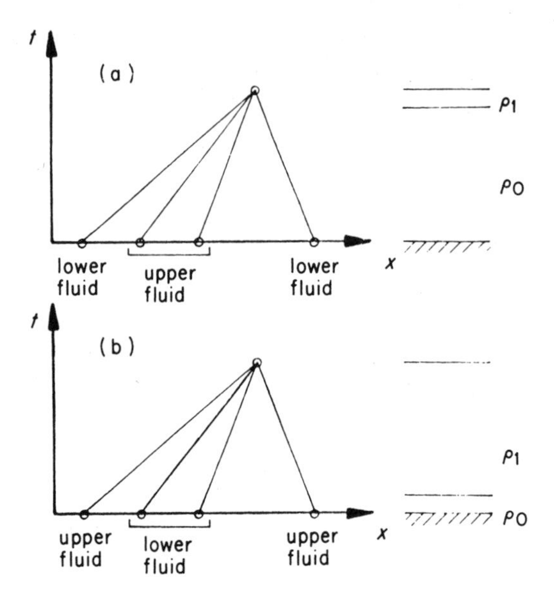

Fig. 3.26. Further subdivision of characteristic structures necessary when these are used as indicators of data structure

Schematizing these results in Fig. 3.26 it is seen that it is not only necessary to present the correct number of data-points at each boundary, but it is also necessary that these be presented in the correct place. Thus in the case of Fig. 3.26a, one-point data should be presented at both left- and right-hand boundaries of the lower fluid while two-point data should be presented at the left-hand boundary and zero-point data at the right-hand boundary of the upper fluid. It is, of course, necessary that such data be properly posed to any solution method of an initial-value problem.

It is easily shown (Abbott, 1961) that the celerities of the 'uncoupled' component representing the total fluid, $u_0 \pm (gh_0)^{\frac{1}{2}}$ in the one case and $u_1 \pm (gh_1)^{\frac{1}{2}}$ in the other case above, always bound the magnitudes of the outer characteristic celerities given by Equation (3.5.9) from below.

The Riemann differential equation in this two-layer case takes the form

$$[(u_1 - \dot{x})^2 - gh_1][\dot{h}(u_0 - \dot{x}) - h_0\dot{u}_0] + \sigma g h_0[\dot{h}(u - \dot{x}) - h\dot{u}] = 0 \qquad (3.5.11)$$

the solution of which has been investigated by Hydén (1974).

3.6 Further worked examples

Problem 1: What relations corresponding to Riemann invariants obtain for a uniform horizontal channel with

(a) a small slope i, measured positive when increasing the fluid velocity

and

(b) resistance τ (force/unit area), when $\tau = \tau(u, h)$?

Solution: The differential equations of motion and their variations can be obtained by writing Equations (2.1.2), (2.2.3) and (2.2.5) in the form of (3.5.4):

$$\begin{bmatrix} 1 & 0 & 0 & 1 \\ 0 & (gh-u^2) & 1 & 2u \\ dt & dx & 0 & 0 \\ 0 & 0 & dt & dx \end{bmatrix} \begin{bmatrix} \partial h/\partial t \\ \partial h/\partial x \\ \partial(uh)/\partial t \\ \partial(uh)/\partial x \end{bmatrix} = \begin{bmatrix} 0 \\ ghi - \dfrac{gu|u|}{C^2} \\ dh \\ d(uh) \end{bmatrix}$$

The characteristics are given by

$$\begin{vmatrix} 0-\dot{x} & 1 \\ (gh-u^2) & 2u-\dot{x} \end{vmatrix} = 0$$

giving again

$$\dot{x} = u \pm (gh)^{\frac{1}{2}}$$

The equation defining the characteristic direction celerities, as it depends only upon the instantaneous values of $u = u(x,t)$ and $h = h(x,t)$, is not changed by the introduction of energy-diffusing terms. Of course, the values of u and h that appear at any (x,t) in the presence of energy diffusion will be different to the values that will appear there without energy diffusion, but the form of the expression for $\dot{x}$ cannot be influenced. The characteristic directions are examples of what are accordingly sometimes called 'point quantities'. The Riemann invariant relations will be changed, however, to

$$\begin{vmatrix} -\dot{x} & -\dot{h} \\ (gh-u^2) & \left(\left(ghi - \dfrac{gu|u|}{C^2}\right) - \dfrac{d(uh)}{dt}\right) \end{vmatrix} = 0$$

or

$$\dot{x}\left(\left(ghi - \frac{gu|u|}{C^2}\right) - \frac{d(uh)}{dt}\right) - (gh-u^2)\dot{h} = 0$$

When $\dot{x} = u + (gh)^{\frac{1}{2}}$ this reduces to

$$\left(ghi - \frac{gu|u|}{C^2}\right) dt - d(uh) + (u - (gh)^{\frac{1}{2}})\, dh = 0$$

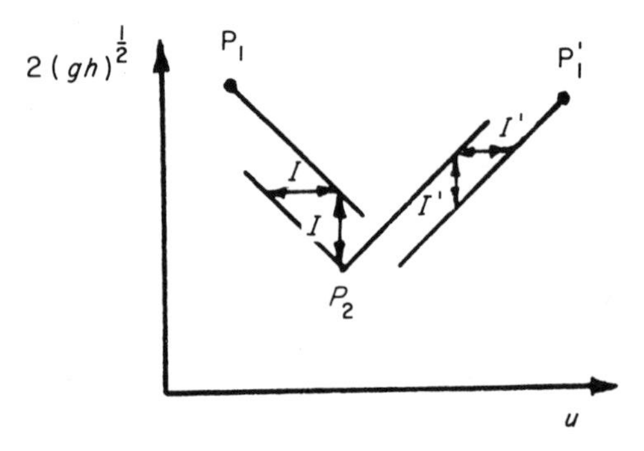

Fig. 3.27. Shifting of characteristics in the hodograph plane to account for the effects of resistance to flow

giving the Riemann differential equation

$$du + \left(\frac{g}{h}\right)^{\frac{1}{2}} dh = \left(gi - \frac{gu|u|}{C^2h}\right) dt$$

or

$$[u + 2(gh)^{\frac{1}{2}}]_1^2 = \int_{t_1}^{t_2} \left(gi - \frac{gu|u|}{C^2h}\right) dt = I \tag{3.6.1}$$

Similarly, when $\dot{x} = u - gh$,

$$[u - 2(gh)^{\frac{1}{2}}]_1^2 = \int_{t_1}^{t_2} \left(gi - \frac{gu|u|}{C^2h}\right) dt = I \tag{3.6.2}$$

When the integrals on the right of Equations (3.6.1) and (3.6.2) are sufficiently small (and they can always be made 'sufficiently small' by choosing a suitable $(t_2 - t_1)$, or physical characteristic net), the quantities $[u \pm 2(gh)^{\frac{1}{2}}]$ are 'nearly invariant' and they are accordingly often referred to as *quasi-invariants*. By use of the hodograph plane, their computation is relatively simple, as schematized in Fig. 3.27. It is seen that if lines $u + 2(gh)^{\frac{1}{2}}$ const. are drawn through P_1 and P_2, then by virtue of Equations (3.6.1) and (3.6.2), the horizontal or vertical distance between these lines is the quantity I, as illustrated in Fig. 3.27. Much the same holds for lines $u - 2(gh)^{\frac{1}{2}}$ = const. through P_1 and P_2, with the integral I', corresponding to conditions at P_1'. Thus a solution in the hodograph plane may be constructed by drawing lines at ±45° through P_1 and P_1' respectively, stepping off distances I and I' as indicated, using approximate values of h and $(t_2 - t_1)$ and approximating the integral by

$$I \approx \frac{\tau}{\rho h}(t_2 - t_1)$$

The intersection of the C_+ and C_- hodograph characteristics gives the required state P_2 and the position of the corresponding point in the hodograph plane.

When this state has been found, a better approximation can be introduced for h and $(t_2 - t_1)$, but such a better approximation is rarely necessary. The direction of stepping is easily checked from whether u is increased or decreased by the step, in relation to the balance of slope and resistance terms.

Problem 2: What relations, corresponding to Riemann invariants, obtain for a horizontal channel with breadth increasing uniformly with length?

Solution: By using exactly the same methods, the quasi-invariants

$$[u + 2(gh)^{\frac{1}{2}}]_{t_1}^{t_2} = -\int_{t_1}^{t_2} \frac{u(gh)^{\frac{1}{2}}}{r}\, dt$$

$$[u - 2(gh)^{\frac{1}{2}}]_{t_1}^{t_2} = +\int_{t_1}^{t_2} \frac{u(gh)^{\frac{1}{2}}}{r}\, dt$$

are determined.

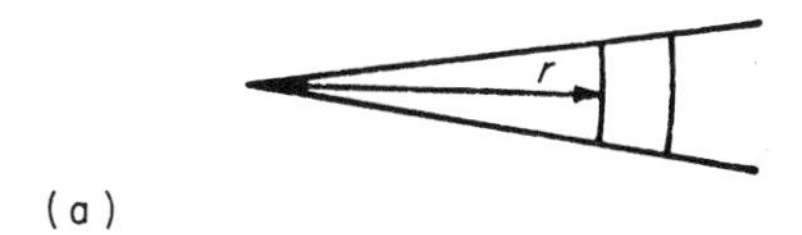

Fig. 3.28. Schematization of the radial flow used at a change of section in the method of characteristics

Expressions of this type are valuable not only for computing flows in gentle changes of section of a simple canal (Fig. 3.28a), but they are essential also for computing through canal junctions, where a condition that h is the same at the junction ends of all the canals can be used only when the velocity remains constant through the junction. (Otherwise it is easy to show, using Equation (1.2.5), that there is an energy falsification.) The equations can also be extended to apply to a rapid and even a sudden change of section, in which although $r \to \infty$, $t_1 \to t_2$ so that the time integrals remain bounded, the integrand tending to a δ-function-like entity. It is possible to make a formal theory of the distributions solution in this case.

Problem 3: The plate A (Fig. 3.29) of mass M Kg per m breadth just fits a very long uniform horizontal canal and is acted upon, for all $t \geqslant 0$ by a force F N/unit breadth. Determine the motion of the plate.

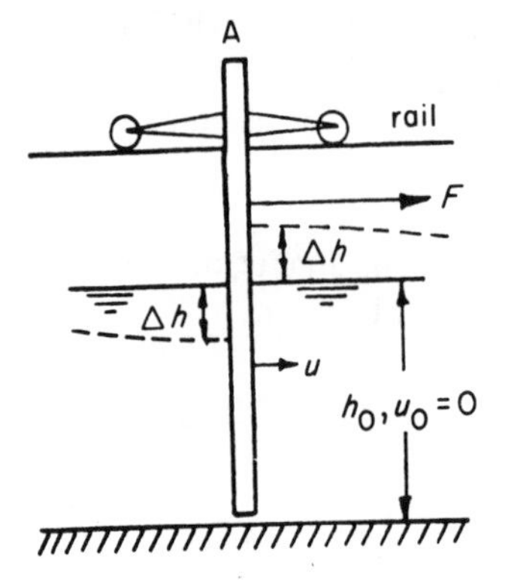

Fig. 3.29. Definition sketch for problem 3

Solution: Denote the velocity of the plate by u. Then the depth variation Δh, neglecting all reflections from the canal ends, is given by

$$u \mp 2(g(h_0 + \Delta h))^{\frac{1}{2}} = \mp 2(gh_0)^{\frac{1}{2}},$$

or for $\Delta h \ll h_0$,

$$u \mp \left(\frac{g}{h_0}\right)^{\frac{1}{2}} \Delta h = 0$$

i.e.

$$\Delta h = \pm \left(\frac{h_0}{g}\right)^{\frac{1}{2}} u$$

Thus, as a result of the motion, a pressure force

$$\rho g h_0 \cdot 2\Delta h = 2\rho g h_0 \left(\frac{h_0}{g}\right)^{\frac{1}{2}} u$$

is induced. The equation of motion of the plate is then

$$M\frac{\mathrm{d}u}{\mathrm{d}t} = F - 2\rho g h_0 \left(\frac{h_0}{g}\right)^{\frac{1}{2}} u$$

or

$$\frac{\mathrm{d}u}{\mathrm{d}t} + \left(\frac{2\rho g h_0}{M}\right)\left(\frac{h_0}{g}\right)^{\frac{1}{2}} u = \frac{F}{M}$$

This has the solution, for $u|_{t=0} = u_0 = 0$;

$$u = \frac{F}{2\rho g h_0 (h_0/g)^{\frac{1}{2}}}\left(1 - \exp\frac{-2\rho g h_0 (h_0/g)^{\frac{1}{2}}}{M} \cdot t\right)$$

This equation provides a first approximation to the low-speed (manoeuvring) motion of a ship under tugs or wind. In particular, it provides a 'lower bound' for the motion, as all other effects (flow under the keel, flow around stem and stern, etc.) will tend to increase the velocity computed above.

Problem 4: A uniform horizontal canal initially contains fluid of depth 5 m travelling with a velocity 1 m s^{-1}. At a certain instant and point, a gate is lowered, cutting off all flow. Assuming that all motion is nearly horizontal, determine the subsequent fluid behaviour in the canal.

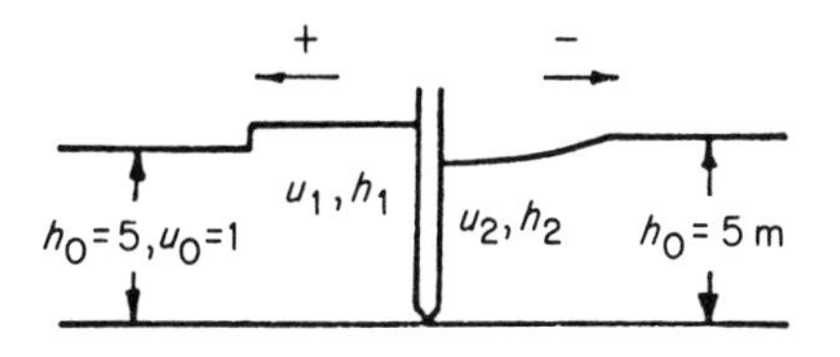

Fig. 3.30. Definition sketch for problem 4

Solution: The situation depicted in Fig. 3.30 will develop: a positive jump propagates to the left and a *negative jump* – decomposing into a long-wave (and short-wave) system – propagates to the right. The positive jump will leave behind it a region with zero velocity and depth given approximately (assuming nearly horizontal flow) by:

$$0 + 2(gh_1)^{\frac{1}{2}} = 1 + (2 \times 7)$$
$$= 15$$

or

$$(gh_1)^{\frac{1}{2}} = 7.5 \text{ m s}^{-1}$$

The positive jump will then, on this assumption, have a celerity between $\dot{x}_- = 1 - 7 = -6$ m s^{-1} and $\dot{x}_- = -7.5$ m s^{-1}. (This result can be controlled using the methods of Chapter 1.) The negative jump may be treated as follows (Fig. 3.31). The leading C_+ characteristic has celerity

$$\dot{x}_+ = 1 + 7 = 8 \text{ m s}^{-1}$$

To the left of this characteristic is a region of constant state. Thus, to the right there will be a simple wave region. A C_- characteristic across this region can be followed to obtain h_2:

$$u_2 - 2(gh_2)^{\frac{1}{2}} = u_0 - 2(gh_0)^{\frac{1}{2}}$$

or

$$0 - 2(gh_2)^{\frac{1}{2}} = 1 - (2 \times 7)$$

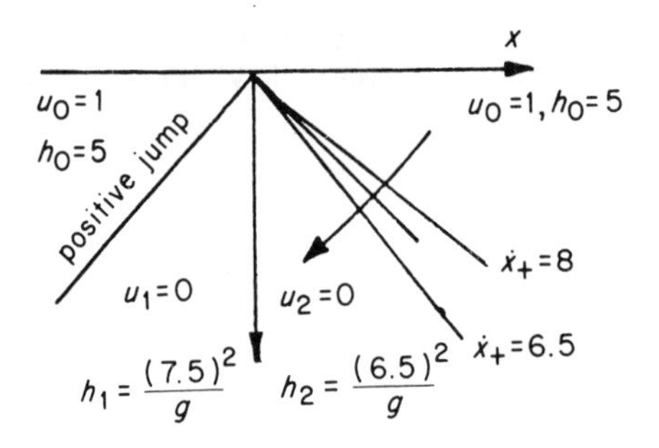

Fig. 3.31. The physical plane for problem 4 showing the positive jump and the centred simple wave realization of the 'negative hydraulic jump'

i.e.

$$(gh_2)^{\frac{1}{2}} = 6.5 \text{ m s}^{-1}$$

The simple wave region is then bounded to the right by another C_+ characteristic with direction

$$\dot{x}_+ = +6.5 \text{ m s}^{-1}$$

To the right of this characteristic is obviously another region of constant state (Fig. 3.31). This completes the solution. The fan-like characteristic structure in the simple wave of this example is a *type structure for the negative jump*. It is called a *centred simple wave*. The limiting case of this wave, as the depth and velocity difference tends to zero, is a single characteristic line. This clearly corresponds as well to the negative jump form of the fundamental solution that was discussed in Section 3.2.

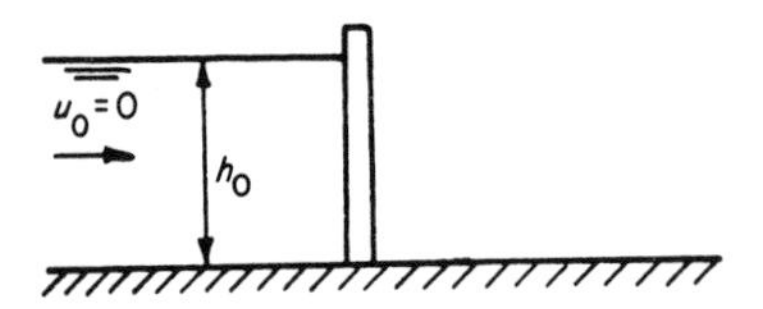

Fig. 3.32. Initial conditions for problem 5

Problem 5: A dam is constructed over a horizontal bed as shown in Fig. 3.32. At time $t = 0$ the dam is removed. Determine the subsequent flow.

Solution: The characteristic 'structure' is constructed from the region of constant state upstream (Fig. 3.33). The initial disturbance C_- characteristic will have a direction

$$\dot{x}_- = -(gh_0)^{\frac{1}{2}}$$

while, from the region of constant state, C_+ characteristics are initiated, along all of which

$$J_+ = +2(gh_0)^{\frac{1}{2}}$$

Accordingly, next to the region of constant state there will be a simple wave region in which all C_- characteristics are straight lines.

Since each of these C_- characteristics carries a different pair of values (h, u), and since characteristics cannot be initiated within the x–t plane (because there are only two characteristics through each point), all of these C_- characteristics must be initiated from the original dam face. (Indeed this is the only place where all the values of h, for example, could obtain initially.) The C_- characteristics thus form a centred simple wave. Writing the slope of a typical (jth) C_- characteristic as ϕ_j, so that

$$\phi_j = \dot{x}_j = u_j - (gh_j)^{\frac{1}{2}}$$

(Fig. 3.33), indicates through

$$u_j + 2(gh_j)^{\frac{1}{2}} = +2(gh_0)^{\frac{1}{2}}$$

that

$$\phi_j + 3(gh_j)^{\frac{1}{2}} = +2(gh_0)^{\frac{1}{2}}$$

i.e.

$$\left.\begin{aligned} h_j &= \frac{(+2(gh_0)^{\frac{1}{2}} - \phi_j)^2}{9g} \\ &\text{and similarly,} \\ u_j &= \frac{u_0 + 2(gh_0)^{\frac{1}{2}} + 2\phi_j}{3} \end{aligned}\right\} \quad (3.6.3)$$

These equations provide a solution until the 'front' $h = 0$, when Equation (3.6.3) gives

$$\phi_{\text{front}} = 2(gh_0)^{\frac{1}{2}} \text{ and thus } u_{\text{front}} = 2(gh_0)^{\frac{1}{2}}$$

In this case the centred simple wave is as shown in Fig. 3.33. It is easily seen,

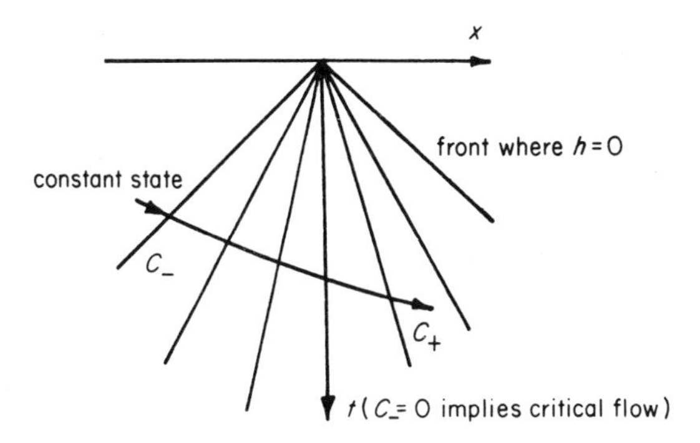

Fig. 3.33. The physical plane for problem 5 composed of a constant-state region and bordered by a centred simple-wave region that continues to the fluid front, where $h = 0$

from Equations (3.6.3), that at the 'front', where $h = 0$, *the C_+ and C_- characteristics coincide.* A front of this type is sometimes called a *St Venant front.* As one advances towards the front, $F, = |u/(gh)^{\frac{1}{2}}|, \rightarrow \infty$. This front, then, usually becomes unstable in practice, so that the St Venant front does not occur. Alternative situations have been discussed by several authors (*see*, for example, Abbott, 1961; Abbott and Torbe, 1963). An alternative solution is provided from the Jeffreys–Vedernikov condition, as described in Section 3.7.

3.7 Uni-directional wave propagation

In the general theory of nearly horizontal one-dimensional flows, information can be transmitted in two directions and this information transmission can be manifested physically as wave motions. A special, degenerate case of this theory arises when wave motion occurs in only one direction so that the corresponding information transmission is effectively restricted to one direction only. If the direction of propagation in the x–t plane is along one of the physical characteristics discussed above, then the uni-directional wave corresponds to a simple wave region. However, as other influences intervene to modify the ideal nearly horizontal flow behaviour – notably, vertical accelerations, resistances and bed slopes – so the direction in the x–t plane of the uni-directional wave may shift away from that of the simple wave discussed so far, introducing other directions of physical wave propagation apart from those defined by the nearly horizontal flow characteristics.

The simple wave of nearly horizontal flow provides a convenient point of reference for all other uni-directional flows of interest in hydraulics. If propagation is restricted to the positive x-direction relative to the fluid, than along each and every C_+ characteristic of the wave

$$(u - 2(gh)^{\frac{1}{2}}) = (u_0 - 2(gh_0)^{\frac{1}{2}}) = J_0 = \text{const.} \tag{3.7.1}$$

Substituting this in the first equation of (3.2.6), namely

$$\frac{\partial}{\partial t}(u + 2(gh)^{\frac{1}{2}}) + (u + (gh)^{\frac{1}{2}})\frac{\partial}{\partial x}(u + 2(gh)^{\frac{1}{2}}) = 0 \tag{3.7.2}$$

gives

$$\frac{\partial}{\partial t}(4(gh)^{\frac{1}{2}} + u_0 - 2(gh_0)^{\frac{1}{2}}) + (3(gh)^{\frac{1}{2}} + u_0 - 2(gh_0)^{\frac{1}{2}}) \times$$

$$\times \frac{\partial}{\partial x}(4(gh)^{\frac{1}{2}} + u_0 - 2(gh_0)^{\frac{1}{2}}) = 0$$

or

$$\frac{\partial}{\partial t}(gh)^{\frac{1}{2}} + (3(gh)^{\frac{1}{2}} + u_0 - 2(gh_0)^{\frac{1}{2}})\frac{\partial}{\partial x}(gh)^{\frac{1}{2}} = 0 \tag{3.7.3}$$

implying that throughout the simple wave region,

$$\frac{\partial h}{\partial t} + (3(gh)^{\frac{1}{2}} + u_0 - 2(gh_0)^{\frac{1}{2}})\frac{\partial h}{\partial x} = 0 \tag{3.7.4}$$

Similarly, throughout the same region,

$$\frac{\partial u}{\partial t} + (3(gh)^{\frac{1}{2}} + u_0 - 2(gh_0)^{\frac{1}{2}})\frac{\partial u}{\partial x} = \frac{\partial u}{\partial t} + \left(\frac{3u}{2} - \frac{u_0}{2} + (gh_0)^{\frac{1}{2}}\right)\frac{\partial u}{\partial x} = 0 \tag{3.7.5}$$

Thus, the condition (3.7.1) provides a relation between the dependent variables that is independent of the relation (3.7.2), so that the system with two characteristic directions can be reduced to a system, (3.7.3) or (3.7.4) or (3.7.5), that has only one characteristic direction $\dot{x}_i = (3(gh_i)^{\frac{1}{2}} + u_0 - 2(gh_0)^{\frac{1}{2}})$. The other characteristic still exists, of course, and its influence is felt through the relation (3.7.1), but it plays no further role in the calculations of depths and velocities through (3.7.4) or (3.7.5).

Referring back to Equation (1.4.4), a general expression for the uni-directional celerity c was found in the form

$$c = \left.\frac{\mathrm{d}g(f)}{\mathrm{d}f}\right|_{x=\text{const.}}$$

and indeed the special case of

$$c = \left.\frac{\mathrm{d}(uh)}{\mathrm{d}h}\right|_{x=\text{const.}} \tag{3.7.6}$$

was used in Section 3.2 to determine the Riemann differential equation. It is now seen that multiplying the volume conservation law

$$\frac{\partial h}{\partial t} + \frac{\partial}{\partial x}(uh) = 0$$

by Equation (3.7.6) in fact gives

$$\frac{\partial}{\partial t}(uh) + c\frac{\partial}{\partial x}(uh) = 0 \tag{3.7.7}$$

This has characteristic directions $\dot{x} = c$ while along the characteristics the dependent variable (uh) is constant. The existence of a uni-directional wave with *scalar wave equation form* for one or more of the dependent variables is thus equivalent to the existence of a relation between the variables, such as (3.7.1).

In exactly the same way, a simple wave travelling in the negative x-direction relative to the fluid will satisfy the scalar wave equations

$$\frac{\partial h}{\partial t} + (-3(gh)^{\frac{1}{2}} + u_0 + 2(gh_0)^{\frac{1}{2}})\frac{\partial h}{\partial x} = 0$$

$$\frac{\partial u}{\partial t} + (-3(gh)^{\frac{1}{2}} + u_0 + 2(gh_0)^{\frac{1}{2}})\frac{\partial u}{\partial x} = \frac{\partial u}{\partial t} + \left(\frac{3u}{2} - \frac{u_0}{2} - (gh_0)^{\frac{1}{2}}\right)\frac{\partial u}{\partial x} = 0$$

In the rest of this section, only the case of waves travelling in the positive x-direction will be considered, the other case being entirely symmetric. In the simple-wave case so far treated, the celerity can be written

$$\dot{x}_i = (3(gh_i)^{\frac{1}{2}} - 3(gh_0)^{\frac{1}{2}}) + (u_0 + (gh_0)^{\frac{1}{2}})$$
$$= (3(gh_i)^{\frac{1}{2}} - 3(gh_0)^{\frac{1}{2}}) + \dot{x}_0$$

so that $h_i > h_0$ implies $\dot{x}_i > \dot{x}_0$ and $h_i < h_0$ implies $\dot{x}_i < \dot{x}_0$. Thus the simple wave is continuously deformed, as its deeper sections overtake its shallow sections, until the vertical accelerations influence the solution, forming up into a hydraulic jump. As has been remarked earlier, the vertical accelerations may have a tendency, in nature, to prevent further deformation locally, so that the system runs further as a simple wave composed of waves with significant vertical accelerations.

The simple-wave region of nearly horizontal flow theory thus leads to a first generalization: to a simple-wave region of vertically accelerated flows. These flows may exist as wave trains which are now, however, dispersive. Radiating energy out behind them, the leading waves are attenuated while the radiation develops and strengthens waves at the end of the train so that the velocity of propagation of the train as a whole (its 'group velocity') is less than the velocities of propagation of its individual component waves. Indeed, from classical hydrodynamics (e.g. Lamb, 1932), uni-directional progressive waves of infinitesimal amplitude propagating in water of depth h are known to have a wave velocity of

$$c = [g/k \tanh kh]^{\frac{1}{2}}$$

and a group velocity of

$$c_{\mathrm{gr}} = \left.\frac{\mathrm{d}(kc)}{\mathrm{d}k}\right|_{x=\mathrm{const}} = \frac{c}{2}\left[1 + \frac{2kh}{\sinh 2kh}\right]$$

where k is the 'wave number', the number of waves over distance 2π (i.e. $k = 2\pi/L$, where L is the wave length). When the water is very deep, then $c_{\mathrm{gr}}/c \to \frac{1}{2}$, while when it is shallow the nearly horizontal flow situation reappears through $c_{\mathrm{gr}}/c \to 1$. Lighthill and Whitham (1955) emphasized the simple-wave nature of such short wave trains and their corresponding description by a scalar wave equation – both inevitable consequences of the existence of hydrodynamic relations between water surface elevation and flow velocity in all such waves.

Computationally satisfactory conservation laws for short shallow-water waves of the above type have been given by Jonsson (1978). The one-dimensional form is

$$\frac{\partial}{\partial t}\left(\frac{E}{\omega_{\mathrm{r}}}\right) + \frac{\partial}{\partial x}\left((u + c_{\mathrm{gr}})\left(\frac{E}{\omega_{\mathrm{r}}}\right)\right) = 0$$

in which E is the mean wave energy level taken as the specific energy that would be calculated on infinitesimal amplitude theory by an observer moving in a Galilean frame with the medium at velocity u, ω_{r} is the relative angular frequency, calculated as for E, and c_{gr} is the corresponding group velocity.

Differentiating out and introducing the uni-directional celerity $(\partial g/\partial f)_{x=\text{const.}}$ again provides the scalar wave equation form.

Expanding upon their generalization of simple wave theory, that Lighthill and Whitham (1955) call *kinematic* wave theory, these authors remark, 'However convenient such devices may be for developing the theory of a number of important phenomena from a simple and unified point of view, one must not forget that in these . . . problems the system is only a kinematic wave system if attenuation be deliberately restricted to waves travelling in one direction only. The methods cannot be used to treat reflexion . . . and, indeed, in a true kinematic wave system no reflexion of any kind is possible (mathematically, there is only one system of characteristics).'

These limitations notwithstanding, the scalar wave formulation of short-wave trains generalizes to two space dimensions to provide a direct application of the methods of geometric optics for calculating wave refraction and diffraction, while extensive trackings of wave energy can be carried out using just such a formulation, for wave 'hindcasting' and forecasting purposes. A proper treatment of this subject, relating it to more general calculations discussed in Chapter 4 and the results shown in Chapter 6, is more suited to a specialized text.

The second, and closely related generalization of nearly horizontal flow simple wave theory is to the propagation of kinematic waves or *monoclinal* waves, in which the effects of resistance compete with (and, in classical theory, dominate over) the dynamic effects, for the periods of time considered in the calculations. In order to introduce this kinematic wave theory, it is germane to reconsider in Fig. 3.34 the initial stages in the construction of the physical and hodograph planes of Fig. 3.18 as reworked with resistance effects included. It is seen at once that $2(gh_{12})^{\frac{1}{2}}$ with resistance effects included is, of necessity, less than $2(gh_{12})^{\frac{1}{2}}$ without such resistance effects. Thus a line of constant h emanating from point 11 in the x–t plane will pass behind point 12. It is similarly remarked that the velocity u_{11} with resistance effects included is, of necessity, less than the velocity u_{11} without resistance effects. Thus the quantity of water entering the fluid system is less with resistance than it is without it and, given the finite celerities of wave propagation, the elevations will be, in consequence, generally less at corresponding points within the rising and advancing elevation system. That this effect is observable as a kinematic wave travelling more slowly than the characteristic of the same sense, even in quite rapidly varying flows, can be seen from the example illustrated in Fig. 3.35. Evidently, the longer the time for which the integrals of the quasi-invariants operate, the greater will be the effect of the resistance.

Consider now any such kinematic wave travelling in the same manner as a simple wave but at a celerity that departs from the characteristic celerity of the same sense in accordance with the resistance effect being augmented. Referring to Fig. 3.36, it is seen that the lines of constant h and constant u of such a simple-wave-analogous system appear as characteristic lines that will now be called the *kinematic characteristics*. They will be identified as mathematical characteristics of a scalar wave equation in the sequel, indeed propagating h or u indifferently.

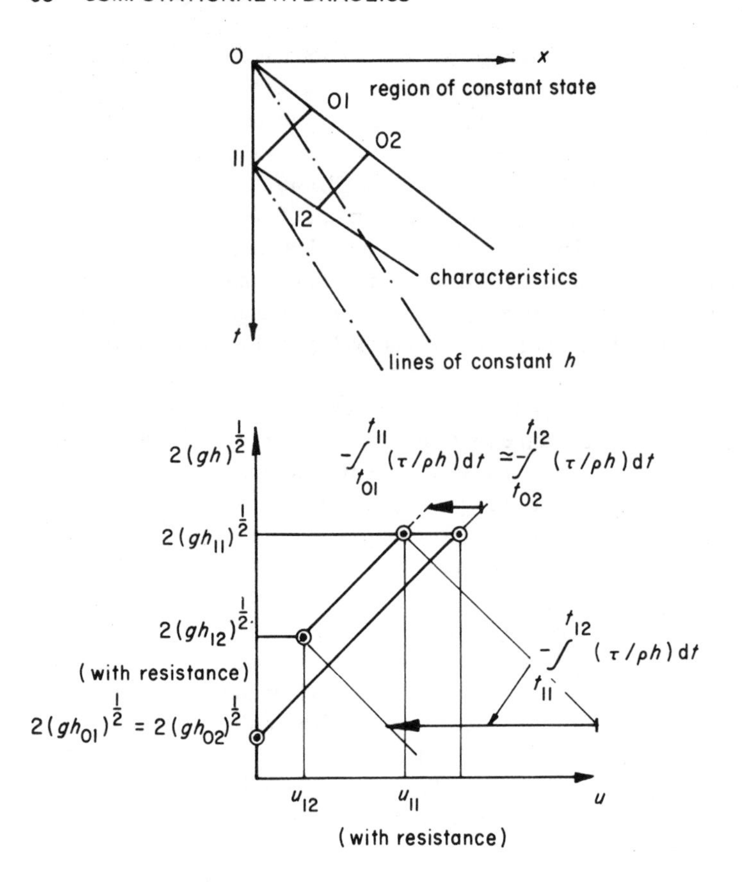

Fig. 3.34. First stages in the reconstruction of the problem of Fig. 3.18 with resistance effects included

The characteristics of the dynamic system will then be called the *dynamic characteristics*. Now consider two dynamic characteristics emanating from any one kinematic characteristic in the physical plane. These meet at a point where h and u take values other than those of the kinematic characteristic at which they are initiated, as shown in the physical plane of Fig. 3.36a. Referring then to the hodograph plane of Fig. 3.36b shows that this is possible if and only if the resistance effect is able to turn the C_+ and C_- characteristics initiated at (h_1, u_1) so that they intersect at one and the same point (h_0, u_0).

Since this construction can be carried out for points as close together as desired on the initiating kinematic characteristics, it is seen that the kinematic simple wave is characterized by a coalescence of dynamic characteristics in the hodograph plane, so as again to map the entire kinematic wave region into a single line in the hodograph plane. This mapping, in turn, implies the existence of a distinct relation between the dependent variables, but one that now depends upon the resistance term. Expressing this in the form of Equation (3.7.2) again

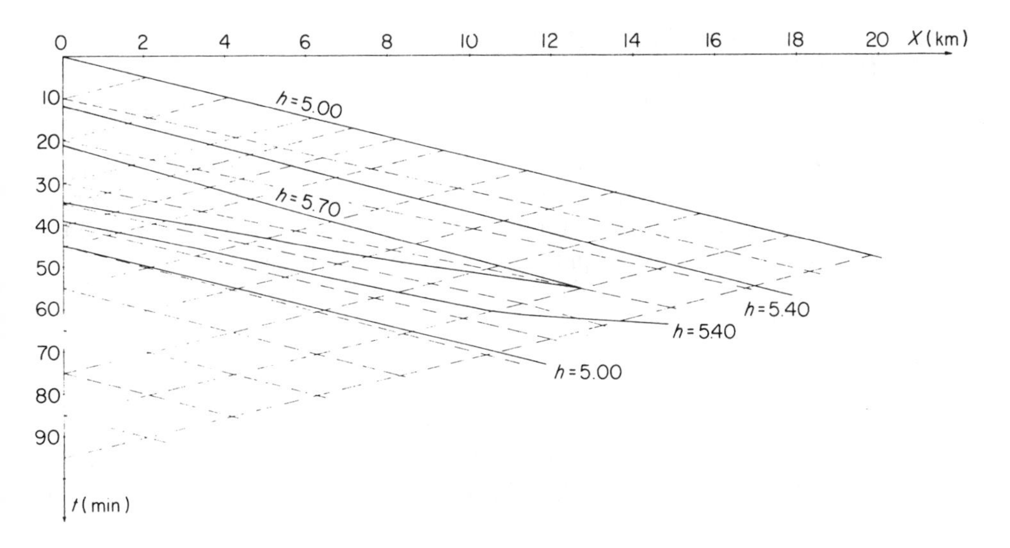

Fig. 3.35. Characteristics and lines of constant elevation for a resisted nearly horizontal flow

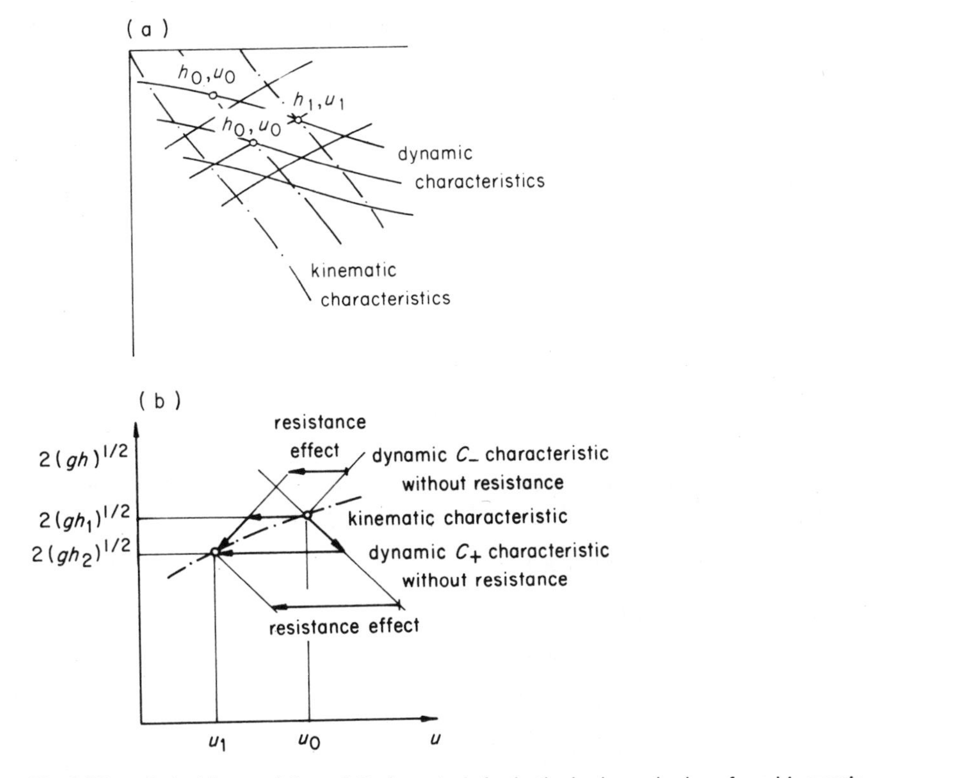

Fig. 3.36. Coincidence of C_+ and C_- characteristics in the hodograph plane for a kinematic simple wave

provides the scalar wave equation (3.7.7), the characteristics of which are now the kinematic characteristics. If a simple wave is defined as one that reduces to a line in the hodograph plane (or has a topological index of 1 – *see* Abbott, 1966) then the kinematic wave is a simple wave.

A kinematic simple-wave theory is then obtained by supposing that the mapping of the simple wave, as a line in the hodograph plane, is determined only by the balance of resistance and bed slope. The dynamic effects are then supposed to occur so slowly, as through taking averages over such long times, that the dynamic equation reduces to a simple resistance law of steady flows. Then the general celerity relation

$$c = \left(\frac{\partial g(f)}{\partial f}\right)_{x=\text{const.}} \qquad (3.7.8, = 1.4.5)$$

depends only on the resistance law relation between $g(f)$ and f. Consider, for example, Equation (3.7.8) in the volume form $c = (\mathrm{d}Q/\mathrm{d}A)_{x=\text{const.}}$, with a resistance law of the form

$$u = \left(\frac{2}{f}\right)^{\frac{1}{2}} (gRi)^{\frac{1}{2}} \qquad (3.7.9)$$

where f is a friction factor and R the resistance radius (Engelund, 1965, 1 and 2). For a rough wall, f can be expressed through

$$\left(\frac{2}{f}\right)^{\frac{1}{2}} = 6 + 2.5 \ln \frac{R}{k}$$

where k is the roughness factor. Substituting Equation (3.7.9) into the volume form of (3.7.8) then gives

$$c = \beta u \text{ with } \beta = 1 + \frac{n}{2}\left(1 + 5(\tfrac{1}{2}f)^{\frac{1}{2}}\right)$$

with

$$n = \frac{A}{R} \cdot \frac{\mathrm{d}R}{\mathrm{d}A} = \frac{\mathrm{d}(\ln R)}{\mathrm{d}(\ln A)}$$

The value of n is thus obtained by constructing a double logarithmic plot of R against A. Taking, as a typical value, $(2/f)^{\frac{1}{2}} = 15$, provides:

(a) for rectangular section channels:

$n = 1, \qquad c = 1.67u \qquad \text{or} \qquad \beta = 1.67$

(b) for parabolic section channels:

$n = \frac{2}{3}, \qquad c = 1.44u \qquad \text{or} \qquad \beta = 1.44$

(c) for triangular section channels:

$n = \frac{1}{2}, \qquad c = 1.33u \qquad \text{or} \qquad \beta = 1.33$

In the event that the simplest Chézy resistance law is used over an infinite width, it is seen that $c = \mathrm{d}(uh)/\mathrm{d}h = 1.5u$.

Equation (3.7.8) then provides the simple-wave solution

$$Q(x, t) = Q(x - c\,\Delta t, t - \Delta t) \tag{3.7.10}$$

and thence, through the resistance law relations for a uniform channel,

$$\left.\begin{aligned} h(x, t) &= h(x - c\,\Delta t, t - \Delta t) \\ u(x, t) &= u(x - c\,\Delta t, t - \Delta t) \end{aligned}\right\} \tag{3.7.11}$$

It is observed, in practice, that the kinematic wave, as it appears in a river flood wave, is not only propagated but also diffused. A simple generalization of the above theory that accounts for this observation has been given by Daubert (1964) (*see also* Cunge, 1969, 1975; Grijsen and Vreugdenhil, 1976). Daubert took the equations

$$\frac{\partial Q}{\partial x} + B\frac{\partial h}{\partial t} = 0$$

$$\frac{\partial h}{\partial x} = -\tau$$

where τ, the resistance term, is a function of Q, h and x. Differentiating the first equation with respect to x and the second with respect to t and eliminating the cross derivatives in h then provides

$$\frac{\partial^2 Q}{\partial x^2} - \left[\left(\frac{\partial \tau}{\partial Q}\right)_x \frac{\partial Q}{\partial t} + \left(\frac{\partial \tau}{\partial h}\right)_x \frac{\partial h}{\partial t}\right] B + \left[\frac{\partial B}{\partial x} + \left(\frac{\partial B}{\partial h}\right)_t \frac{\partial h}{\partial x}\right] \frac{\partial h}{\partial t} = 0$$

Introducing the original equations, this reduces to

$$\frac{\partial Q}{\partial t} + \frac{1}{B}\left[-\frac{(\partial \tau/\partial h)_x}{(\partial \tau/\partial Q)_x} + \frac{\partial B/\partial x - \tau(\partial B/\partial h)_t}{B(\partial \tau/\partial Q)_x}\right]\frac{\partial Q}{\partial x} - \frac{1}{B(\partial \tau/\partial Q)_x}\frac{\partial^2 Q}{\partial x^2} = 0$$

In the case that the section has parallel vertical walls, the celerity reduces to $c = -(\mathrm{d}\tau/\mathrm{d}h)_x/B(\mathrm{d}\tau/\mathrm{d}Q)_x$. When, further, τ remains constant, as in a nearly steady flow, then

$$\left(\frac{\partial \tau}{\partial h}\right)_x \mathrm{d}h + \left(\frac{\partial \tau}{\partial Q}\right)_x \mathrm{d}Q = 0$$

or

$$c = \frac{1}{B}\left(\frac{\partial Q}{\partial h}\right)_x$$

exactly as computed earlier.

Consider now the kinematic wave situation when the flow becomes increasingly supercritical through a reduction in depth at constant velocity, as schematized in Fig. 3.37 for a Chézy resistance law. It is seen that, as the depth decreases to raise the Froude number above unity, so the time taken for the C_+ dynamic

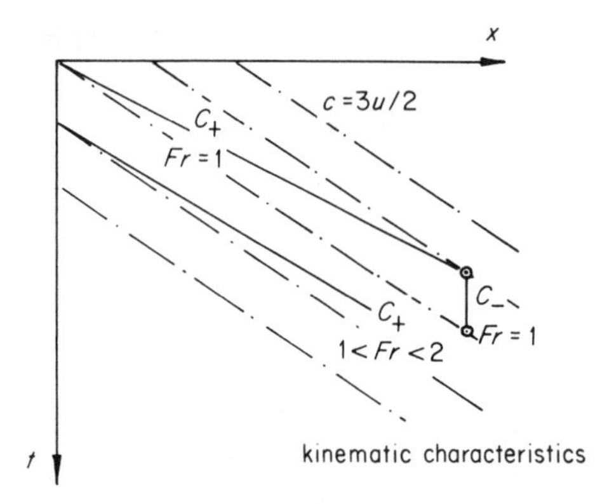

Fig. 3.37. As the dynamic and kinematic characteristics come to coincide, so the time over which the resistance integrals of the quasi-invariants are calculated increases without bound. As a result, conditions at adjacent kinematic characteristics become increasingly separated and the nearly horizontal flow breaks down to a flow with large vertical accelerations

characteristic to travel from one kinematic characteristic to any adjacent characteristic increases rapidly. The integrals that intervene in the quasi-invariants (3.6.1), (3.6.2) increase correspondingly, so that variations in conditions from one kinematic characteristic to its neighbours become rapidly more intense. Vertical accelerations are induced accordingly and the flow develops into a 'roll-wave' pattern, with a greatly augmented irreversible energy flow from out of the nearly horizontal system. Clearly, this situation occurs as

$$\dot{x}_{+\ \text{dynamic}} \rightarrow \dot{x}_{\text{kinematic}}$$

i.e. as

$$u + (gh)^{\frac{1}{2}} \rightarrow \beta u$$

or

$$Fr = \left| \frac{u}{(gh)^{\frac{1}{2}}} \right| \rightarrow \frac{1}{\beta - 1}$$

The condition

$$Fr = \frac{1}{\beta - 1}$$

which is the limit of this process, is called the Jeffreys–Vedernikov condition. The Jeffreys–Vedernikov condition then gives a value of $Fr = 2$ for the simplest flow (uniform channel, including uniform bed slope) governed by a Chézy resistance law. It is sometimes used to modify the dam break profile of the classical St Venant solution, on the basis of the argument that the flow will be so arrested by the energy dissipation induced by the roll wave formation that a front will be formed, characterized by its Froude number that is

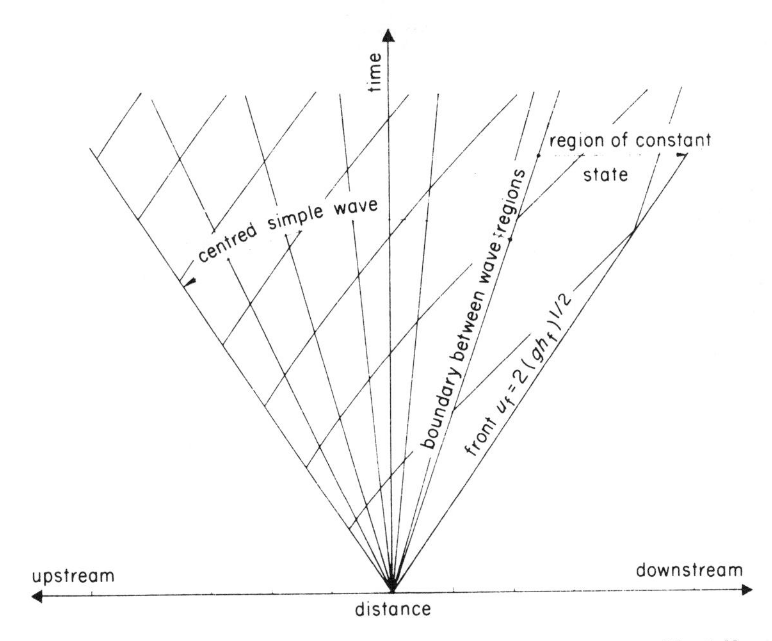

Fig. 3.38. The physical characteristics of the centred simple wave and its Jeffreys–Vedernikov-fronted region of constant state for a wave front computation

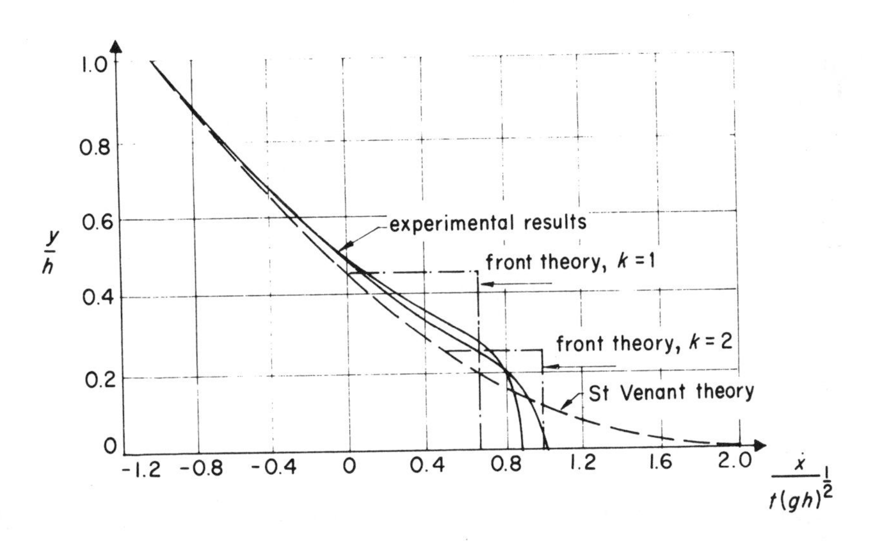

Fig. 3.39. Comparison of wave front theories with the experimental results of Schoklitsch (*see* Rouse, 1938)

then given in turn by the Jeffreys–Vedernikov condition (e.g. Abbott, 1961). The corresponding physical plane characteristics and dimensionless flow profile then appear as in Fig. 3.38 and 3.39. The flow profile is seen in Fig. 3.39 to agree quite well with the experimental profiles of Schoklitsch (*see* Rouse, 1938), obtained using a uniform channel.

3.8 The linearized method of characteristics: uni-directional waves in hydrology

The linearized theory of uni-directional waves provides a convenient representation of the linear systems used so widely in hydrology (e.g. Dooge, 1973). The celerity c can be supposed constant for all x under a pure transport, but the 'catchment' can vary in width in any way, $\psi = \psi(x)$, as illustrated in Fig. 3.40a. Consider the discharge I produced at $x = 0$ when water is placed instantaneously with unit depth over the entire catchment, as calculated from Fig. 3.40a to Fig. 3.40b. The curve of $I = I(t)$ is simply the curve of $\psi = \psi(x)$ scaled by the celerity c, since, geometrically,

$$I(\tau)\,\mathrm{d}\tau = \psi(c\tau)\,c\cdot\mathrm{d}\tau$$

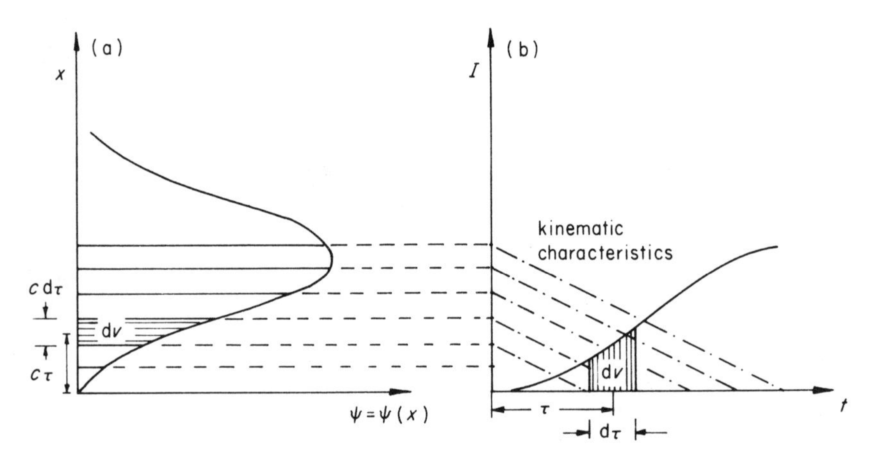

Fig. 3.40. The variation of the width of the idealized catchment with distance from a point of outflow is translated linearly through the straight parallel kinematic characteristics to provide the instantaneous unit hydrograph

whence

$$I(\tau) = \psi(c\tau)\cdot c \tag{3.8.1}$$

Consider next a net precipitation rate $w = w(t)$ in measures of depth per unit time, uniformly distributed over the entire catchment. Then the contribution $\mathrm{d}Q(t)$ to the total outflow $Q(t)$ from the catchment in unit time of the element of length $c\,\mathrm{d}\tau$ in x and distance $c\tau$ removed from $x = 0$ is

$$\mathrm{d}Q(t) = w(t-\tau)\,\psi(c\tau)\cdot c\,\mathrm{d}\tau$$

Thence the total outflow $Q(t)$ in unit time – i.e. the total rate of outflow – is the sum over all such elements:

$$Q(t) = \int_0^t w(t-\tau)\,\psi(c\tau) \cdot c\, d\tau$$

$$= \int_0^t w(t-\tau) I(\tau)\, d\tau \tag{3.8.2}$$

The function $I(\tau)$ defined in this simple model by Equation (3.8.1) is called the *instantaneous* unit hydrograph while Equation (3.8.2) gives the rule for any such linear system that the catchment outflow at any time is the convolution over elapsed time of the instantaneous unit hydrograph and the time series of the net precipitation.

It should be remarked that although the above is one representation of a linear system, it is not at all necessary to restrict linear systems to such models. This is easily seen just by introducing curved, but time-invariant, kinematic characteristics in Fig. 3.40 to obtain the same results. The representation does indicate, however, that a linear system algorithm of the form of (3.8.2) can as well be formulated through a transport algorithm: the two algorithms are essentially equivalent.

3.9 The linearized method of characteristics: periodic waves

One of the essential features of the characteristic method for general nearly horizontal flows is that it describes processes of *interaction*, as opposed to simple *superposition*. This is to say that the meeting of two waves with height Δh does not give rise to a wave of height $2\,\Delta h$, but to some other height, computed through the Riemann invariants. Now in certain cases it is possible to approximate the flow situation by one in which the principle of superposition applies. The approximating theory is then called a 'linearized theory'. One line of reasoning leading to the description 'linearized' appears at once from the forms of the Riemann invariants, which go over from expressions non-linear in depth variation h':

$$u + 2(gh)^{\frac{1}{2}} = u + 2(g(h_0 + h'))^{\frac{1}{2}} = J_+$$

to expressions linear in depth variation:

$$u + \left(\frac{g}{h_0}\right)^{\frac{1}{2}} h' = J_+ - 2(gh_0)^{\frac{1}{2}} = K_+, \text{ another constant} \tag{3.9.1}$$

according to the binomial theorem – or the original equations (3.2.2), (3.2.3) – when $h' \ll h_0$. The linearized theory is thus a 'small-disturbance' or 'small-perturbation' theory.

Clearly, if $h' \ll h_0$ throughout a flow system, then u will also change but little in the system, and in fact $u \ll (gh_0)^{\frac{1}{2}}$. In this case, however, the characteristic directions

$$\dot{x}_{\pm} = u \pm (gh)^{\frac{1}{2}}$$

will go over to

$$\dot{x}_{\pm} = \pm(gh_0)^{\frac{1}{2}} \tag{3.9.2}$$

so that in such a linearized theory the entire characteristic structure is approximated by the structure of a region of constant state. The theory is only sketched in its simplest outlines here, as a first-approximation aid to more detailed numerical study and as a means of introducing some material for subsequent linear analyses. The theory is discussed in detail by Dronkers (1964).

The occurrence of periodic flows is closely associated with two canal boundary conditions, corresponding to an 'open end', where h = constant, or $h' = 0$,

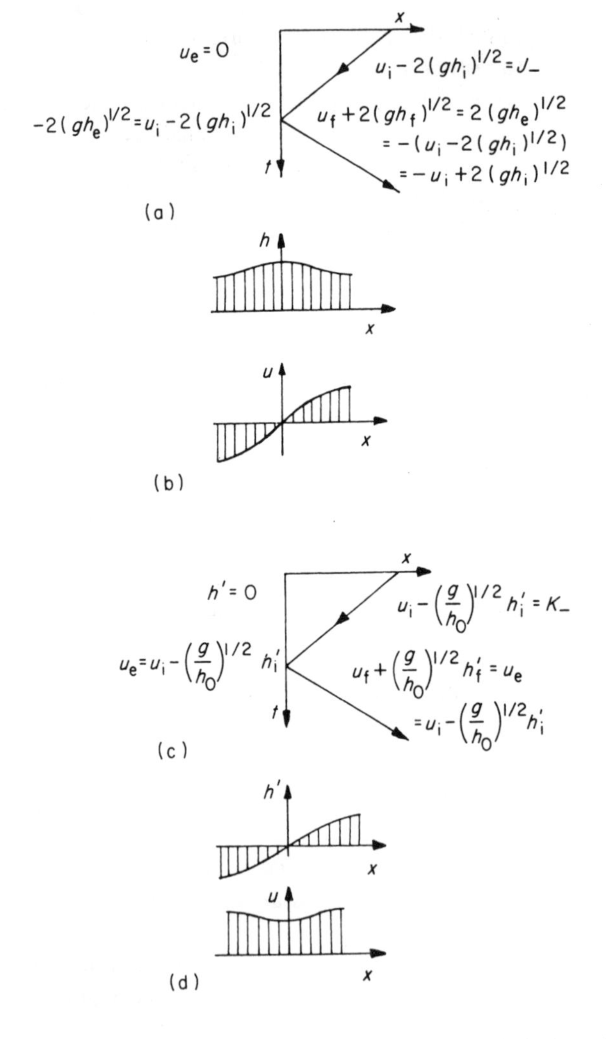

Fig. 3.41. Schematization of the manner in which the boundary conditions of zero velocity and zero depth variation can be replaced by reflection conditions

and to a 'closed end', where $u = 0$. Their essential properties can be demonstrated using Fig. 3.41.

For $u = 0$, the property concerned follows from Fig. 3.41a, where it is seen that following a reflection:

$$u_f + 2(gh_f)^{\frac{1}{2}} = -u_i + 2(gh_i)^{\frac{1}{2}}$$

(subscript i for 'initial', f for 'final'). This is then to say that if instead of the canal being closed, it is extended, at the same time introducing a fluid system to the left whose depths are normal reflections of those on the right and whose velocities are also reflections of those on the right but with opposite sign, the subsequent situation within the canal itself would be entirely unaltered (Fig. 3.41b). This is a general situation for all nearly horizontal flows. For $h' = 0$ and *for the special situations of the linearized theory*, the situation depicted in Fig. 3.41c occurs whereby

$$u_f + \left(\frac{g}{h_0}\right)^{\frac{1}{2}} h_f' = u_i - \left(\frac{g}{h_0}\right)^{\frac{1}{2}} h_i'$$

This corresponds to the situation where the depth variation h' is reversed in sign under reflection while the velocity is reflected normally (Fig. 3.41d).

These concepts are generalized in terms of mirror images in Fig. 3.42.

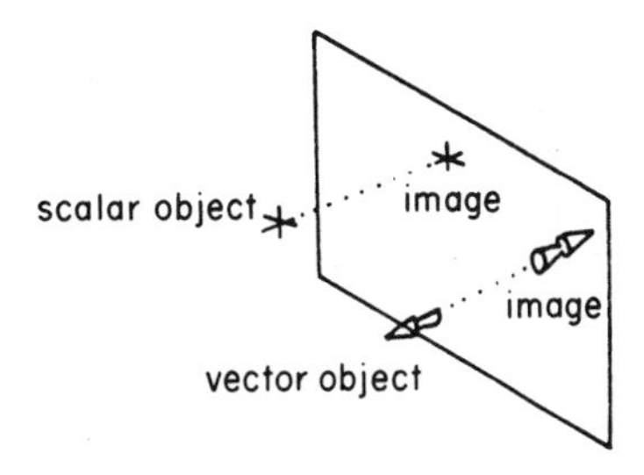

Fig. 3.42. A 'scalar' object appears normally when viewed in a mirror, but a vector orthogonal to the mirror surface appears to be reversed in direction

Functions of the form of $h = h(x)$ in Fig. 3.41b and $u = u(x)$ in Fig. 3.41d are called *even functions*, while functions of the form $u = u(x)$ in Fig. 3.41b and $h' = h'(x)$ in Fig. 3.41d are called *odd functions*.

The linearized theory is often connected to the above boundary conditions, essentially for the reason adumbrated in Fig. 3.43 for $u(0) = u(l) = 0$. It is seen at once that, by virtue of the reflection properties and the constant directions of the $\dot{x}_{\pm}$,

$$u_f + \left(\frac{g}{h_0}\right)^{\frac{1}{2}} h_f' = u_i + \left(\frac{g}{h_0}\right)^{\frac{1}{2}} h_i' \tag{3.9.3}$$

$$u_f - \left(\frac{g}{h_0}\right)^{\frac{1}{2}} h_f' = u_i - \left(\frac{g}{h_0}\right)^{\frac{1}{2}} h_i'$$

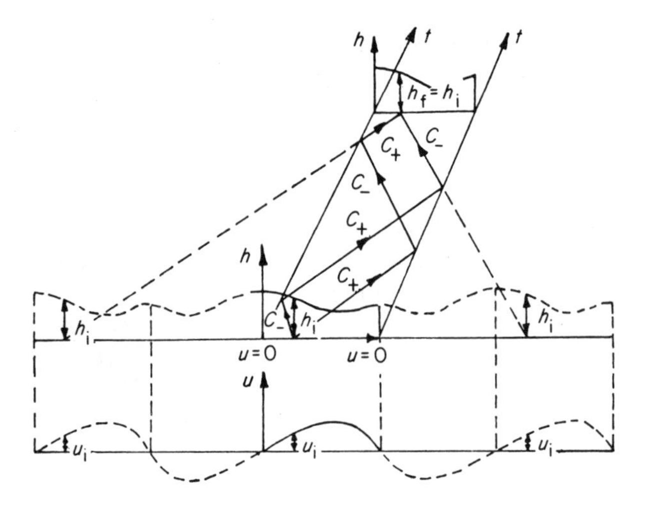

Fig. 3.43. Construction showing the formation of a period solution in a closed region

so that

$$u_f = u_i \quad \text{and} \quad h'_f = h'_i$$

for all $u_i, u_f; h_i, h_f$ associated with the same x and separated in time by a quantity T, called the *natural period* of the system. Any function $f = f(t)$ with the property $f(t + T) = f(t)$ is said to be *periodic* with period T. It is then seen that any long-wave system with boundary conditions $u(0) = u(l) = 0$ is periodic with a period defined as the time taken for a disturbance to traverse the canal once in each direction:

$$T = \frac{2l}{(gh_0)^{\frac{1}{2}}} \tag{3.9.4}$$

The same property of periodicity will evidently obtain for a canal open at both ends to a 'sea', i.e. open to a region so great in extent that it is quite unaffected by the flow in the canal.

A further property of linearized systems arises from their periodicity. It is readily seen that in general, during time, any initial profile transforms into itself through a sequence of profiles of other 'type', i.e. of other function form. One may then ask if there exist functions which retain their type in time. From Fig. 3.43 it is clear that such functions must have the property that when one is superimposed on its translated self, the result is always a function of the same type. Clearly the most common functions with this property are *sine* and *cosine*:

$$\sin \omega t + \sin \omega(t + \tau) = 2 \sin \tfrac{1}{2}(\omega t + \omega(t + \tau)) \cos \tfrac{1}{2}(\omega t - \omega(t + \tau))$$

$$= 2 \cos\left(-\frac{\omega\tau}{2}\right) \cdot \sin\left(\omega t + \frac{\omega\tau}{2}\right)$$

$$\cos \omega t + \cos \omega(t + \tau) = 2 \cos \tfrac{1}{2}(\omega t + \omega(t + \tau)) \cos \tfrac{1}{2}(\omega t - \omega(t + \tau))$$

$$= 2 \cos\left(-\frac{\omega\tau}{2}\right) \cdot \cos\left(\omega t + \frac{\omega\tau}{2}\right)$$

Here ω is the *frequency*, defined by $\omega = 2\pi/T$. (3.9.5)

The above behaviour, described as 'conservation of type', is illustrated in Fig. 3.44 where it is seen to allow an infinity of realizations. The first of these,

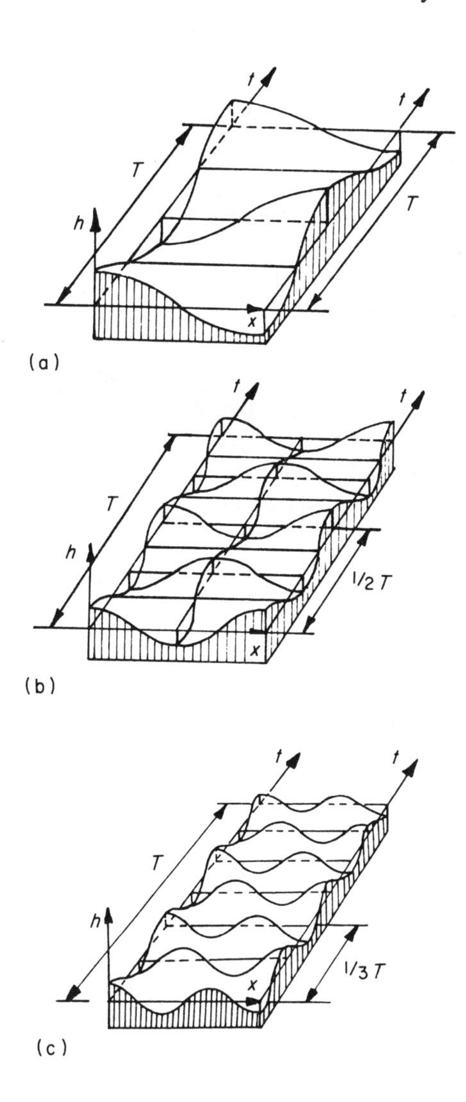

Fig. 3.44. The fundamental or first harmonic (a) and the second harmonic (b) and the third harmonic (c) for simple harmonic oscillations of a fluid. These illustrate the components of the Fourier series used so much in the next chapter

illustrated in Fig. 3.44a, corresponds to motion with the natural period T: in this case the motion is described as *motion in the fundamental mode*. The next interpretation, shown in Fig. 3.44b, has not only a period T, but also a period of $T/2$ for it reappears after $T/2$ as well as T. It is called the *second harmonic*. In Fig. 3.44c the third harmonic is illustrated, with period $T/3$. Clearly, by taking successive aliquot parts of T, further successive harmonics can be introduced. Now the linearized theory has the property that it is always possible to superimpose one solution on another. Since the sine and cosine terms do not 'disperse' in time, it is often convenient to analyse the function of interest into a series in sine and cosine waves, hence into a sine–cosine *Fourier series,* in order to investigate the form of these waves at a later time, which is again of sine or cosine form, and then to build the result at that time by superposition. Thus the profile of Fig. 3.45a can be analysed approximately into the cosine waves of Fig. 3.45b and 3.45c.

The form of the latter at a later time (taken as $T/2$) is easily determined (Figs 3.45d and 3.45c) and then these can be superimposed to give the result at that time (Fig. 3.47f). This procedure, often referred to as the 'linear-periodic' or 'linear-harmonic' approach, is much used in tidal theory (e.g. Dronkers, 1964). It should be remarked that, in all harmonic motions, energy is being transformed from all potential to all kinetic and vice versa, just as in a simple pendulum. The process of decomposition of flows into Fourier components will be used extensively in Chapter 4 in connection with the analysis of numerical stability.

One further simplification should be introduced; it arises when the initial profile $h = h(x)$ is chosen such that $u = 0$. In this case the initial energy is entirely potential. Then Equations (3.9.3) become generally

$$u_{\mathrm{f}} + \left(\frac{g}{h_0}\right)^{\frac{1}{2}} h'_{\mathrm{f}} = \left(\frac{g}{h_0}\right)^{\frac{1}{2}} h'_{\mathrm{i1}}$$

$$u_{\mathrm{f}} - \left(\frac{g}{h_0}\right)^{\frac{1}{2}} h'_{\mathrm{f}} = -\left(\frac{g}{h_0}\right)^{\frac{1}{2}} h'_{\mathrm{i2}}$$

i.e.

$$2\left(\frac{g}{h_0}\right)^{\frac{1}{2}} h'_{\mathrm{f}} = \left(\frac{g}{h_0}\right)^{\frac{1}{2}} (h'_{\mathrm{i1}} + h'_{\mathrm{i2}})$$

or

$$h'_{\mathrm{f}} = \frac{h'_{\mathrm{i1}} + h'_{\mathrm{i2}}}{2}$$

$$u_{\mathrm{f}} = \left(\frac{g}{h_0}\right)^{\frac{1}{2}} \left(\frac{h'_{\mathrm{i1}} - h'_{\mathrm{i2}}}{2}\right) \qquad (3.9.6)$$

A solution in this case is illustrated in Fig. 3.46.

In the flow systems so far considered, with closed ends or with constant level

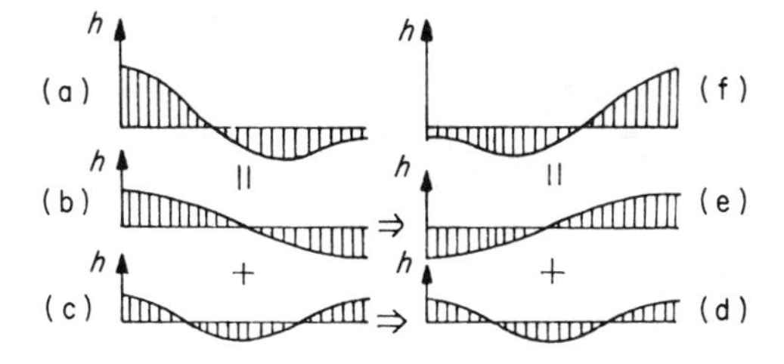

Fig. 3.45. Schematization of a Fourier-series decomposition followed by the transformation in time of each Fourier component in turn and the subsequent recomposition of these components

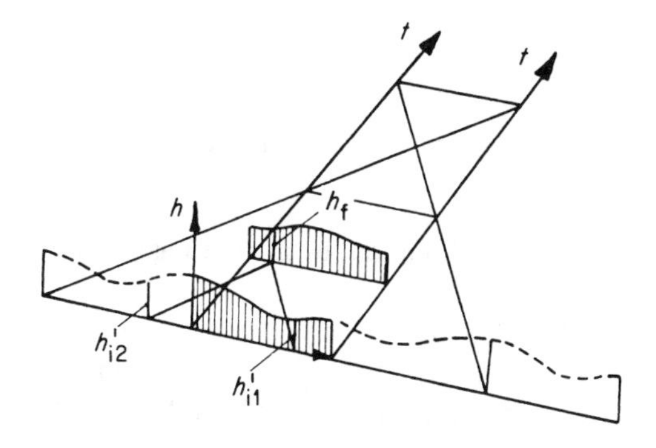

Fig. 3.46. A special linearized form of the method of characteristics obtained when all energy is initially potential

at the ends, all motion occurs with a certain period, determined by the system itself. The motion may be induced by meteorological effects (wind, change of pressure), by surrounding mechanical effects (landslide, ice calving from a glacier, earthquake) or even by certain external hydraulic influences. A motion of this type is often called a *seiche*. It is now seen that the natural period of a seiche is given by

$$T = \frac{2l}{(gh)^{\frac{1}{2}}} \tag{3.9.4}$$

for both ends open or both ends closed (as above), or by

$$T = \frac{4l}{(gh)^{\frac{1}{2}}} \tag{3.9.7}$$

with one end open and one end closed (Fig. 3.47). The phenomenon of seiching occurs, of course, even though the end conditions are not ideally 'closed' or 'open' – although T is still generally approximated by one of the above formulae. If h varies in the system, the period must again be defined as the time for a

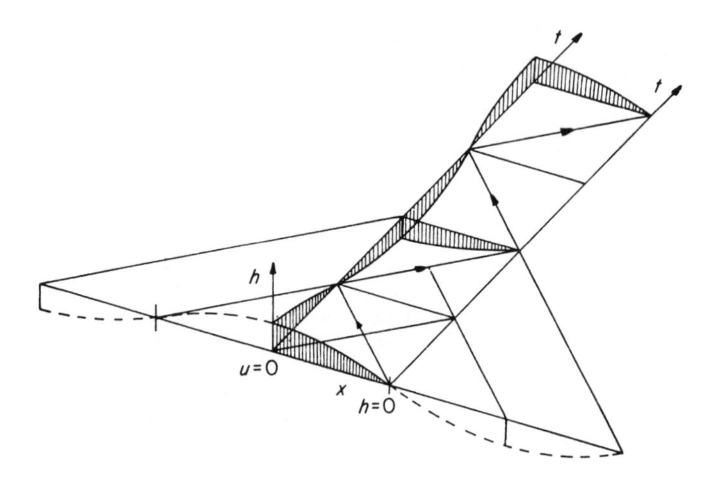

Fig. 3.47. Free oscillations in a basin with one end open and one end closed, so effectively halving the construction of Fig. 3.43

disturbance to traverse the system once in each direction, by integrating celerity over distance.

The above properties of *free oscillation*, or 'pendulation', refer to a system that is in no way forced by its surroundings. Indeed, in the above connection, one often refers to the 'own-frequencies' (eigenfrequencies) and 'own-functions' (eigenfunctions) of a system.

A contrary but related case arises when the fluid system is connected to a much larger one which is in motion, and which accordingly completely impresses its motion upon the much smaller system. In this latter case *forced oscillations* occur, following the motion, or taking the frequency (if any) of the much larger system. The situation in this case is illustrated in Fig. 3.48. It is seen, that, by virtue of the invariant relations, the depth at the open end at zero velocity is transferred to the closed end (where the velocity is always zero) along the corresponding $C_{\pm}$ characteristics. Then a sinusoidal variation at the open end must provide a sinusoidal variation at the closed end and, in the case illustrated, there always occurs an *amplification* of the wave along the canal. It is seen, moreover, that as the canal length l tends to $(gh)^{\frac{1}{2}} \cdot T/4$, so this amplification becomes infinite, i.e. $h(x = 0) \to \infty$ as $l \to (gh)^{\frac{1}{2}} \cdot T/4$ or as $T \to 4l/(gh)^{\frac{1}{2}}$, and this for arbitrary disturbance amplitude at the open end.

This phenomenon is called *resonance*. It is not, of course, restricted to a canal open at only one end. In a pendulum analogy it corresponds to the application of a periodic force with the same period as the natural period of oscillation of the pendulum.

The phenomenon of resonance explains also the 'filtering' effects of certain systems, which amplify those incoming disturbances with the system's natural frequency much more than they amplify disturbances with other periods.

For this reason systems with a distinct 'eigenfrequency' are not directly suit-

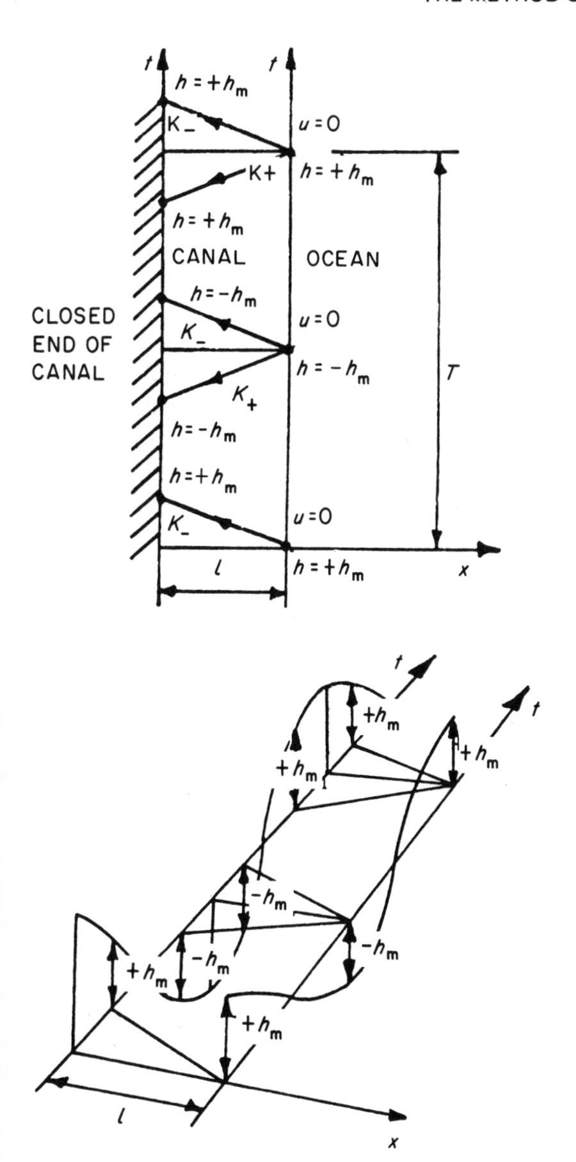

Fig. 3.48. Construction of the amplified wave induced by a periodic forcing surface elevation variation at one end of a canal closed at its other end. Characteristics are set back into the canal from its open end at each time of zero velocity at the end, or, effectively, at each time when the flow is all potential

able for measuring disturbances in the open ocean. If, for example, an attempt is made to make long-wave observations from an island with a wave-cut platform, generally only distorted results are obtained, with disturbances that are aliquot parts of the period of the platform greatly amplified over other disturbances.

The above phenomena of seiching and resonance have been discussed only in terms of two idealized conditions: in the one case a complete closed system, and in the other case a system connected to an ocean of infinite depth (i.e. celerity) and infinite extent. It can happen, however, that connected systems of basins, reservoirs, or channels have to be considered, all of which are of about the same extent so that a variation in any one always gives rise to considerable variations in the others. Such systems are said to be *coupled*. The mechanics of coupled systems is a considerable subject, requiring the introduction of *Rayleigh's principle* (e.g. Lamb, 1932; Temple and Bickley, 1933).

Related to the mechanics of coupled systems is the theory of schematization, whereby complicated flow sections and systems of canals in series and parallel are reduced to simple rectangular sections providing the same attenuation and phase difference. This theory is also outside the scope of the present work. (*See*, for example, Dronkers, 1964.) For rigorous analytical formulations of linear-periodic wave theories, reference is made to Riesz and Nagy (1955, pp. 247–260) and Mikusinski (1959, pp. 201–218).

3.10 The characteristic varieties for two-dimensional nearly horizontal flows

It is useful, especially for the subsequent study of numerical methods for two-dimensional nearly horizontal flows, to derive the characteristic varieties that then correspond to the characteristic lines of the one-dimensional case. For this purpose, the methodology of Daubert and Graffe (1967) is followed, and this methodology is best introduced by applying it to the one-dimensional case already treated at such length above.

Consider, then, the one-dimensional nearly horizontal flow equations in the form

$$\frac{\partial f}{\partial t} + A\frac{\partial f}{\partial x} = 0 \qquad (3.10.1, = 2.6.6)$$

Next define a local relation between t and x, that can later be conditioned to be a characteristic relation through a local origin, in the form $t = \phi(x)$. Then for all f on the variety $t = \phi(x)$,

$$f(x,t) = f(x,\phi(x)) = f_0(x)$$

so that

$$\frac{\partial f_0}{\partial x} = \frac{\partial f}{\partial x} + \frac{\partial \phi}{\partial x}\frac{\partial f}{\partial t} \qquad (3.10.2)$$

Then Equation (3.10.1) can be written as

$$\frac{\partial f}{\partial t} - A\frac{\partial \phi}{\partial x}\frac{\partial f}{\partial t} = -A\frac{\partial f_0}{\partial x}$$

or

$$\left(I - A\frac{\partial \phi}{\partial x}\right)\frac{\partial f}{\partial t} = -A\frac{\partial f_0}{\partial x} \tag{3.10.3}$$

When $t = \phi(x)$ is a characteristic variety, then the derivatives $\partial f/\partial t$ are indeterminate and, so long as they differ from zero, then

$$\text{Det}\left(I - A\frac{\partial \phi}{\partial x}\right) = 0 \tag{3.10.4}$$

Writing $\partial\phi/\partial x = p$ and using the f and the corresponding A of Equation (2.6.7) in Equations (3.10.3), Equation (3.10.4) reads

$$\text{Det}\begin{bmatrix}(1-up) & gp \\ hp & (1-up)\end{bmatrix} = 0$$

i.e.

$$(1-up)^2 - ghp^2 = 0$$

or

$$p = \frac{1}{u \pm (gh)^{\frac{1}{2}}} \tag{3.10.5}$$

Thus the characteristics are defined locally by

$$t = \frac{x}{u \pm (gh)^{\frac{1}{2}}}$$

This corresponds, of course, exactly to the result obtained earlier, and the method used is transparently equivalent to that used in Section 3.5.

In order to extend this method to the two-dimensional equations,

$$\frac{\partial f}{\partial t} + A_1\frac{\partial f}{\partial x} + A_2\frac{\partial f}{\partial y} = 0 \tag{3.10.6, = 2.6.1}$$

a relation $t = \phi(x,y)$ is defined so that for all f on this variety,

$$f(x,y,t) = f(x,y,\phi(x,y)) = f_0(x,y)$$

and

$$\frac{\partial f_0}{\partial x} = \frac{\partial f}{\partial x} + \frac{\partial \phi}{\partial x}\frac{\partial f}{\partial t}$$

$$\frac{\partial f_0}{\partial y} = \frac{\partial f}{\partial y} + \frac{\partial \phi}{\partial y}\frac{\partial f}{\partial t}$$

Then Equation (3.10.6) can be written as

$$\left(I - A_1\frac{\partial \phi}{\partial x} - A_2\frac{\partial \phi}{\partial y}\right)\frac{\partial f}{\partial t} = -A_1\frac{\partial f_0}{\partial x} - A_2\frac{\partial f_0}{\partial y}$$

entirely analogously to Equation (3.10.3), and the indeterminacy of the $\partial f/\partial t$ on the characteristic variety implies that

$$\text{Det}\left(I - A_1 \frac{\partial \phi}{\partial x} - A_2 \frac{\partial \phi}{\partial y}\right) = 0 \tag{3.10.7}$$

which is analogous to Equation (3.10.4).

Using the notation $\partial\phi/\partial x = p$ and $\partial\phi/\partial y = q$ and the f, A_1 and A_2 of Equations (2.6.11), (2.6.12) gives (3.10.7) as

$$\text{Det}\begin{bmatrix} (1 - up - vq) & -gp & 0 \\ -hp & (1 - up - vq) & -hq \\ 0 & -gq & (1 - up - vq) \end{bmatrix} = 0$$

or

$$(1 - up - vq)((1 - up - vq)^2 - gh(p^2 + q^2)) = 0$$

Thus, *either*

$$(1 - up - vq)^2 - gh(p^2 + q^2) = 0 \tag{3.10.8}$$

or

$$(1 - up - vq) = 0 \tag{3.10.9}$$

These two equations define two different families of characteristics. The first, Equation (3.10.8), is simplified by writing

$$p = \rho \cos\theta, \quad q = \rho \sin\theta$$

so that Equation (3.10.8) reduces to

$$1 - 2\rho(u \cos\theta + v \sin\theta) + \rho^2((u \cos\theta + v \sin\theta)^2 - gh) = 0$$

giving

$$\rho = \frac{1}{(u \cos\theta + v \sin\theta \pm (gh)^{\frac{1}{2}})} \tag{3.10.10}$$

Now, locally,

$$t = \frac{\partial \phi}{\partial x} \cdot x + \frac{\partial \phi}{\partial y} \cdot y = px + qy$$

so that the characteristic variety can be expressed locally as

$$t(u \cos\theta + v \sin\theta \pm (gh)^{\frac{1}{2}}) = x \cos\theta + y \sin\theta$$

or

$$\cos\theta(x - ut) + \sin\theta(y - vt) = \pm(gh)^{\frac{1}{2}} t \tag{3.10.11}$$

Differentiating this with respect to θ, it becomes

$$-\sin\theta(x - ut) + \cos\theta(y - vt) = 0 \tag{3.10.12}$$

Then squaring Equations (3.10.11) and (3.10.12) in turn and adding provides

$$(x - ut)^2 + (y - vt)^2 = ght^2 \tag{3.10.13}$$

Equation (3.10.13) defines the *characteristic cone* of two-dimensional nearly horizontal flow. It corresponds to the radial spreading of a disturbance with celerity $(gh)^{\frac{1}{2}}$ in a Galilean frame moving with the convective velocity (u,v). It is schematized in Fig. 3.49. This clearly generalizes the one-dimensional situation of Fig. 3.4. The two-dimensional case is complicated, however, through the exigencies of *Huygens' principle*, the most relevant premise of which may be stated as follows (e.g. Hadamard, 1923; Abbott, 1966): 'If, at an instant $t = 0$ – or more precisely throughout a short interval $\epsilon \leqslant t \leqslant 0$ – a disturbance is produced that is localized in the immediate neighbourhood of P, the effect of it will be, for $t = t'$, localized in the immediate neighbourhood of the surface of the sphere with centre P and radius ct': that is, it will be localized in a very thin spherical shell with centre P including the aforesaid sphere'.

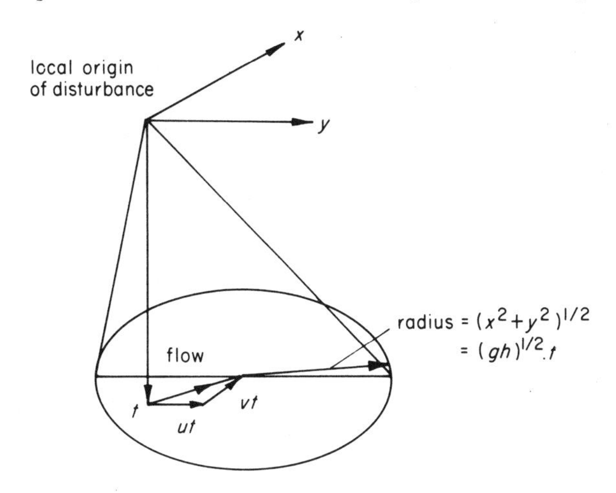

Fig. 3.49. The characteristic cone of two-dimensional nearly horizontal flow

The very validity of Huygens' principle for three-dimensional propagation necessitates that it cannot hold for two-dimensional propagation, although it again reforms for one-dimensional propagation (Hadamard, 1923; Abbott, 1966). Thus, although in ideal one-dimensional propagation the flow conditions at a point are determined completely from characteristically connected conditions at two discrete points of an earlier time level, in even the most ideal two-dimensional propagation the flow conditions at a point P are determined by conditions at all points within the lower-time-level plane intercepted by the characteristic cone emanating from P. On the basis of this principle it will later be seen that the exact solutions of one-dimensional flows can have no exact equivalents in two-dimensional flows.

The other characteristic variety, corresponding to Equation (3.10.9), clearly corresponds to the axis of the cone in Fig. 3.49 and follows, in effect, the convective velocity. In the linear case it obviously follows a line (x = const., y = const.).

In one-dimensional flow, this 'particle path characteristic', with celerity $\dot{x} = u$, is always enclosed within the dynamic characteristics, with celerities $\dot{x} = u \pm (gh)^{\frac{1}{2}}$, while, by virtue of Huygens' principle, it does not introduce further information from outside of these characteristics during their propagation. In two-dimensional flow, on the other hand, it follows from the failure of Huygens' principle that there may be an interaction between the influences following the particle path characteristic variety, so essentially the convective influences, and the influences following the dynamic characteristic variety, so essentially the impulse influences.

On the basis of the properties of the characteristic varieties and their Riemann differential equations, Daubert and Graffe argued that both elevations and velocities should be given at boundaries with inflow, while at boundaries with outflow the elevation alone sufficed. Abbott, Damsgaard and Rodenhuis (1973) pointed out that the two-point inflow conditions could be relaxed, and elevations alone given, when the balance of head loss and resistance effectively determined the magnitudes and directions of the boundary velocities over the area considered. The application of these notions in modelling practice is discussed further in Section 4.11.

4 Numerical methods

4.1 The description problem: first remark

For the purposes of engineering practice, a computation has to describe a particular phenomenon with enough accuracy and at a low enough price to make the entire exercise worth while. Accuracy and price are usually opposing constraints in the design of any numerical model. In principle, at least, the accuracy of the model can be increased indefinitely by spending more, whether upon its field measuring program, its numerics or investigations into its physical foundations. In practice, the accuracy of the model has to be balanced, on the one hand against the accuracy of description of the physical system as derived from field surveys and the available knowledge of basic physical laws and processes and, on the other hand, against the requirements of the job in hand. The field study and computing expenditure for modelling a $100 000 extension to a small fishing harbour, for seiching effects for example, will generally be much less than the modelling of salinities between and around the intakes and outfalls of a $4 000 000 000 desalination plant. Both applications necessitate the use of computational hydraulic methods, but in the first case a few runs of a rather simple and coarse model, based on the available limited data, will probably have to suffice, while in the second case a very advanced and sophisticated modelling exercise is justified, backed up by an extensive field program, closely coordinated with the modelling effort.*

One consequence of this diversity in the needs and possibilities of numerical modelling is that a very large number of alternative numerical methods have been and still are employed in practice. The objective of this chapter is to introduce the main principles underlying these methods.

The methods used with digital computers are, of necessity, discrete methods,

* It unfortunately happens all too often that a critical investigation for expensive works has to be made without adequate (sufficiently redundant) field studies. Worse still, it is then sometimes argued that in this situation only a minor modelling effort is justified. Quite the opposite is in fact true: in just such a case a very extensive modelling study is needed, precisely to investigate the gaps in the field data and to demonstrate the uncertainty of outcome of the works when designed on such an inadequate data base.

in which quantities are computed at a finite number of points in distance and time. These points at which numbers are computed are called 'grid points' or 'nodal points'. Between these grid points nothing is given directly by the computation, but quantities must be determined by a process of interpolation on the quantities given at the grid points. The type of interpolation process used influences not only how quantities are determined from values at the grid points, but also, profoundly, the way in which the computation proceeds, from values of dependent variables at grid points at one time level to their values at the next time level. In principle, the accuracy of the description increases as the accuracy of the interpolation increases, both as concerns the recovery of values between grid-defined values and as concerns the description of the process of change in time.

It follows that the choice of a grid, or net of points, to describe the flow system is intimately related to the methods used to describe the laws determining the behaviour of the flow system. In this chapter, the choice of grid and the choice of governing laws will be considered together.

4.2 The three-point method of characteristics

A natural starting point for this review is the method described in the previous chapter. Given values of u and c, $= (gh)^{\frac{1}{2}}$, at two points in (x,t), the values at a third point, situated at the intersection of characteristics through the two given points, can be determined. When used with a number or digital description, as opposed to the graphical description of the previous chapter, the (x,t) coordinates of the points of computation must be stored as number-pairs. These coordinates are the grid points of the three-point method of characteristics. Evidently the method requires iterations for determining its third points as well as for interpolations at boundaries, all entirely analogous to those used in the graphical method. The accuracy of the method is then limited by the speed of

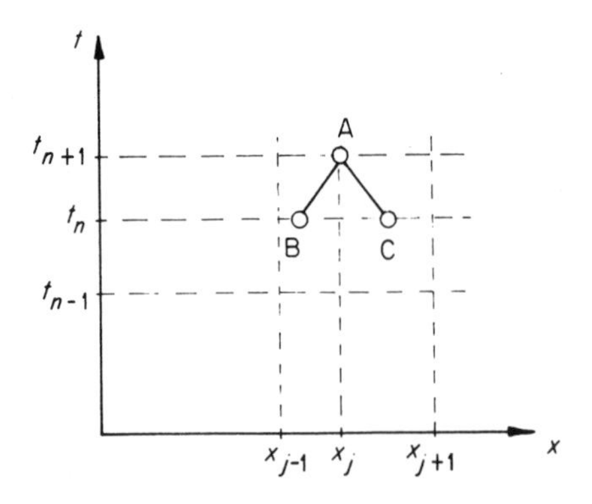

Fig. 4.1. The characteristics in a fixed grid system: explicit formulation

the iterations, which will themselves be functions of the level of refinement and corresponding complexity of the interpolation processes.

The three-point method provides a correspondence of values of (u,c) pairs with values of (x,t) pairs of a type that is quite convenient for plotting contours of u and c in (x,t), using a digital–analog plotting table and its associated software.

However, the use of contours of dependent variables in the space of the independent variables still does not endear itself to most practising engineers, who prefer to look at a length profile of h against x and u against x at specific times or the variation of these quantities in time at a particular place. The user prefers, or the market requires, the use of a grid of the type illustrated in Fig. 4.1. Here the fluid system is defined at distances $x_1, x_2, \ldots, x_{j-1}, x_j, x_{j+1}, \ldots x_{jj}$ written as $\{x_j\}$, and at time $t_1, t_2, \ldots, t_{n-1}, t_n, t_{n+1}, \ldots, t_{nn}$, written as $\{t_n\}$. A total description of the grid then requires the storage of only $(jj \times nn)$ numbers. If $x_j - x_{j-1} = \Delta x$ for all j and/or $t_n - t_{n-1} = \Delta t$ for all n, then the description is simplified further.

The values of u and c at any point A, written as u_a and c_a, can be related to the grid coordinates of A through the notation $u_a = u_j^{n+1}$ and $c_a = c_j^{n+1}$. The three-point method of characteristics says that if lines AB and BC in Fig. 4.1 are characteristic then

$$u_a + 2c_a = u_b + 2c_b$$

$$u_a - 2c_a = u_c - 2c_c$$

or

$$u_a = \frac{u_b + u_c}{2} + (c_b - c_c) \tag{4.2.1}$$

$$c_a = \frac{u_b - u_c}{4} + \frac{(c_b + c_c)}{2} \tag{4.2.2}$$

The problem of finding u_a and c_a is then reduced to one of determining $(u,c)_b$ and $(u,c)_c$, which is essentially a problem of interpolating for these B and C values from values already computed at the $\{x_j\}$ at time level t_n. The following is the simplest procedure:

(1) Determine the positions of x_b and x_c using the characteristic directions known at x_j at t_n:

$$x_b = x_a - \dot{x}_+(t_{n+1} - t_n)$$

$$\approx x_j - (u_j^n + c_j^n)(t_{n+1} - t_n)$$

$$x_c = x_a - \dot{x}_-(t_{n+1} - t_n)$$

$$\approx x_j - (u_j^n - c_j^n)(t_{n+1} - t_n)$$

(2) Interpolate linearly for u and c between x_{j-1} and x_j for values at B and

between x_j and x_{j+1} for values at C.

$$u_b \approx \left(\frac{x_b - x_{j-1}}{x_j - x_{j-1}}\right) u_j^n + \left(\frac{x_j - x_b}{x_j - x_{j-1}}\right) u_{j-1}^n$$

$$c_b \approx \left(\frac{x_b - x_{j-1}}{x_j - x_{j-1}}\right) c_j^n + \left(\frac{x_j - x_b}{x_j - x_{j-1}}\right) c_{j-1}^n$$

$$u_c \approx \left(\frac{x_c - x_j}{x_{j+1} - x_j}\right) u_{j+1}^n + \left(\frac{x_{j+1} - x_c}{x_{j+1} - x_j}\right) u_j^n$$

$$c_c \approx \left(\frac{x_c - x_j}{x_{j+1} - x_j}\right) c_{j+1}^n + \left(\frac{x_{j+1} - x_c}{x_{j+1} - x_j}\right) c_j^n$$

(3) Then apply Equations (4.2.1) and (4.2.2).

(4) The procedure must obviously be modified at a boundary so as, for example, to compute further data-points additional to the one that is given for subcritical flows. The modification is easily made using the same interpolation process.

The above method uses the characteristic directions at (x_j, t_n) as approximations for the mean directions along AB and AC and it uses a linear interpolation that corresponds to passing a first-order algebraic equation between values of the dependent variable at adjacent grid points. Because of this property, the method is said to have only 'first-order accuracy' or to be only 'first-order accurate'. As can be anticipated, its accuracy is very poor but can be improved by making successively better approximations for the characteristic directions and by using a higher-order interpolation scheme. Clearly, in this case, it would not be consistent to improve on the one approximation without simultaneously improving on the other. The above method has a variation that is schematized in Fig. 4.2.

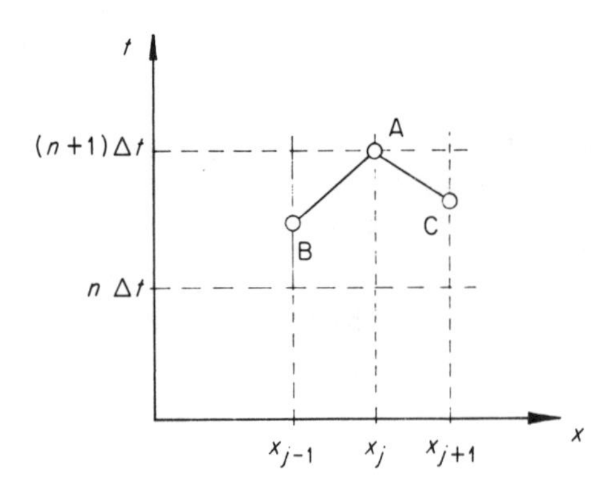

Fig. 4.2. The characteristics in a fixed grid system: implicit formulation

Here the characteristics are taken to lines of constant x. The values of $u_a = u_j^{n+1}$ are then expressed in terms of unknowns $(u,c)_{j+1}^{n+1}$, $(u,c)_j^{n+1}$ and $(u,c)_{j-1}^{n+1}$ as well as the known values $(u,c)_{j+1}^n$, $(u,c)_j^n$ and $(u,c)_{j-1}^n$. The result is a computational

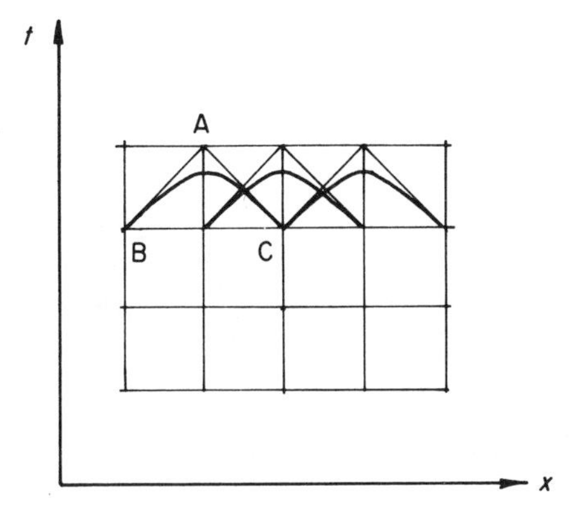

Fig. 4.3. In a linearized scheme the characteristics can always be fitted with a fixed grid system

scheme that involves more than one unknown at the forward time at every j, so that the solution of the scheme necessitates the solution of a set of simultaneous equations. This variant of the three-point method then has to use algorithms of the type normally associated with the implicit schemes discussed in Section 4.7.

Like most methods, the three-point method becomes much simpler when applied to the linearized equations. In this case the grid can be conveniently set to fit the characteristic net, as schematized in Fig. 4.3 with $\Delta x/\Delta t = c = (gh_0)^{\frac{1}{2}}$. The numerical scheme schematized across the points A, B and C in Fig. 4.3 then reduces to Equations (4.2.1) and (4.2.2) without any need for interpolation. Evidently, when c varies as h_0 varies, a grid non-uniform in x can still be found to provide an equivalent simplification.

4.3 The four-point method of characteristics

This method appears to have been originated, in the theory of plasticity, by Hansen (1965), although certain of its properties, especially those relating to linearized equations, were known very much earlier. The remarkable symmetry of the method appears, at its most elementary level, when an element of the (x,t) space is considered that is bounded by two C_+ and two C_- characteristics, as shown in Fig. 4.4. The points of intersection of these bounding characteristics are denoted by the integers 1, 2, 3 and 4, ordered so that they increase as one passes, clockwise or anticlockwise, around the element, and so that the first two (1, 2) span a + characteristic. In the system so defined, although conditions at any two opposite points are sufficient to determine conditions at a third point, conditions at the remaining point are also introduced. However, so long as exact relations are used between the dependent variables at 1, 2, 3 and 4, the system is not overdetermined and the existence of solutions is not thereby compromised. The fourth point thus functions as a 'memory', in the manner familiar from the

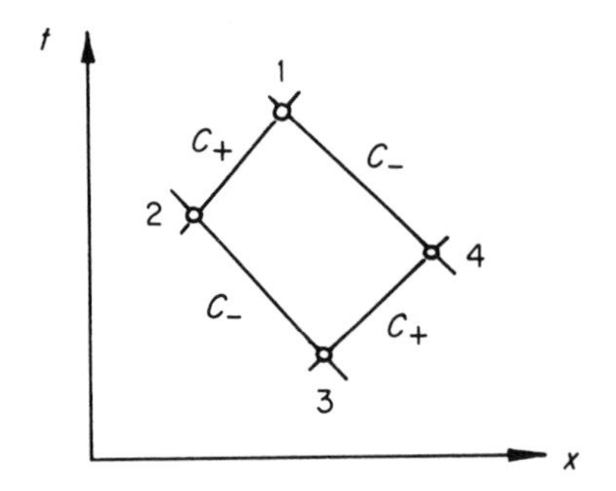

Fig. 4.4. Addressing sequence for the four-point method of characteristics

numerical analysis of ordinary differential equations (e.g. Kopal, 1955). In order to notate the symmetry of the four-point method it is convenient to introduce the concept of 'the set of natural numbers modulo four'. The properties of this set are that:

(a) it consists of the four integers, 1, 2, 3 and 4. Thus, if (j) is any element of the set

$$(j) + 1 = (j + 1)$$

(b) it satisfies the condition that for any element (j)

$$(j) + 4 = (j)$$

For example:

$$1 + 1 = 2, \quad 2 + 1 = 3, \quad 2 + 3 = 1 + 4 = 1$$

It is easily seen that this set, with the law of composition +, forms a group (Lederman, 1957). With this convention, the path from point j to point $j + 1$, is taken along a $(-1)^{j+1}$ characteristic.

The Riemann invariants for the element of Fig. 4.4 appear as follows:

$$\left.\begin{aligned} u_1 + 2c_1 &= u_2 + 2c_2 \\ u_3 + 2c_3 &= u_4 + 2c_4 \\ u_1 - 2c_1 &= u_4 - 2c_4 \\ u_3 - 2c_3 &= u_2 - 2c_2 \end{aligned}\right\} \tag{4.3.1}$$

where u_j is the velocity at point j and $c_j = (gh_j)^{\frac{1}{2}}$ with h_j the fluid depth at point j. If j is an element of the set of integers modulo four, the Riemann equations (4.3.1) can be written as

$$(u_j - u_{j+1}) = (-1)^j \, 2(c_j - c_{j+1}) \tag{4.3.2}$$

Adding, with the $(-1)^j$ successively to the right and to the left, provides the relations

$$\sum_j (-1)^j u_j = 0 \tag{4.3.3}$$

$$\sum_j (-1)^j c_j = 0 \tag{4.3.4}$$

Thus, for example, for explicit computation of point 1 in terms of points 2, 3 and 4:

$$u_1 = u_2 - u_3 + u_4$$

$$c_1 = c_2 - c_3 + c_4$$

It follows at once from Equations (4.3.3) and (4.3.4) that

$$\sum_j (-1)^j (\underset{+}{\dot{x}})_j = 0, \quad \sum_j (-1)^j (\underset{-}{\dot{x}})_j = 0 \tag{4.3.5}$$

$$\sum_j (-1)^j (\underset{+}{J})_j = 0, \quad \sum_j (-1)^j (\underset{-}{J})_j = 0 \tag{4.3.6}$$

where the $\underset{(\pm)_j}{\dot{x}}$, $(= u_j \pm c_j)$, are the characteristic directions and $\underset{(\pm)_j}{J}$, $(= u_j \pm 2c_j)$ are the Riemann invariants. Stronger statements of Equations (4.3.5) and (4.3.6) can be formed by further introducing the notation

$$\left(\underset{(-)^k}{\dot{x}}\right)_j \overset{\text{def}}{=} u_j + (-1)^k c_j \qquad \left(\underset{(-)^k}{J}\right)_j \overset{\text{def}}{=} u_j + (-1)^k 2c_j \tag{4.3.7}$$

Then Equations (4.3.5) and (4.3.6) contract to:

$$\sum_j (-1)^j \left(\underset{(-)^k}{\dot{x}}\right)_j = 0, \quad k = 1, 2$$

$$\sum_j (-1)^j \left(\underset{(-)^k}{J}\right)_j = 0, \quad k = 1, 2$$

Higher-order symmetries can be built from these relations. Thus Equation (4.3.3) and (4.3.4) continue to apply to the system illustrated in Fig. 4.5, except that the set $\{j\}$ then becomes the set of integers modulo six. It may be remarked that relations (4.3.3) onwards are no longer subject to the restriction that the path from point j to point $j + 1$ must be along a $(-1)^{j+1}$ characteristic. They are, so to speak, 'more symmetric' than the Riemann invariant relations.

Relations (4.3.3) and (4.3.4) may be used to determine arc and chord length relations, as exemplified by Hansen (1965) for the case of plastic deformations, where the characteristics are approximated by circular arcs. In hydraulics, however, these relations are physically meaningless and explicit relations for x–t coordinates are required, with a view to realizing results with an x–t plotter.

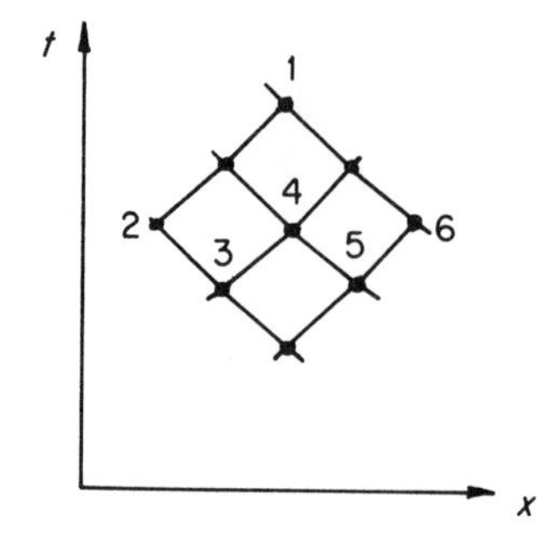

Fig. 4.5. Addressing sequence for a higher-order method of characteristics

The coordinates of descriptive points can be found from the relations

$$\left[\underset{(+)1}{\dot{x}} + \underset{(+)2}{\dot{x}}\right]\left[t_1 - t_2\right] \approx 2\left[x_1 - x_2\right]$$

$$\left[\underset{(-)2}{\dot{x}} + \underset{(-)3}{\dot{x}}\right]\left[t_2 - t_3\right] \approx 2\left[x_2 - x_3\right]$$

$$\left[\underset{(+)3}{\dot{x}} + \underset{(+)4}{\dot{x}}\right]\left[t_3 - t_4\right] \approx 2\left[x_3 - x_4\right]$$

$$\left[\underset{(-)4}{\dot{x}} + \underset{(-)1}{\dot{x}}\right]\left[t_4 - t_1\right] \approx 2\left[x_4 - x_1\right]$$

where t_j and x_j are the time and distance coordinates respectively of the point j. This equation can be written more generally, using the notation of Equation (4.3.7), as

$$\left[\left(\underset{(-)}{\dot{x}}{}^{j}\right)_{j-1} + \left(\underset{(-)}{\dot{x}}{}^{j}\right)_{j}\right]\left[t_{j-1} - t_j\right] \approx 2\left[x_{j-1} - x_j\right] \tag{4.3.8}$$

or, through Equation (4.3.7) again, as

$$[(u_{j-1} + u_j) + (-1)^j(c_{j-1} + c_j)]\,[t_{j-1} - t_j] \approx 2[x_{j-1} - x_j] \tag{4.3.9}$$

Eliminating x_j in Equation (4.3.8) or (4.3.9) by summing over all j provides

$$\sum_j \left[\left(\underset{(-)}{\dot{x}}{}^{j}\right)_{j-1} + \left(\underset{(-)}{\dot{x}}{}^{j}\right)_{j}\right]\left[t_{j-1} - t_j\right] \approx 0 \tag{4.3.10}$$

or

$$\sum_j\ [(u_{j-1} + u_j) + (-1)^j(c_{j-1} + c_j)]\,[t_{j-1} - t_j] \approx 0 \tag{4.3.11}$$

Collecting terms in t_j in Equation (4.3.10)

$$\sum_j \left[\left(\underset{(-)}{\dot{x}}{}^{j+1}\right)_{j+1} + \left(\underset{(-)}{\dot{x}}{}^{j+1}\right)_{j} - \left(\underset{(-)}{\dot{x}}{}^{j}\right)_{j} - \left(\underset{(-)}{\dot{x}}{}^{j}\right)_{j-1}\right] t_j \approx 0 \tag{4.3.12}$$

which expresses the set $\{t_j\}$ entirely in terms of the set $\{\underset{\pm}{\dot{x}_j}\}$.

Collecting terms in t_j in Equation (4.3.11), however, shows that

$$\sum_j [(u_{j+1} - u_{j-1}) + (-1)^{j+1}(c_{j+1} + 2c_j + c_{j-1})]\, t_j \approx 0$$

It follows from (4.3.2) that,

$$\begin{aligned}(u_{j+1} - u_{j-1}) &= (u_{j+1} - u_j) + (u_j - u_{j-1}) \\ &= (-1)^j \cdot 2(c_{j+1} - c_j) + (-1)^{j-1} \cdot 2(c_j - c_{j-1}) \\ &= (-1)^{j-1} 2(-c_{j+1} + 2c_j - c_{j-1})\end{aligned}$$

Hence

$$\sum_j (-1)^{j+1}(-c_{j+1} + 6c_j - c_{j-1})t_j \approx 0$$

or, by virtue of Equation (4.3.4):

$$\sum_j (-1)^{j+1}(5c_j - c_{j+2})t_j \approx 0 \tag{4.3.13}$$

which expresses the set $\{t_j\}$ entirely in terms of the set $\{c_j\}$.

For example, if t_1 is to be found:

$$t_1(5c_1 - c_3) - t_2(5c_2 - c_4) + t_3(5c_3 - c_1) - t_4(5c_4 - c_2) \approx 0$$

gives explicitly

$$t_1 \approx \frac{+t_2(5c_2 - c_4) - t_3(5c_3 - c_1) + t_4(5c_4 - c_2)}{(5c_1 - c_3)}$$

Equation (4.3.13) is invariant under interchanges of its even and odd components, so that it is again not subject to the restriction that the path j to $j + 1$ be along a $(-1)^j$ characteristic. Clearly none of the $\{t_j\}$ becomes singular in any physically realistic problem.

The above elimination cannot be repeated for the x_j, since in the present case the independent variables x and t, being of different dimension, provide an essentially non-isotropic space, with the result that $1/\dot{x}$ cannot have the same symmetry as $\dot{x}$ itself. (In the theory of plasticity, on the other hand, where the independent variables are distances, x and y, the space is isotropic and the elimination is always possible.) From Equation (4.3.8) however, it follows that

$$\sum_j \begin{pmatrix}\dot{x}\\(+)\end{pmatrix}_j \approx 2\,\frac{[x_1 - x_2]}{[t_1 - t_2]} + 2\,\frac{[x_3 - x_4]}{[t_3 - t_4]}$$

and

$$\sum_j \begin{pmatrix}\dot{x}\\(-)\end{pmatrix}_j \approx 2\,\frac{[x_4 - x_1]}{[t_4 - t_1]} + 2\,\frac{[x_2 - x_3]}{[t_2 - t_3]}$$

whence, adding,

$$\sum_j u_j \approx \sum_j \left(\frac{x_{j-1} - x_j}{t_{j-1} - t_j}\right)$$

or, subtracting:

$$\sum_j c_j \approx \sum_j (-1)^j \left(\frac{x_{j-1} - x_j}{t_{j-1} - t_j}\right)$$

Collecting terms, provides

$$\sum_j u_j \approx \sum_j \left(\frac{1}{t_j - t_{j+1}} - \frac{1}{t_{j-1} - t_j}\right) x_j \tag{4.3.14}$$

$$\sum_j c_j \approx \sum_j (-1)^{j+1} \left(\frac{1}{t_j - t_{j+1}} + \frac{1}{t_{j-1} - t_j}\right) x_j \tag{4.3.15}$$

It should be noted that Equations (4.3.14) and (4.3.15) do not provide entirely independent expressions connecting the $\{x_j\}$, as it is possible that one (or two) of the $\{x_j\}$ become singular. Thus, if x_1 is determined from Equation (4.3.14):

$$\sum_j u_j \approx \left(\frac{1}{t_1 - t_2} - \frac{1}{t_4 - t_1}\right) x_1 + \left(\frac{1}{t_2 - t_3} - \frac{1}{t_1 - t_2}\right) x_2 + \\ + \left(\frac{1}{t_3 - t_4} - \frac{1}{t_2 - t_3}\right) x_3 + \left(\frac{1}{t_4 - t_1} - \frac{1}{t_3 - t_4}\right) x_4$$

provides

$$x_1 \approx \left\{ \frac{\sum_j u_j - \left[\left(\frac{1}{t_2 - t_3} - \frac{1}{t_1 - t_2}\right) x_2 + \left(\frac{1}{t_3 - t_4} - \frac{1}{t_2 - t_3}\right) x_3 + \left(\frac{1}{t_4 - t_1} - \frac{1}{t_3 - t_4}\right) x_4\right.}{\frac{1}{t_1 - t_2} - \frac{1}{t_4 - t_1}} \right.$$

This becomes singular when

$$(t_1 - t_2) = (t_4 - t_1)$$

which is a situation that can easily occur, as for example in a region of constant state, as depicted in Fig. 4.6. However, whenever

$$\left(\frac{1}{t_j - t_{j+1}} - \frac{1}{t_{j-1} - t_j}\right) \approx 0,$$

it is clear that

$$\left(\frac{1}{t_j - t_{j+1}} + \frac{1}{t_{j-1} - t_j}\right)$$

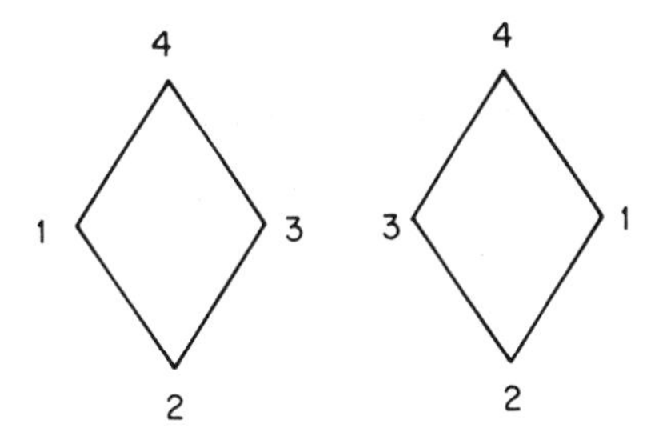

Fig. 4.6. Alternative addressing used in the four-point method in order to avoid a singularity

can never be zero, so that whenever Equation (4.3.14) leads to a singular value for x_1, then Equation (4.3.15) will provide the correct value and, of course conversely. This leads to the use of Equation (4.3.14) for computing x_j when the sense of computation is as shown in Fig. 4.7a, and Equation (4.3.15) when the sense of computation appears as in Fig. 4.7b. This is to say that Equation (4.3.14)

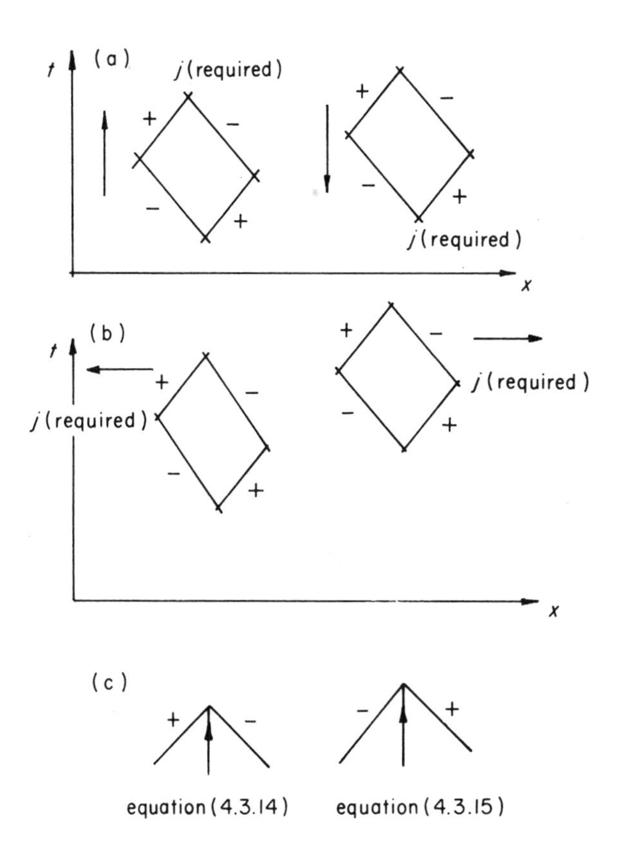

Fig. 4.7. Procedures for avoiding singularities

is used whenever the + characteristic is to the left of the direction of computation and Equation (4.3.15) is used whenever the + characteristic is to the right of this direction (Fig. 4.7c), when t is in the direction shown.

The method is easily extended to account for boundary conditions and a typical output is shown in Fig. 4.8.

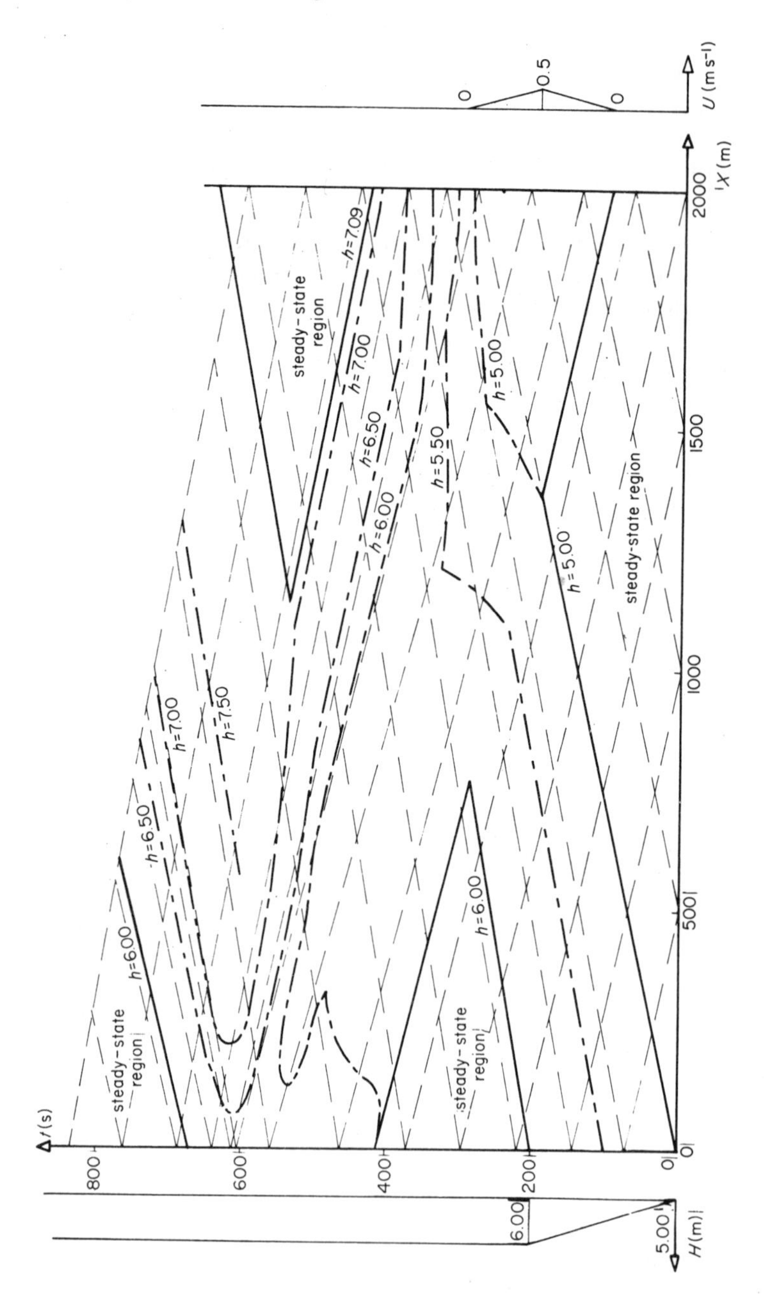

Fig. 4.8. Physical plane characteristics and isolines of water depth and velocity plotted from four-point method output

The four-point method of characteristics appears to be generally superior to the three-point method for machine computation, and indeed it provides a method of exceptional accuracy. However, its widest field of application is to linearized schemes. In Fig. 4.9, it is schematized for the one-dimensional situation, when it reads, for the primitive situation,

$$u_j^{n+1} = u_{j+1}^n + u_{j-1}^n - u_j^{n-1} \tag{4.3.16}$$

$$c_j^{n+1} = c_{j+1}^n + c_{j-1}^n - c_j^{n-1} \tag{4.3.17}$$

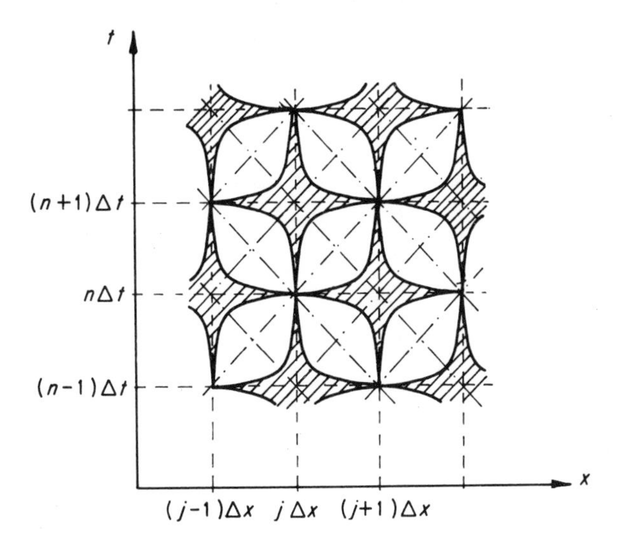

Fig. 4.9. In one dimension the linearized four-point operator coincides with the leap-frog operator of numerical analysis under linearization

Evidently, under the linearizing assumptions, Equation (4.3.17) can be expanded in a binomial series in the variation in h, written h', which, when truncated at its first variation, gives

$$h_j'^{n+1} = h_{j+1}'^n + h_{j-1}'^n - h_j'^{n-1} \tag{4.3.18}$$

These expressions hold for $j = 1, 2 \ldots jj - 1$ with values at $j = 0$ and $j = jj$ supplied from the boundary conditions.

Referring to Fig. 4.9, it is seen that Equation (4.3.18), for example, can be used to construct a solution in h' from two-point initial data $h' = h'(x,0)$ and $h' = h(x,\Delta t)$ and subject to one-point boundary data $h' = h'(0,t)$ at the left-hand end and $h' = h'(jj\,\Delta x,t)$ at the right-hand boundary. When the initial data is given in other forms (as, for example, h' and u), these data can always be transformed to the h' form by using the three-point method, in effect as a starting sub-routine. Similarly, when the boundary data are given in other forms (such as in u) then either the reflection conditions of Section 3.7 can be introduced or the three-point method can be used. In the sequel, h' will be written as h.

When viewed as numerical schemes, Equations (4.3.16) to (4.3.18) advance values given at $(n-1)\,\Delta t$ to values at $(n+1)\,\Delta t$ using the intermediate values at $n\,\Delta t$. A method of this type is accordingly popularly called a 'leapfrog' scheme. Since the above one-dimensional four-point method derives from an exact solution of the primitive nearly horizontal flow equations without the intervention of numerical approximations, Equations (4.3.16)–(4.3.18) constitute exact solutions of the linearized wave equation (2.6.17). Evidently, this method is easily extended to non-uniform mean depths by the use of non-equidistant grid spacings, as introduced above for the three-point scheme.

The linear four-point method can be modified very easily to allow for energy-diffusing terms, of which the most important practically is resistance. It is supposed that τ can always be calculated locally as a function of u (and, in principle, h), i.e. $\tau_j^n = \tau(u_j^n, h_j^n)$ or in practice $\tau_j^n = \tau(u_j^n)$. Use can then be made of the relations (Fig. 4.10):

$$\left(u_1 + \left(\frac{g}{h_0}\right)^{\frac{1}{2}} h_1\right) - \left(u_4 + \left(\frac{g}{h_0}\right)^{\frac{1}{2}} h_4\right) = -\int_{t_4}^{t_1} \frac{\tau}{\rho h}\,\mathrm{d}t$$

$$\left(u_4 - \left(\frac{g}{h_0}\right)^{\frac{1}{2}} h_4\right) - \left(u_3 - \left(\frac{g}{h_0}\right)^{\frac{1}{2}} h_3\right) = -\int_{t_3}^{t_4} \frac{\tau}{\rho h}\,\mathrm{d}t \tag{4.3.19}$$

$$\left(u_1 - \left(\frac{g}{h_0}\right)^{\frac{1}{2}} h_1\right) - \left(u_2 - \left(\frac{g}{h_0}\right)^{\frac{1}{2}} h_2\right) = -\int_{t_2}^{t_1} \frac{\tau}{\rho h}\,\mathrm{d}t$$

$$\left(u_2 + \left(\frac{g}{h_0}\right)^{\frac{1}{2}} h_2\right) - \left(u_3 + \left(\frac{g}{h_0}\right)^{\frac{1}{2}} h_3\right) = -\int_{t_3}^{t_2} \frac{\tau}{\rho h}\,\mathrm{d}t \tag{4.3.20}$$

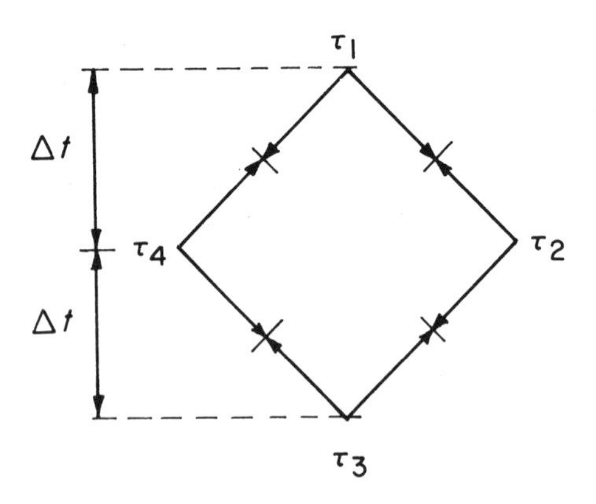

Fig. 4.10. Distribution of resistances in the four-point method

From these, it is found, by eliminating u_1, u_2, u_3 and u_4 in Equations (4.3.17) and (4.3.20), that

$$2\left(\frac{g}{h_0}\right)^{\frac{1}{2}}(h_1 + h_3 - h_2 - h_4) = \int_{t_2}^{t_1} \frac{\tau}{\rho h}\,\mathrm{d}t + \int_{t_3}^{t_2} \frac{\tau}{\rho h}\,\mathrm{d}t - \int_{t_4}^{t_1} \frac{\tau}{\rho h}\,\mathrm{d}t - \int_{t_3}^{t_4} \frac{\tau}{\rho h}\,\mathrm{d}t$$

$$= \int_{321} \frac{\tau}{\rho h}\,\mathrm{d}t - \int_{341} \frac{\tau}{\rho h}\,\mathrm{d}t$$

Taking $\tau_{321} = \frac{\tau_3}{4} + \frac{\tau_2}{2} + \frac{\tau_1}{4}$ and $\tau_{341} = \frac{\tau_3}{4} + \frac{\tau_4}{2} + \frac{\tau_1}{4}$, as schematized in Fig. 4.10, the integrals can be approximated by linear interpolations to obtain

$$2\left(\frac{g}{h_0}\right)^{\frac{1}{2}}(h_1 + h_3 - h_2 - h_4) = \frac{1}{\rho h_0}\left(\frac{\tau_2 - \tau_4}{2}\right) \cdot 2\Delta t$$

This may finally be written in difference form (Section 4.5) as

$$(h_j^{n+1} + h_j^{n-1} - h_{j+1}^n - h_{j-1}^n) - \frac{1}{\rho(gh_0)^{\frac{1}{2}}}\left(\frac{\tau_{j+1}^n - \tau_{j-1}^n}{2}\right)\Delta t = 0 \tag{4.3.21}$$

The following form is derived in a similar manner:

$$h_j^{n+1} = h_j^{n-1} + \left(\frac{h_0}{g}\right)^{\frac{1}{2}}(u_{j+1}^n - u_{j-1}^n) \tag{4.3.22}$$

and, since it is necessary to find u_j^n in order to obtain τ_j^n, this is rather more convenient. Similarly, one can derive

$$u_j^{n+1} + u_j^{n-1} - u_{j+1}^n - u_{j-1}^n - \frac{\Delta t}{\rho h_0}(\tau_j^{n+1} - \tau_j^{n-1}) = 0$$

which is inconvenient, in that it contains both u_j^{n+1} and τ_j^{n+1}, $= \tau(u_j^{n+1})$. It then seems best to use

$$u_j^{n+1} = u_j^{n-1} - \left(\frac{g}{h_0}\right)^{\frac{1}{2}}(h_{j+1}^n - h_{j-1}^n) - \left(\frac{\tau_{j+1}^n + \tau_{j-1}^n}{\rho h_0}\right)\Delta t \tag{4.3.23}$$

and then $\tau_j^{n+1} = \tau(u_j^{n+1})$. If required, the value of τ_j^{n+1} so found can be used to obtain a somewhat better approximation to the dissipation integrals:

$$u_j^{n+1} = u_j^{n-1} - \left(\frac{g}{h_0}\right)^{\frac{1}{2}}(h_{j+1}^n - h_{j-1}^n) - \left(\frac{\tau_j^{n+1} + \tau_{j+1}^n + \tau_j^{n-1} + \tau_{j-1}^n}{2\rho h_0}\right)\Delta t$$

Such an iterative approach is rarely necessary, however.

The four-point method can be extended as well to two-dimensional nearly horizontal flows under the linearizing assumptions. Consider the case of the primitive equations, putting aside all bathymetric complications, resistances, Coriolis accelerations, etc. Then divide up the x–y plane into a system of parallel canals in x and a similar system in y, as shown in Fig. 4.11. Then, in the x-direction,

$$h_{j,k}^{n+1}\Big|_x + h_{j,k}^{n-1}\Big|_x - h_{j+1,k}^n - h_{j-1,k}^n = 0 \tag{4.3.24}$$

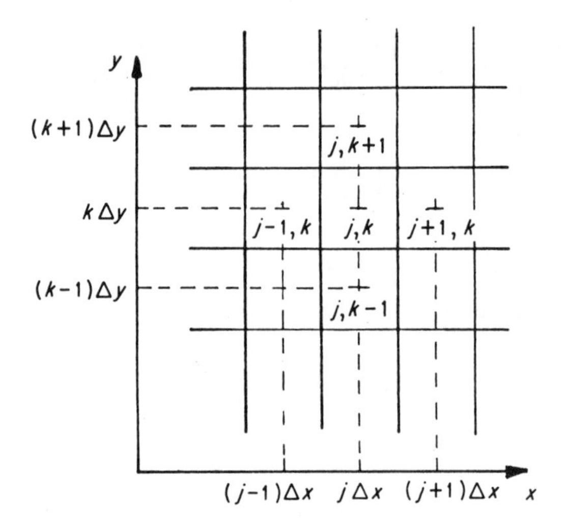

Fig. 4.11. Canal schematization for the use of the linearized four-point method for two-dimensional linearized flows

from Equation (4.3.18) applied in the x-direction, and

$$h_{j,k}^{n+1}\Big|_y + h_{j,k}^{n-1}\Big|_y - h_{j,k+1}^n - h_{j,k-1}^n = 0 \tag{4.3.25}$$

from the same equation applied in the y-direction. If it is now supposed that the x-canals and y-canals do not interact, the following averaging process can be introduced:

$$(h_{j,k}^{n\pm1} - h_{j,k}^n) = (h_{j,k}^{n\pm1}\Big|_x - h_{j,k}^n) + (h_{j,k}^{n\pm1}\Big|_y - h_{j,k}^n) \tag{4.3.26}$$

i.e. the *change* in h over Δt is set as the sum of the *change* due to x influences and the *change* due to y influences. Adding Equations (4.3.24) and (4.3.25) and introducing (4.3.26) provides

$$h_{j,k}^{n+1} + h_{j,k}^{n-1} = h_{j+1,k}^n + h_{j-1,k}^n + h_{j,k+1}^n + h_{j,k-1}^n - 2h_{j,k}^n \tag{4.3.27}$$

This expression appears to allow the construction of solutions for nearly horizontal flows in x- and y-directions, so long, of course, as $u \ll (gh)^{\frac{1}{2}}$. The operator (4.3.27) is schematized in Fig. 4.12.

It should be remarked that whereas Equation (4.3.18) constitutes an exact solution for linearized nearly horizontal flows, Equation (4.3.27) constitutes only an approximation, owing, on the one hand, to the approximation inherent in Equation (4.3.26), and, on the other hand, to the diffusive nature of two-dimensional flow, as introduced in Section 3.10.

By making Δx and Δt sufficiently small, however (and, of course, always such that $\Delta x/\Delta t = (gh_0)^{\frac{1}{2}}$), the error involved can be made arbitrarily small.

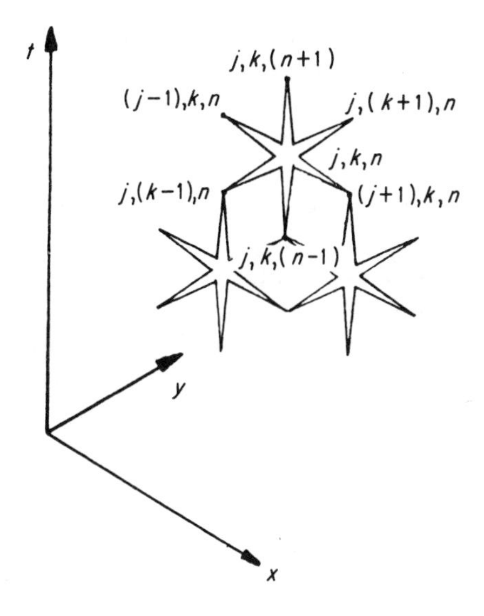

Fig. 4.12. The four-point operator also coincides with the leap-frog operator under linearization in two-dimensional flows

This aspect of the scheme (4.3.27) will be considered in more detail in Section 4.10, where, indeed, it will be shown to be *unstable* and therefore unusable.

4.4 The description problem: second remark

There are, in applied mathematics, two ways of viewing the world: the continuous and the discrete. The one leads to differential equations and the other to difference equations. The earlier physicists and mathematicians of the Enlightenment, such as Galileo, Huygens, Newton, Leibnitz and Euler, used both continuum and discrete mathematics dualistically, but after Euler the tendency was increasingly towards the continuum representation, with its prestigious superstructure of mathematical analysis, while the discrete point of view was comparatively neglected and had no comparable prestige. However, the very development of analysis showed that it contained serious internal contradictions, especially as concerned the treatment of infinities, and this led to a concentrated study of the relations between countable sets and sets of the power of the continuum. It fell to Cantor, in 1873, to show that the infinite countable sets were not equivalent to sets of the power of the continuum; that is to say, that the elements of the one could not be put into a one-to-one correspondence with the elements of the other. So long as methods of analysis are used to describe continuous flow problems, such as the method of characteristics is used to describe flows that are nearly horizontal everywhere, this fundamental non-equivalence of the two points of view is of no great practical consequence. However, as the centre of interest turns to discrete representations, through difference methods, so this non-equivalence becomes of increasing importance and, as will appear, it underlies the entire non-equivalence of momentum and energy formulations in difference methods, and much else besides. It is this non-equivalence that makes numerical instability possible – and unstable schemes are just as common as stable schemes. It is this non-equivalence that also makes possible the subsuming of continuous and discontinuous flows into a single theory, based on the discrete formulation. Clearly this non-equivalence constitutes one of the foundations of computational hydraulics, but as its more general formulation necessitates some preparation, it is better first to consider difference methods in some detail and then to return to this fundamental aspect in a separate chapter. The simplest of the difference methods are those appertaining to the degenerate case of monoclinal wave propagation, as described by the scalar wave equation, and these will be described first, by way of an introduction to difference methods.

4.5 Introduction to direct difference methods using the example of the scalar wave equation

In the differential formulation of Equation (3.7.7), the scalar wave equation was seen to be an equation that could be expressed in its simplest form as

$$\frac{\partial z}{\partial t} + c(z) \cdot \frac{\partial z}{\partial x} = 0 \tag{4.5.1}$$

This equation expresses the transport of any property z with celerity $c(z)$ in the region under consideration. Any such differential equation is a statement about an infinitesimal region, or about conditions at a point. In difference methods this statement is extended to comprehend a finite region. Consider the solution surface of an equation of the type (4.5.1), drawing upon this surface a finite grid (Fig. 4.13).

The projection of this grid on the x–t plane for the case of constant intervals in x and t is shown in Fig. 4.14a. One approach to the difference formulation is to consider any point P and to proceed to relate the function values at the grid points at and around P so as to form approximations to partial derivatives. In order to illustrate the approximation process, a section at $n\,\Delta t$, i.e. parallel to the z–x plane, is sketched in Fig. 4.14b.

The derivative $\partial z/\partial x$ at P is the slope of the tangent TT in Fig. 4.14. Now it is possible to approximate this derivative at P in several different ways, for example:

(1) Using the slope of the chord AP, giving the 'backward difference' approximation:

$$\left(\frac{\partial z}{\partial x}\right)_j^n \approx \frac{z_j^n - z_{j-1}^n}{\Delta x}$$

(2) Using the slope of the chord PB, giving the 'forward difference' approximation:

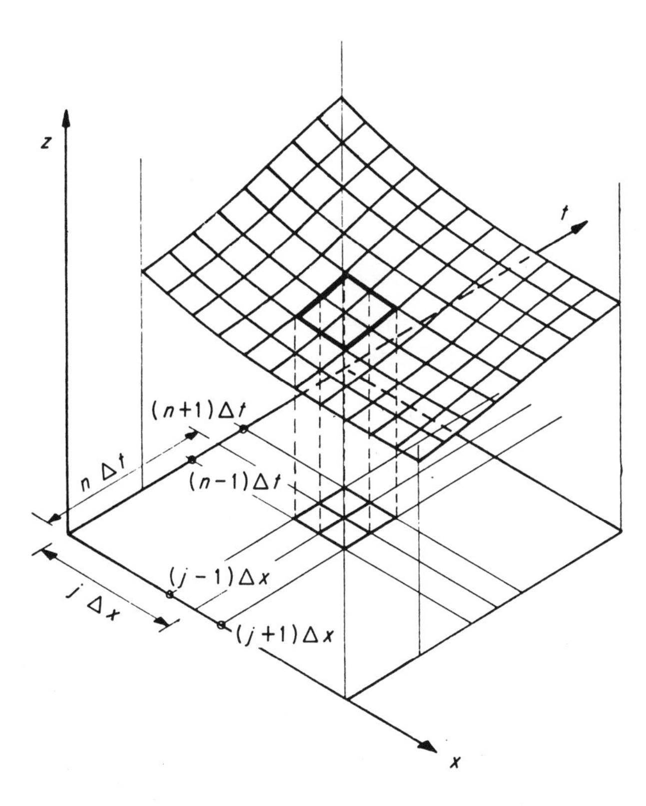

Fig. 4.13. Three-dimensional visualization of the discretization of the 'hydraulic surface' in (z, x, t) space

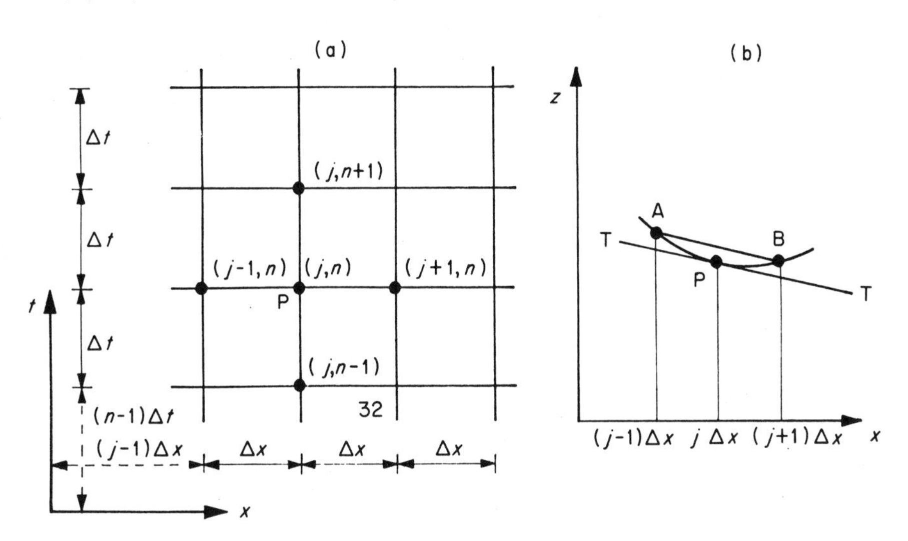

Fig. 4.14. (a) A plan view on Fig. 4.13 introducing the grid notations. (b) A corresponding view in a z–x plane

$$\left(\frac{\partial z}{\partial x}\right)_j^n \approx \frac{z_{j+1}^n - z_j^n}{\Delta x}$$

(3) Using the slope of the chord AB, giving the 'central difference' approximation:

$$\left(\frac{\partial z}{\partial x}\right)_j^n \approx \frac{z_{j+1}^n - z_{j-1}^n}{2\Delta x}$$

Other approximations can be made by using weighted combinations of (1), (2) and/or (3), by introducing values at points further from $j\,\Delta x$ and by introducing time levels other than $n\,\Delta t$.

For the purpose of approximating $(\partial z/\partial t)_j^n$, the same range of possibilities exists and, in particular, the approximations

(a) $$\left(\frac{\partial z}{\partial t}\right)_j^n \approx \frac{z_j^n - z_j^{n-1}}{\Delta t}$$

(b) $$\left(\frac{\partial z}{\partial t}\right)_j^n \approx \frac{z_j^{n+1} - z_j^n}{\Delta t}$$

(c) $$\left(\frac{\partial z}{\partial t}\right)_j^n \approx \frac{z_j^{n+1} - z_j^{n-1}}{2\Delta t}$$

correspond to the approximations (1), (2) and (3) of the x-derivative case.

By using these differences it is then possible to approximate a first-order partial differential equation. Consider, as a first example, Equation (4.5.1) approximated by a backward difference in the x-direction and a forward difference in the t-direction, so that

$$\frac{\partial z}{\partial t} + c\frac{\partial z}{\partial x} = 0, \text{ with } c \text{ a positive real number}$$

is approximated by

$$\frac{z_j^{n+1} - z_j^n}{\Delta t} + c\frac{z_j^n - z_{j-1}^n}{\Delta x} \approx 0$$

or

$$z_j^{n+1} \approx z_j^n - \left(\frac{c\,\Delta t}{\Delta x}\right)(z_j^n - z_{j-1}^n)$$

or

$$z_j^{n+1} \approx \left(1 - \frac{c\,\Delta t}{\Delta x}\right) z_j^n + \left(\frac{c\,\Delta t}{\Delta x}\right) z_{j-1}^n \qquad (4.5.2)$$

When the relation (4.5.2) is treated as an equality it provides a 'difference scheme', giving values of a function at one time level from its values at one or more lower time levels, essentially through a modelling, at finite scale, of the differential equation. Thus, if the set $\{z_1^n, z_2^n, z_3^n, \ldots z_{jj}^n\}$ is known, then the set $\{z_2^{n+1}, z_3^{n+1}, \ldots, z_{jj}^{n+1}\}$ can be determined by using relation (4.5.2).

One then seeks solutions to the equality

$$z_j^{n+1} = \left(1 - \frac{c\,\Delta t}{\Delta x}\right) z_j^n + \left(\frac{c\,\Delta t}{\Delta x}\right) z_{j-1}^n \tag{4.5.3}$$

and although it is hoped that these solutions will approximate solutions of Equation (4.5.1) sufficiently well, it is accepted that they will not usually be identical to those solutions. Whenever difference equations are used to derive approximate solutions to differential equations one speaks of *approximate solutions* generated by the former as compared with *true solutions* generated by the latter.

In this case it is seen that when $c\,\Delta t/\Delta x = 1$, Equation (4.5.3) reduces to $z_j^{n+1} = z_{j-1}^n$, which in fact is seen to correspond to the exact solution found in Section 3.7. In Fig. 4.15 some solutions generated with Equation (4.5.3) are calculated out for various values of $c\,\Delta t/\Delta x$. The various results at $T = nn\,\Delta t = 4$ are illustrated in Fig. 4.16. It is seen that the approximate solution with $c\,\Delta t/\Delta x = \frac{1}{2}$ is much diffused compared with the true solution: it would be regarded as a very poor approximation for most practical purposes. The 'approximate solution' with $\Delta t/\Delta x = 1$ gives in this case, the exact solution: it is certainly the best possible approximation in this case! The 'approximate solution' with $\Delta t/\Delta x = 2$ bears little or no resemblance to the true solution: oscillations set in, which make complete nonsense of the calculation. Such an approximation is evidently completely useless for practical purposes. In such a case it is said that the difference scheme with $\Delta t/\Delta x$ so chosen is *unstable*.

Clearly it is desirable to have the 'best' approximation to the differential equation. It is then necessary:

(a) to choose the 'best' difference approximation out of a very great number of possible difference schemes;

(b) to run this scheme at optimal accuracy conditions, such as an optimal ratio of $\Delta t : \Delta x$;

(c) to run it under stable conditions.

The scheme should also, of course, use the smallest possible amount of time and the smallest amount of core storage on a digital machine, commensurable with its accuracy and stability requirements.

Up to this point, difference equations have been considered as approximations to a differential equation, using the example of modelling the differential equation (4.5.1) with a difference equation, such as (4.5.2). It is now necessary, however, to describe this aspect of the 'modelling process' more precisely.

This is often done by remarking that every difference equation may be written as a differential equation simply by a term-by-term expansion in a Taylor series. Keeping to the example equation of (4.5.2) provides

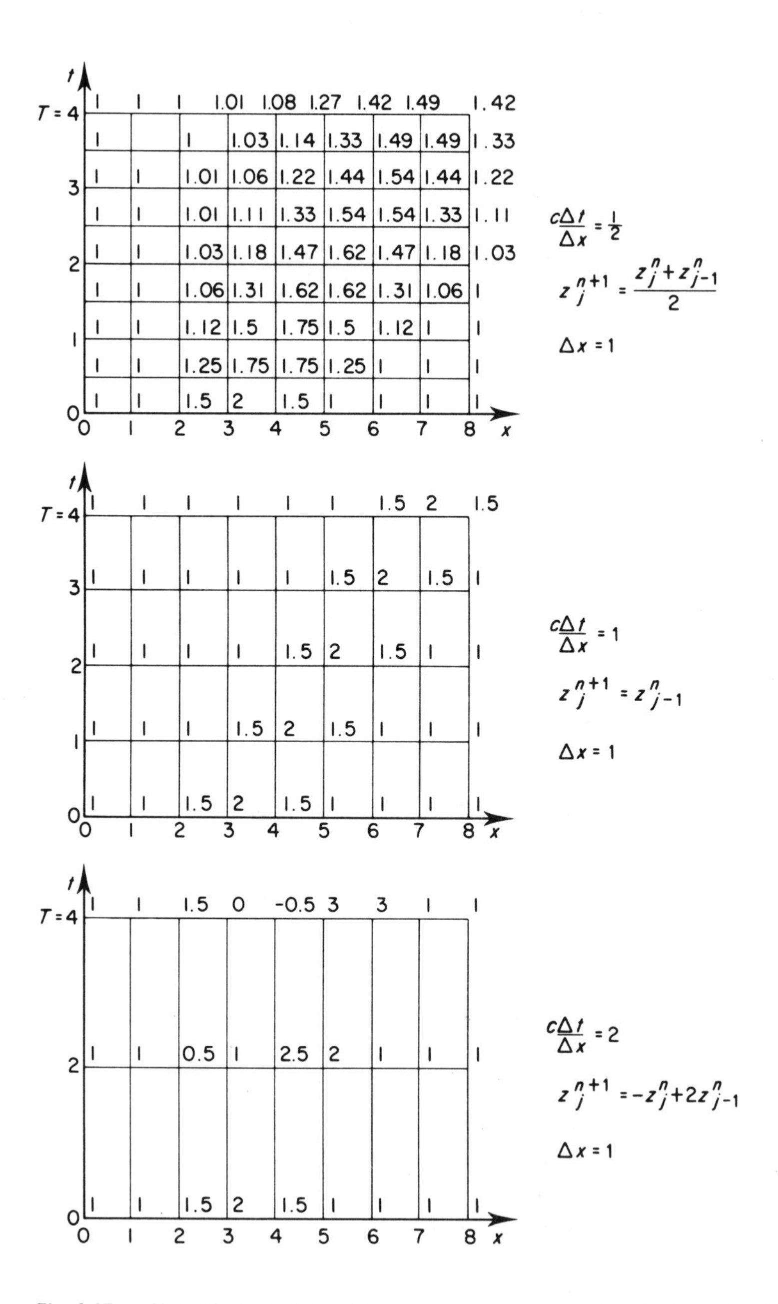

Fig. 4.15. Numerical solutions of the scalar wave equation given by Equation (4.5.3) run at three values of $c\,\Delta t/\Delta x$

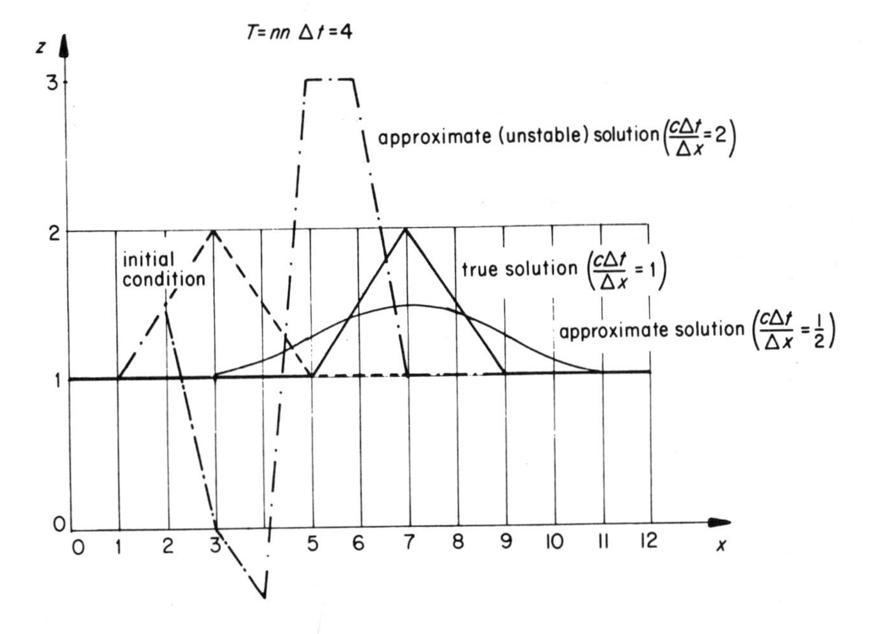

Fig. 4.16. Comparison of solutions of the scalar wave equation at one real time

$$z_j^{n+1} - (1-r)z_j^n - rz_{j-1}^n = 0, \qquad r = \frac{c\,\Delta t}{\Delta x} \tag{4.5.4, = 4.5.3}$$

But

$$z_j^{n+1} = z_j^n + \left(\frac{\partial z}{\partial t}\right)_j^n \frac{\Delta t}{1!} + \left(\frac{\partial^2 z}{\partial t^2}\right)_j^n \frac{\Delta t^2}{2!} + \left(\frac{\partial^3 z}{\partial t^3}\right)_j^n \frac{\Delta t^3}{3!} + \ldots$$

and

$$z_{j-1}^n = z_j^n - \left(\frac{\partial z}{\partial x}\right)_j^n \frac{\Delta x}{1!} + \left(\frac{\partial^2 z}{\partial x^2}\right)_j^n \frac{\Delta x^2}{2!} - \left(\frac{\partial^3 z}{\partial x^3}\right)_j^n \frac{\Delta x^3}{3!} + \ldots$$

so that Equation (4.5.4) becomes

$$\left[\left(\frac{\partial z}{\partial t}\right)_j^n \frac{\Delta t}{1!} + \left(\frac{\partial^2 z}{\partial t^2}\right)_j^n \frac{\Delta t^2}{2!} + \left(\frac{\partial^3 z}{\partial t^3}\right)_j^n \frac{\Delta t^3}{3!} + \ldots\right] - r\left[-\left(\frac{\partial z}{\partial x}\right)_j^n \frac{\Delta x}{1!} + \right.$$
$$\left. + \left(\frac{\partial^2 z}{\partial x^2}\right)_j^n \frac{\Delta x^2}{2!} - \left(\frac{\partial^3 z}{\partial x^3}\right)_j^n \frac{\Delta x^3}{3!} + \ldots\right] = 0$$

or, dividing by Δt and cancelling r, $= c\,\Delta t/\Delta x$,

$$\left[\left(\frac{\partial z}{\partial t}\right)_j^n + c\left(\frac{\partial z}{\partial x}\right)_j^n\right] + \left[\left(\frac{\partial^2 z}{\partial t^2}\right)_j^n \frac{\Delta t}{2!} - \left(\frac{\partial^2 z}{\partial x^2}\right)_j^n \frac{c\Delta x}{2!} + \left(\frac{\partial^3 z}{\partial t^3}\right)_j^n \frac{\Delta t^2}{3!} + \right.$$
$$\left. + \left(\frac{\partial^3 z}{\partial x^3}\right)_j^n \frac{c\Delta x^2}{3!} + \ldots\right] = 0 \tag{4.5.5}$$

Equation (4.5.5) is simply another way of writing (4.5.4), but from (4.5.5) it is clear that, except for the second bracketed term in (4.5.5), Equation (4.5.4) corresponds to Equation (4.5.1) at all points ($j\ \Delta x, n\ \Delta t$). This second bracketed term, which is the 'amount' by which a difference equation D(z) differs from a differential equation C(z), is called the *truncation error* of D(z) relative to C(z). (D stands for 'discrete', C for 'continuous'.) To be more precise, the truncation error of a D(z), of the form D(z) = $\{z_j^{n+1}$ + terms in other grid-dependent variables}, relative to a C(z), of the form C(z) = $\{\partial z/\partial t$ + terms in other derivatives of dependent variables}, is defined as the quantity

$$\left|\frac{\mathrm{D}(z)}{\Delta t} - \mathrm{C}(z)\right|$$

If this truncation error tends to zero as $\Delta x, \Delta t \to 0$ it is said that D(z) is *consistent* with C(z). In the event that the truncation error between D(z) and C(z) tends to zero regardless of how $\Delta x, \Delta t \to 0$ it is further said that D(z) is *unconditionally consistent* with C(z). In the above example it is readily seen from Equation (4.5.5) that (4.5.4) is unconditionally consistent with (4.5.1), but examples of conditional consistency also exist, where the truncation error between a D(z) and a C(z) tends to zero only when Δx and Δt are taken to zero in a particular way.

In general, the truncation error of a D(z) with a C(z) is considered only when the D(z) is consistent in some way with the C(z).

Evidently, the notion of 'consistency' should be conditioned by the notion of 'characteristic', since the extension of functions by Taylor-series expansions is delimited by the characteristics. In this elementary hydraulics text, these difficulties will be put to one side and it will be assumed that solution surfaces are, in some way, 'locally analytic'. The characteristic conditions will assert themselves, anyhow, in the later discussion of stability.

The relation of consistency is a relation between discrete (difference) and continuous (differential) *equations*. However, the question of relations between discrete and continuous formulations can also be considered in terms of their respective *solutions*. It has been seen that Equation (4.5.4) differs from (4.5.1) by the terms in the second square bracket of (4.5.5), which terms have now been described collectively as 'truncation error'. As $\Delta x, \Delta t \to 0$ in any way whatsoever (e.g. with $\Delta t/\Delta x$ fixed, or $\Delta t/\Delta x^2$ fixed, etc.) the truncation error of Equation (4.5.4) relative to (4.5.1) tended also to zero and accordingly it was said that Equation (4.5.4) was unconditionally consistent with (4.5.1). The question must now be posed of whether, during this process, the solutions of the difference schemes will converge towards the true solution as $\Delta x, \Delta t \to 0$, also.

Denote now by $z(x, T)$ the true solution at some *fixed time* T and represent by $z(j\ \Delta x, nn\ \Delta t)$ the approximate solution at the time T, so that $nn\ \Delta t = T$. Thus, in Fig. 4.16, the true solution and two approximate solutions have been compared at $T = 4$. Now it must suffice for this elementary hydraulics text to say that the sequence of approximate solutions $\{z(j\Delta x, n\ \Delta t); j = 1, 2 \ldots jj,$ $n = 1, 2 \ldots nn\}$ ***converges*** to the true solution $z(x, T)$ if

$$\Delta x \sum_{j=1}^{jj} | z(j\,\Delta x, nn\,\Delta t) - z(x = j\,\Delta x, T) | \to 0$$

for every choice of $nn\,\Delta t = T$ as $\Delta x \to 0$ and $\Delta t \to 0$ (i.e. as $jj \to \infty$ and $nn \to \infty$)

The convergence of Equation (4.5.4) towards (4.5.1) can be investigated by

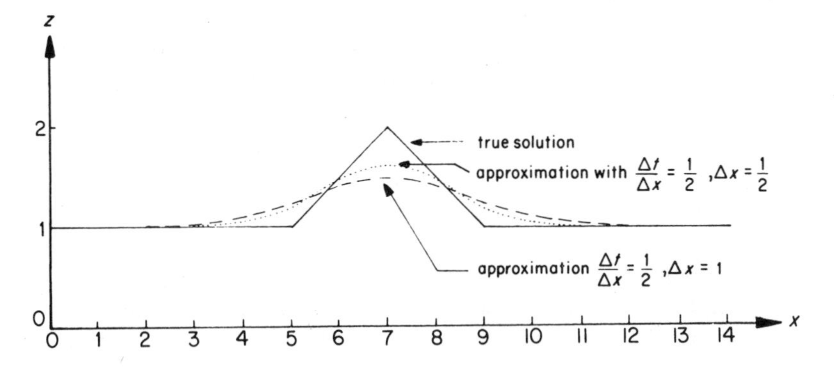

Fig. 4.17. The effect of reducing the distance and time steps is shown for a stable scheme

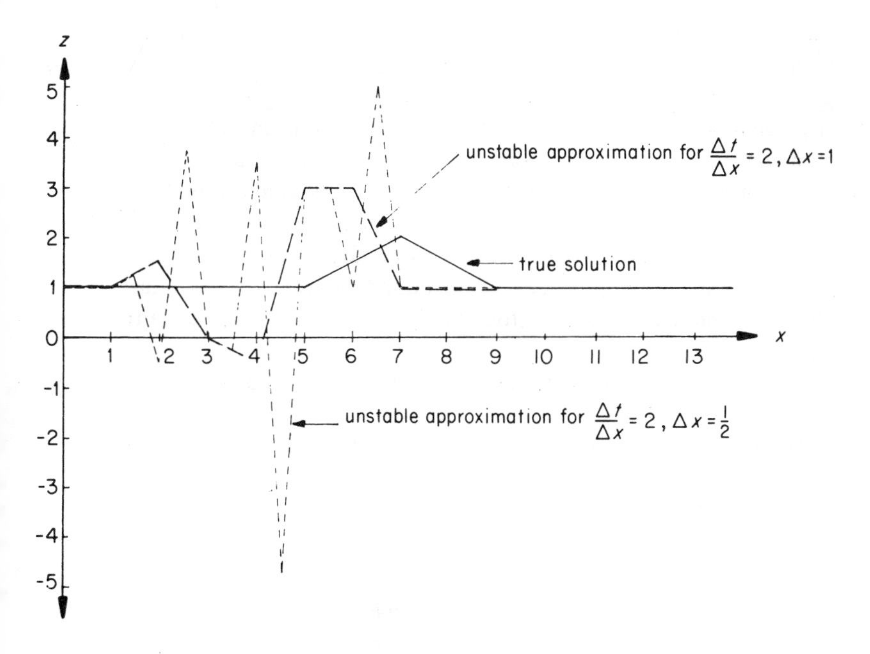

Fig. 4.18. The effect of reducing the Δx and Δt in an unstable scheme is to make the corresponding solutions at any one real time depart further and further from the true solution

actually constructing some approximate solutions for descending Δx and Δt The ratio of $\Delta t/\Delta x$ is kept fixed during this process, as a matter of convenience. Results are first shown for $c = 1$ and $\Delta t/\Delta x = \frac{1}{2}$, with $\Delta t = \frac{1}{2}$ (so $\Delta x = 1$) and $\Delta t = \frac{1}{4}$ (so $\Delta x = \frac{1}{2}$), which clearly converge towards the true solution as $\Delta t, \Delta x \to 0$ (Fig. 4.17). Figure 4.18 then shows results for $\Delta t/\Delta x = 2$ with $\Delta t = 2$ (so $\Delta x = 1$) and $\Delta t = 1$ (so $\Delta x = \frac{1}{2}$) in which it is seen that the approximation becomes progressively worse in this situation. These results suggest that, for those $\Delta t \sim \Delta x$ relations for which the system is stable, the sequence of approximate solutions converges to the true solution as $\Delta t, \Delta x \to 0$, while for those $\Delta t \sim \Delta x$ relations for which the system is unstable the sequence of approximate solutions departs further and further from the true solution as $\Delta t, \Delta x \to 0$. This possibility can be followed up by the study of *stability*.

Clearly, as $\Delta t, \Delta x \to 0$, the truncation error goes to zero. But only when the computation is stable can the sequence of approximation *solutions* $\{z(j\,\Delta x, nn\,\Delta t)\}$ tend to the true *solution* as $\Delta t, \Delta x \to 0$. When the computation is unstable, these solutions differ more and more as $\Delta t, \Delta x \to 0$, even though the truncation error does tend to zero. This observation suggests that stability should be defined in terms of the sequence of solutions of the difference equation, $\{z(j\,\Delta x, nn\,\Delta t)\}$ as $\Delta t, \Delta x \to 0$. In order to do this, consider a set of calculations taking the solution from time $t = 0$ to time $t = T$ using ever-decreasing values of Δt, so that nn, in $T = nn\,\Delta t$, is ever-increasing. It is supposed that during all these calculations, some relation between Δt and Δx is fixed, so that it is possible to write $\Delta x = \text{function}\,(\Delta t) = f(\Delta t)$. Three grids from such a set of calculations are drawn in Fig. 4.19. Then, if the scheme is unstable it may be anticipated that, as the grid is refined in this way, the deviation will increase without bound. Thus for any number that one likes to think of as a bound, the unstable 'approximate' solution will always exceed this bound for some $\Delta t, \Delta x\ (= f(\Delta t))$ sufficiently small, i.e. for some nn sufficiently large. For the stable solution, on the other hand, it should always be possible to find some bound such that the solution always stays within that bound. Note that this bound must be fixed (a 'uniform bound') during the refining process: if the bound increased at every stage, even to infinity, as the refining process took place, it would not be possible to make this distinction.

This construction leads to the following definition of stability:

A difference scheme is stable if the infinite set of computed solutions

$$\{z(j\,\Delta x, n\,\Delta t)\} \qquad 0 \leqslant \Delta t < \tau$$
$$0 \leqslant n\,\Delta t < T$$

is uniformly bounded. Otherwise it is unstable.

For linear equations with constant coefficients operating on uniformly continuous initial and boundary data, the following equivalence theorem of Lax (1954) can be proved (Richtmyer and Morton, 1967, p. 45):

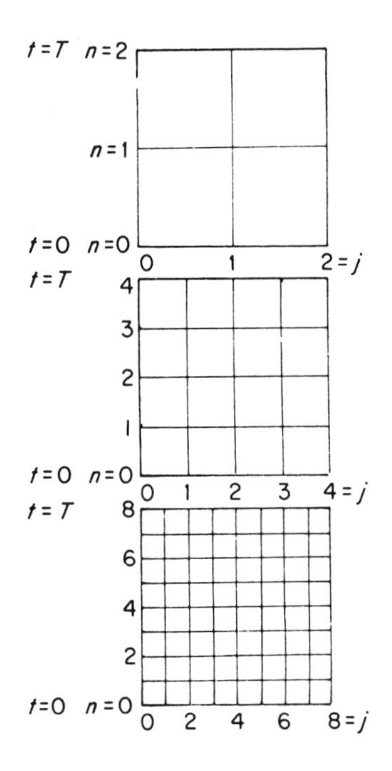

Fig. 4.19. Schematization of the grid-refining process through which stability is defined

> *Given a properly posed initial-value problem and a finite-difference approximation to it that satisfies the consistency condition, stability is the necessary and sufficient condition for convergence.*

This equivalence theorem can be extended to certain more general conditions, as also described by Richtmyer and Morton (1967). It may not, however, always hold in the more general flow situations considered in computational hydraulics, that are characterized by the formation of discontinuities and turbulence, and especially by turbulence at the scale of the numerical discretization. In these cases there may be no equivalence in the above sense, in that sequences of stable solutions may then converge, in some sense, to a solution that differs essentially from the solution of the differential equation, as described in detail in Chapter 5. As will then be seen, this breakdown of the elementary equivalence theorem is of central importance in computational hydraulics; it may even be regarded as distinguishing computational hydraulics from classical hydraulics and hydrodynamics.

It should be remarked, following Richtmyer and Morton (1967), that 'the concept of stability makes no reference to the differential equation that one wishes to solve, but is a property solely of the sequence of difference equations'.

In the case of Equation (4.5.4) it is easy to show that, when $c\,\Delta t/\Delta x = r$ is a constant $\leqslant 1$ the equation is stable or: 'a sufficient condition for the stability of

Equation (4.5.4) is that r is a constant $\leqslant 1$'. Write:

$$z_j^{n+1} = (1-r)z_j^n + rz_{j-1}^n \quad \text{as} \quad z_j^{n+1} = Az_j^n + Bz_{j-1}^n$$

where $r = c\,\Delta t/\Delta x$. Since r is a constant $\leqslant 1$, then both A and B are non-negative, while $A + B = 1$. But then one can as well write

$$\begin{aligned} z_j^n &\leqslant (A+B)\,\text{Max}\,|z_p^{n-1}|, && p = j, j-1 \\ &= \text{Max}\,|z_p^{n-1}|, && p = j, j-1 \\ &\leqslant \text{Max}\,|z_p^{n-2}|, && p = j, j-1, j-2 \\ &\;\;\vdots \\ &\leqslant \text{Max}\,|z_p^0|, && p = j, j-1, \ldots j-n \end{aligned}$$

so that the sequence $z(j\,\Delta x, n\,\Delta t)$ is uniformly bounded by the number

$$\text{Max}\,|z_p^0|, \qquad p = j, j-1, \ldots j-n$$

This sufficient condition for stability, that $r \leqslant 1$, has a simple geometrical interpretation in the x–t plane (Fig. 4.20). It is seen that when the quantity z_j^{n+1} is computed at a point A below the characteristics through $((j-1)\Delta x, n\,\Delta t)$, it

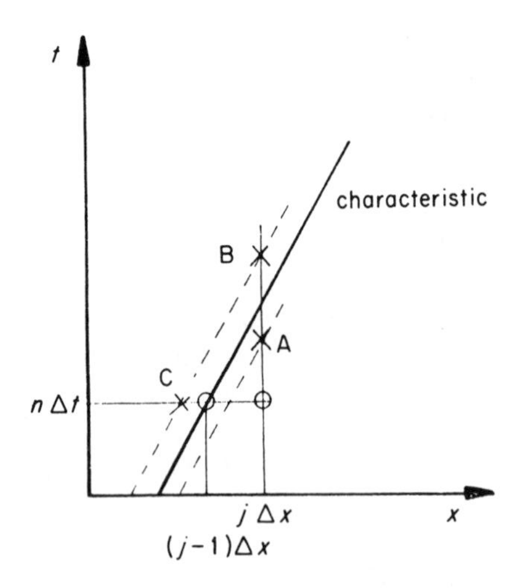

Fig. 4.20. Schematization of the Courant–Friedrichs–Lewy or 'CFL' stability condition for Equation (4.5.3). Conditions at a point A under the characteristic through $(j-1)\,\Delta x$ are determined by the given conditions between $(j-1)\,\Delta x$ and $j\,\Delta x$ while conditions at a point B above the characteristic are not so determined. It is accordingly impossible for the numerical operator to compute any realistic conditions at B, and if given such a time step it will only produce that noise that is associated with 'instability'

should be determined by information given between $(j-1)\Delta x$ and $(j\,\Delta x)$, which information appears as an interpolation in the computation of z_j^{n+1} 1 in Equation (4.5.4) when $r = c\,\Delta t/\Delta x, \leqslant 1$. On the other hand, when z_j^{n+1} is computed at a point B above the characteristic through $((j-1)\,\Delta x, n\,\Delta t)$ it is not determined by the data between $(j-1)\,\Delta x$ and $j\,\Delta x$ that are fed into Equation (4.5.4). In this latter case, to ask Equation (4.5.4) to provide an approximate value of z_j^{n+1} is to ask it to do the impossible, since this value depends on conditions at points such as C that may differ by any amount from z_{j-1}^n and z_j^n. In any such case, the difference scheme may well become unstable. It is then natural to expect that the condition that $r \leqslant 1$ that was seen above to be a ***sufficient*** condition for stability is also a *necessary* condition for stability.

A stability condition of this last type, determined by an 'information argument', is called a 'Courant–Friedrichs–Lewy condition' or simply 'CFL condition' (Courant, Friedrichs and Lewy, 1928).

One can as well obtain necessary and sufficient stability conditions of the above type by using a linear stability analysis, based upon Fourier-series solutions of the difference scheme of the form

$$z_j^n = \sum_k \xi_k^n \exp(ikmj\,\Delta x) \tag{4.5.6}$$

Let $jj\,\Delta x$ again represent the extent of the computed domain, so that $jj\,\Delta x = l$. Then write $m = \pi/l$, so that, for various k, components of the sum (4.5.6) can be represented in the real plane as shown in Fig. 4.21. It is seen that, as k increases, so more and more complicated functions can be represented by higher and higher

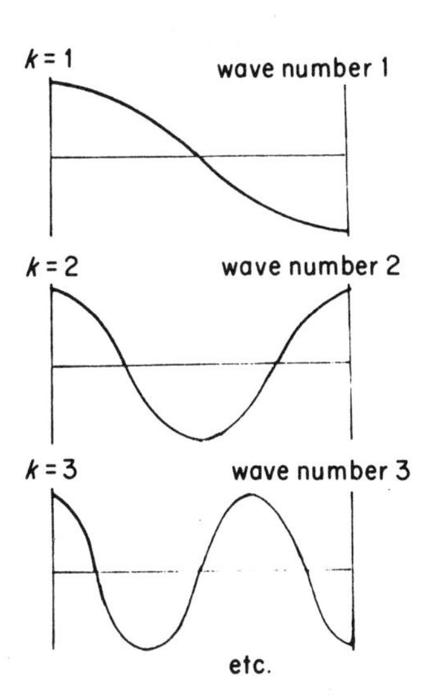

Fig. 4.21. Schematizations of wave number components in a Fourier decomposition (*see also* Fig. 3.44)

'wave numbers' k. In general, for a domain of jj points, up to jj wave numbers can be used, so as, ultimately, to pass the resultant curve through each and every grid point at $n\,\Delta t$ by a suitable choice of the *Fourier coefficients* ξ_k. A more thorough analysis of Fourier series is given in Chapter 5.

The wave number shown in Fig. 4.22a, corresponding to $k = jj$, is called the *highest resolvable wave number*. When $k > jj$, the corresponding function can no longer be resolved, and it is interpreted as a component of lower wave number, as schematized for $k = 3jj/2$ in Fig. 4.22b. In this last case, one speaks of an *aliasing* into the lower wave numbers.

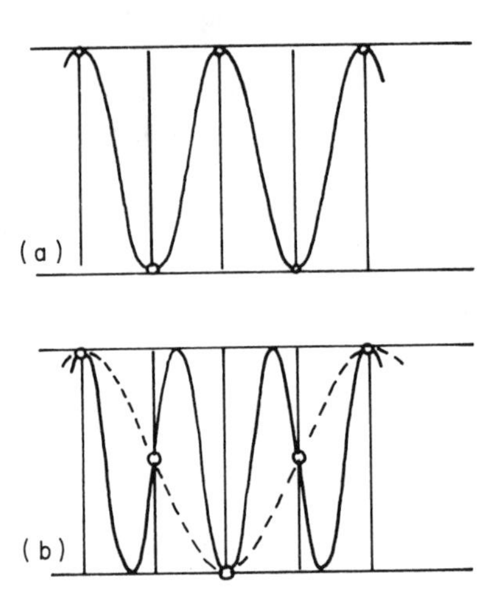

Fig. 4.22. The component (a) of highest resolvable wave number and the 'aliasing' of an even higher wave number into a low wave number component

Now consider the behaviour of any one component of Equation (4.5.6) between $n\,\Delta t$ and $(n+1)\,\Delta t$. Substituting in Equation (4.5.4), as the simplest example, it is found that

$$\xi_k^{n+1}\,\mathrm{e}^{ikmj\,\Delta x} = (1-r)\,\xi_k^n\,\mathrm{e}^{ikmj\,\Delta x} + r\xi_k^n\,\mathrm{e}^{ikm(j-1)\,\Delta x}$$

or, cancelling $\mathrm{e}^{ikmj\,\Delta x}$,

$$\xi_k^{n+1} = (1-r)\xi_k^n + r\mathrm{e}^{-ikm\,\Delta x}\xi_k^n$$

i.e.

$$\xi_k^{n+1} = A\xi_k^n, \text{ where } A = [1 + (\mathrm{e}^{-ikm\,\Delta x} - 1)r]$$

The complex number A is called the *amplification factor* of the difference scheme. In Fig. 4.23 the amplification factor is traced in the Argand diagram for $r = \frac{1}{4}, \frac{1}{2}, \frac{3}{4}$, 1 and $\frac{5}{4}$.

It is seen that for $r \leqslant 1$ the modulus of the amplification factor $|A| \leqslant 1$ so

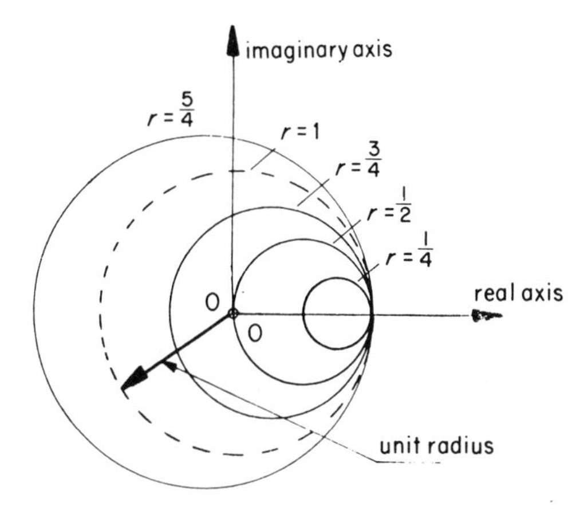

Fig. 4.23. The Argand diagram representation of the amplification factor of Equation (4.5.3)

that, for any component k, the magnitude or 'modulus' of the Fourier coefficient of the kth wave component is not increasing in time. When $km\,\Delta x, = k\pi\,\Delta x/l$, is small, A is near to 1. This is to say that when there are a large number of grid-points per wave length at wave number k, this kth component is very little attenuated. When there are only a very few points per wave length, on the other hand, the attenuation for the corresponding wave number may be very large for $r < 1$.

It is also seen, on the other hand, that when $r > 1$, all components are amplified at each and every time step, with amplification increasing as the number of points per wave length decreases. Although it is possible to obtain more points per wave length by increasing the number of distance steps, so reducing the amplification per time step, the number of time steps and thus the number of amplifications will increase correspondingly (for r fixed) and the instability in fact becomes ever more pronounced upon arriving at the fixed $T = nn\,\Delta t$. Thus, although the truncation error may tend to zero as $\Delta x, \Delta t \to 0$, regardless of the stability of the scheme, the solutions of the scheme in the unstable case move further and further from the true solution as $\Delta x, \Delta t \to 0$. Richtmyer expressed the matter in the form of a pseudo-paradox, as follows: 'If the mesh is refined, but in such a way as to violate the stability condition, the exact solution of the differential equation comes closer and closer to satisfying the difference equations, but the exact solution of the difference equations departs, in general, more and more from the true solution of the initial-value problem.'

In the present case, the stability condition $r < 1$ of Equation (4.5.4) can as well be derived heuristically from the Taylor-series expansion of Equation (4.5.4). The leading terms of approximated equation and first-order truncation error read

$$\left[\left(\frac{\partial z}{\partial t}\right)_j + c\left(\frac{\partial z}{\partial x}\right)_j\right] + \left[\left(\frac{\partial^2 z}{\partial t^2}\right)_j \frac{\Delta t}{2!} - c\left(\frac{\partial^2 z}{\partial x^2}\right)\frac{\Delta x}{2!}\right] = \text{(higher-order terms)}$$

But

$$\frac{\partial^2 z}{\partial t} = \frac{\partial}{\partial t}\left(\frac{\partial z}{\partial t}\right) = \frac{\partial}{\partial t}\left(-c\frac{\partial z}{\partial x}\right) = -c\frac{\partial}{\partial x}\left(\frac{\partial z}{\partial t}\right) = -c\frac{\partial}{\partial x}\left(\frac{-c\,\partial z}{\partial x}\right) = c^2\frac{\partial^2 z}{\partial x^2}$$

so that Equation (4.5.4) can be written as

$$\left(\frac{\partial z}{\partial t}\right)_j + c\left(\frac{\partial z}{\partial x}\right)_j + \left(\frac{\partial^2 z}{\partial x^2}\right)_j \frac{(r-1)c\,\Delta x}{2} = \text{(higher-order terms)}$$

When $r < 1$, the coefficient of the third term is negative with c positive so that the dominant part of the differential equation equivalent to Equation (4.5.4) is a transport–diffusion equation with a positive diffusion coefficient. As soon as $r > 1$, on the other hand, the diffusion coefficient becomes negative and Equation (4.5.4) corresponds to a physically unstable system.

The above notions can be extended to the non-linear case. Consider, for example, the equation

$$\frac{\partial z}{\partial t} + (c + az)\frac{\partial z}{\partial x} = 0 \tag{4.5.7}$$

The solution evidently consists of constant values of z carried along characteristic lines with slope, or celerity $dx/dt = c + az$. Some characteristics and a solution at a time T are sketched in Fig.4.24. It is seen that the true solution becomes three-

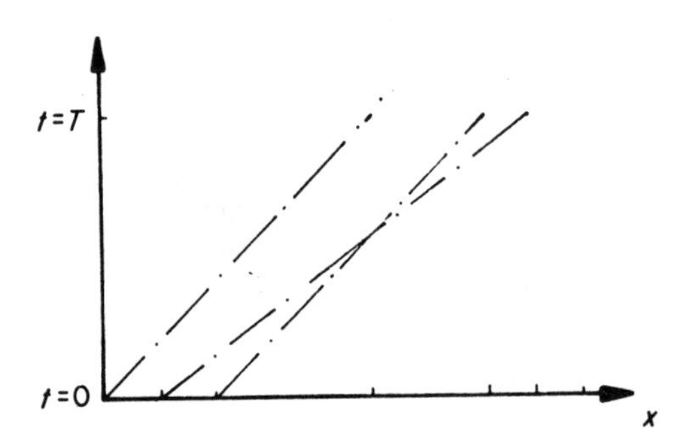

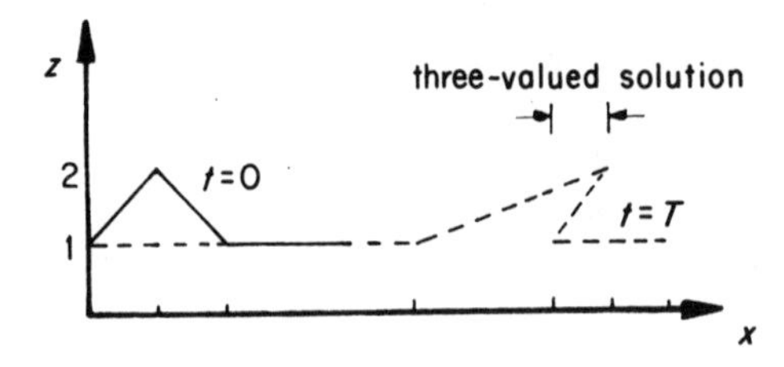

Fig. 4.24. The formation of a three-valued solution of Equation (4.5.7) after intersection of its characteristics

valued after some time for any non-trivial (not everywhere constant) initial profile. Now, of course, three-valuedness cannot occur in any difference scheme on a fixed grid: a scheme such as Equation (4.5.4) gives one and only one value of z at each grid point. What happens in fact in the difference scheme (4.5.4) with $r = [(c + az_j^n)\Delta t]/\Delta x$ is that oscillations occur in the region where three-valuedness is predicted by the exact solution even though $r \leqslant 1$ and even $r \approx 1$ and the solution is otherwise good. As time continues, these oscillations may increase in magnitude and the sequence of solutions may appear to have no uniform bound so that the difference scheme appears to be unstable (Fig. 4.25). It may already be anticipated that this behaviour corresponds, as a scalar simplification, to that observed during the formation of a hydraulic jump, where irreversible physical processes dissipate enough energy to ensure a 'single-valued solution'.

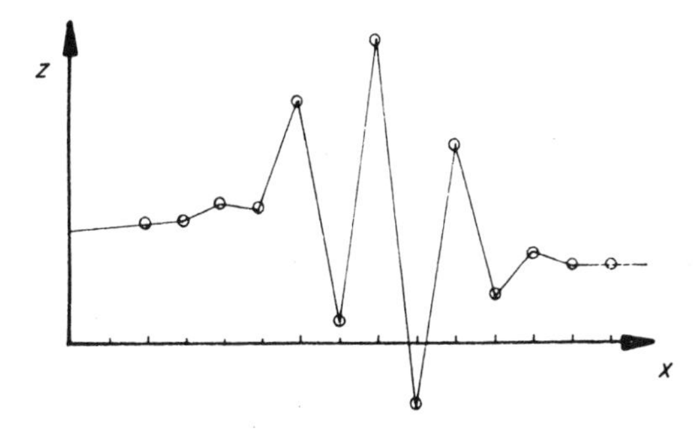

Fig. 4.25. Parasitic waves that are formed by some difference approximations to Equation (4.5.7) in place of the three-valued exact solution

Such a behaviour can also be anticipated mathematically by considering the behaviour of non-linear terms in the difference scheme

$$z_j^{n+1} = (1 - r)z_j^n + rz_{j-1}^n \text{ with } r = [(c + az_j^n)\,\Delta t]/\Delta x$$

in the Fourier-series representation. Product terms

$$z_j^n \cdot z_j^n \text{ and } z_j^n \cdot z_{j-1}^n$$

then occur, which transform under the Fourier substitution to product terms of the type

$$e^{ikmj\Delta x}\, e^{ikmj\Delta x} = e^{i2kmj\Delta x}$$

Thus it is seen that components of wave number k 'breed' components of higher wave number through the product terms, and it is these components of higher wave number that are observed in Fig. 4.25.

Now when discussing the analysis of linear stability, it was remarked that for values of r less than unity, the higher-wave-number components would be more attenuated than the lower-wave-number components. A situation may then be envisaged where, by a suitable choice of r, parasitic waves and thus non-linear instability can be suppressed, while the longer-wave components may be very little influenced. This notion is quantified by constructing the 'amplitude portrait' of the difference scheme, as shown in Fig. 4.26a. The figure is easily constructed from Fig. 4.23. It is seen that components of highest resolvable wave number suffer an attentuation of magnitude $(1 - r)$ at each time step. This suggests that running the scheme

$$z_j^{n+1} = (1 - r)z_j^n + rz_{j-1}^n$$

with $r = \frac{3}{4}$ should provide sufficient damping at the head of the wave while maintaining an acceptable accuracy elsewhere.

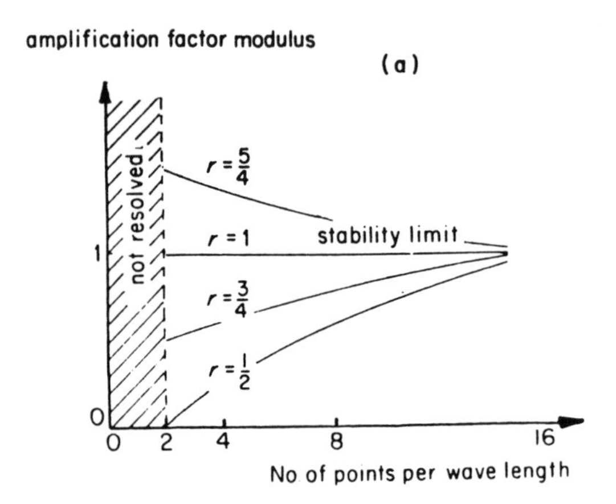

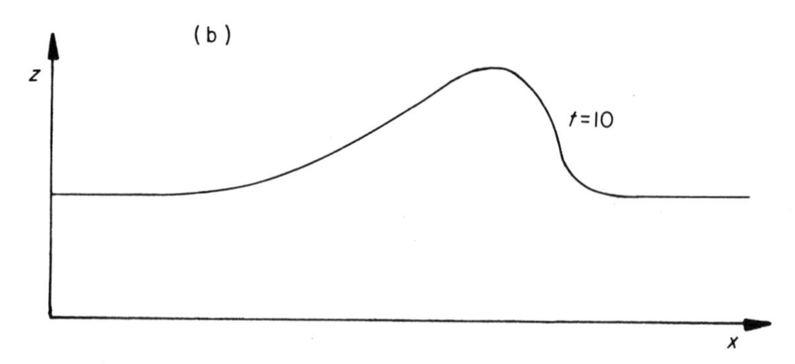

Fig. 4.26. (a) Amplitude portrait of a scheme consistent with Equation (4.5.7). For values of $r > 1$, indefinite amplification or instability of all wave components occurs while for values of $r < 1$, the wave components of the solution are damped. (b) By using a value of r sufficiently below unity, the damping or 'dissipation' predicted in Fig. 4.26a can be used to eliminate the parasitic waves of Fig. 4.25

In Fig. 4.26b it is seen that these expectations are realized. A difference scheme that dissipates components of higher wave number in this way is called a *dissipative* difference scheme. As will be seen later, this dissipative behaviour can sometimes be very useful, but most of the time it appears only as a considerable nuisance.

4.6 Explicit difference methods for one-dimensional nearly horizontal flows

In the difference scheme (4.5.3) that was used as a prototype in the previous section, the value of the dependent variable at one time could be expressed as an explicit function of the values of the dependent variable at an earlier time. Such a scheme is traditionally called an *explicit difference scheme*. In the prototype scheme (4.5.3) the information transfer was along one characteristic so that the use of a forward time difference and a backward space difference was indicated, as schematized in Fig. 4.27a. In the simplest nearly horizontal flow case of the conservation law

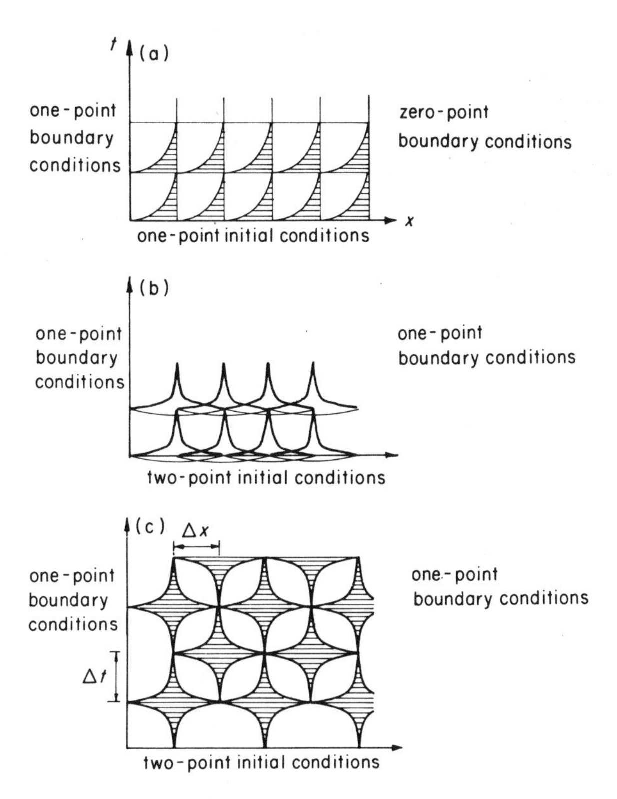

Fig. 4.27. Schematizations (sometimes called 'tilings') of (a) the transport or scalar wave equation operator (4.5.3), and (b), (c) the nearly horizontal flow operators (4.6.2) and (4.6.3)

$$\frac{\partial f}{\partial t} + \frac{\partial g(f)}{\partial x} = 0 \qquad (4.6.1, = 2.1.8)$$

there are two dependent variable components of f and two characteristics, so that other structures are indicated, as schematized in Fig. 4.27b and 4.27c. The scheme of Fig. 4.27b could be written as

$$\frac{f_j^{n+1} - f_j^n}{\Delta t} + \frac{g_{j+1}^n - g_{j-1}^n}{2\Delta x} = 0 \qquad (4.6.2)$$

while the scheme of Fig. 4.27c would appear as

$$\frac{f_j^{n+1} - f_j^{n-1}}{\Delta t} + \frac{g_{j+1}^n - g_{j-1}^n}{\Delta x} = 0 \qquad (4.6.3)$$

The property that f_j^{n+1} can be calculated explicitly as a function of the f values at lower time levels makes Equations (4.6.2) and (4.6.3) 'explicit schemes'. Although in the next section, a somewhat modified terminology will be introduced, these schemes will remain 'explicit'. Equations (4.6.2) and (4.6.3) both immediately present problems at the boundaries, where two values of f would be needed to provide $g = g(f)$ whereas only one-point data are in fact provided. Some special boundary element is then required, such as one based upon the three-point or four-point method of characteristics.

By expanding Equations (4.6.2) and (4.6.3) in Taylor series about $(j\,\Delta x, n\,\Delta t)$, it is seen that the first is of first order in Δt and second order in Δx while the second is of the second order in Δt and Δx.

For the purposes of linear stability analysis, Equation (4.6.1) must be written in the form of (2.6.6):

$$\frac{\partial f}{\partial t} + A\,\frac{\partial f}{\partial x} = 0 \qquad (4.6.4)$$

Then Equation (4.6.2) becomes, to the corresponding local linearization,

$$\frac{f_j^{n+1} - f_j^n}{\Delta t} + A\left(\frac{f_{j+1}^n - f_{j-1}^n}{2\Delta x}\right) = 0 \qquad (4.6.5)$$

while Equation (4.6.3) becomes

$$\frac{f_j^{n+1} - f_j^{n-1}}{\Delta t} + A\left(\frac{f_{j+1}^n - f_{j-1}^n}{\Delta x}\right) = 0 \qquad (4.6.6)$$

Consider now the behaviour of any one component of the Fourier series

$$f_j^n = \sum_k \xi_k^n \exp(\mathrm{i}2\pi kj\,\Delta x/2l) \qquad (4.6.7)$$

under the operation (4.6.5). Just as in (4.5.6), k is a dimensionless wave number. Substituting the general, kth, component of Equation (4.6.7) in (4.6.5) and simplifying provides an *amplification matrix* G, multiplying ξ_k^n to give ξ_k^{n+1} and given by

$$G = I - \mathrm{i}\left(\frac{\Delta t}{\Delta x} \cdot A\right) \sin \gamma \tag{4.6.8}$$

where $\gamma = 2\pi k\ \Delta x/2l$. When $f = \{u, h\}$, this can be written out as

$$\begin{bmatrix} 1 & 0 \\ 0 & 1 \end{bmatrix} - \frac{\mathrm{i}\ \Delta t}{\Delta x} \sin \gamma \begin{bmatrix} u & g \\ h & u \end{bmatrix} = \begin{bmatrix} 1 - \dfrac{\mathrm{i}\ \Delta t}{\Delta x} \sin \gamma \cdot u & -\dfrac{\mathrm{i}\ \Delta t}{\Delta x} \sin \gamma \cdot g \\ -\dfrac{\mathrm{i}\ \Delta t}{\Delta x} \sin \gamma \cdot h & 1 - \dfrac{\mathrm{i}\ \Delta t}{\Delta x} \sin \gamma \cdot u \end{bmatrix}$$

This matrix amplifies the values of the kth Fourier coefficients at $n\ \Delta t$ into their values at $(n + 1)\ \Delta t$. The problem is then posed of whether these coefficients are amplified, so that the component chosen (still identified by the k in the γ of Equation (4.6.8)) is unstable or whether it is diminished, so that the component is stable. In general any such component Fourier coefficients will not only be amplified, or changed in magnitude, but they will also be changed in direction in the Argand diagram of complex numbers.

However, there will usually exist certain vectors whose directions in the space of complex numbers $\{\xi_1, \xi_2\}$ do not change under multiplication by G. These vectors are called ***eigenvectors*** of G, while the amounts by which they amplify under multiplication by G are called the ***eigenvalues*** of G. Thus, if X is an eigenvector of G and g a corresponding eigenvalue,

$$GX = gX \tag{4.6.9}$$

It is relatively easy to determine whether the eigenvectors increase or decrease in magnitude by solving G for its eigenvalues using Equation (4.6.9) in the form

$$(G - gI)X = 0$$

which implies, if $X \neq 0$, that

$$\mathrm{Det}(G - gI) = 0 \tag{4.6.10}$$

If all such g satisfy $|g| \leqslant 1$ then there is a hope that the scheme is stable, whereas if $|g| \geqslant 1$ for even one g then the corresponding eigenvector is amplified at every Δt and the scheme is unstable. The condition $|g| \leqslant 1$ is called the *von Neumann necessary condition* for stability of a difference scheme of a primitive form. It obviously has to be generalized for forms with mass, momentum and/or energy sources and sinks, where the Fourier coefficients may be amplified by physical influences.

In the case of (4.6.8), it is more convenient to develop Equation (4.6.10) by introducing an X' and a λ that are, respectively, an eigenvector and a corresponding eigenvalue of A. Then

$$GX' = \left(1 - \frac{\mathrm{i}\,\Delta t}{\Delta x} \cdot \lambda \cdot \sin\gamma\right) X' \qquad (4.6.11)$$

so that $X' = X$ and the eigenvalues $\{g\}$ of G are given by

$$g = 1 - \mathrm{i}\left(\frac{\Delta t}{\Delta x}\right) \cdot \lambda \cdot \sin\gamma \qquad (4.6.12)$$

where the λ, being eigenvalues of A, are given by $\lambda = u \pm (gh)^{\frac{1}{2}}$ (e.g. Equation 3.5.7). Equation (4.6.12) is plotted graphically in the Argand diagram of Fig. 4.28, from which it is readily seen that

$$|g| > 1 \text{ for all } \Delta t/\Delta x > 0$$

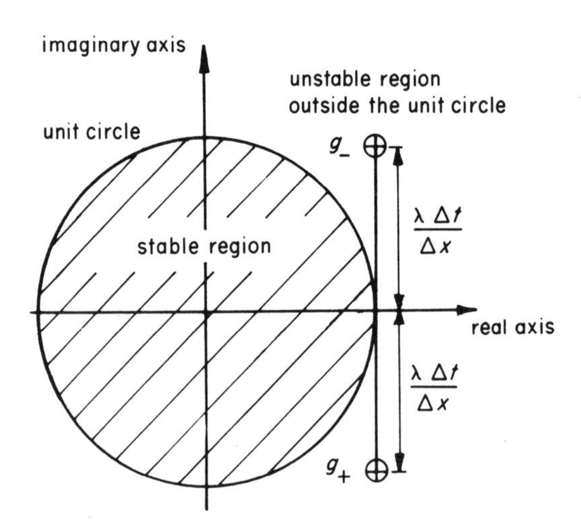

Fig. 4.28. Argand diagram representation of the eigenvalues (4.6.12) of the amplification matrix (4.6.8) of the scheme (4.6.2)

Thus the scheme (4.6.2) is linearly *unstable*. Of course only the behaviour of the eigenvectors has been considered in this analysis and the possibility exists that, in one or more particular calculations, vectors are considered that are so far removed from these eigenvectors that the calculation still survives. Such special cases are occasionally observed, in fact. However, a scheme that functions for only some choices of data but not for others is a very unreliable instrument, and it is impractical to employ such a scheme. It should be remembered, in this connection, that stability is defined through the boundedness of an operation over all possible operands, so that a scheme of the above type is unstable, by this definition, for all $\Delta t/\Delta x$, and in this sense it is even 'unconditionally unstable'. Moreover, an analysis of this type provides only necessary conditions for stability: there is no certainty that the conditions are also sufficient. If, however, the eigenvalues of the eigenvectors bound amplification factors of all other vectors, as the principal axes of an ellipse (e.g. as described in the theory of quadratic forms), then the necessary condition will also be sufficient. For a large class of amplification matrices (primarily *normal* matrices: *see* Richtmyer and Morton, 1967) it can be shown that the necessary and sufficient conditions in fact coincide, but the discussion of these matters is outside the scope of the present introductory work. The maximum of the moduli of the eigenvalues is called the *spectral radius*.

Now it is possible to *stabilize* any unstable scheme, such as Equation (4.6.2), by the introduction of a *dissipative interface*. The general theory of these devices is discussed in the next chapter, but for the moment it will suffice to consider only their most elementary properties. What is done, in effect, is to average out the values of one or both dependent variables $\{f\}$ obtained after every time step of the nearly horizontal flow computation, before using these in the next time difference, as for example through application of the following operators:

$$f_j^* = \alpha f_{j+1} + (1-2\alpha) f_j + \alpha f_{j-1} \qquad 0 < \alpha \leqslant \tfrac{1}{2} \text{ (centred)} \qquad (4.6.13)$$

$$f_j^* = \alpha f_{j+1} + (1-\alpha) f_j \qquad 0 < \alpha < 1 \text{ (forward)} \qquad (4.6.14)$$

$$f_j^* = (1-\alpha) f_j + \alpha f_{j-1} \qquad 0 < \alpha < 1 \text{ (backward)} \qquad (4.6.15)$$

Equation (4.6.13) can be used throughout the domain of the computation except at the ultimate left-hand boundary point, $j = 0$, where Equation (4.6.14) is used, and the ultimate right-hand boundary point, $j = jj$, where (4.6.15) is used. The following sum may then be constructed:

$$\begin{array}{rcccccc}
f_0^* = & (1-\alpha) f_0 + & \alpha f_1 & & & & \\
f_1^* = & \alpha f_0 + & (1-2\alpha) f_1 + & \alpha f_2 & & & \\
f_2^* = & & \alpha f_1 + & (1-2\alpha) f_2 + & \alpha f_3 & & \\
\cdot & & & & & & \\
\cdot & & & & & & \\
\cdot & & & & & & \\
ff_{jj-1}^* = & & & & \alpha f_{jj-2} + & (1-\alpha) f_{jj-1} + & \alpha f_{jj} \\
f_{jj}^* = & & & & & \alpha f_{jj-1} + & \alpha) f_{jj} \\
\hline
\sum_{j=0}^{jj} f_j^* = & f_0 + & f_1 + & f_2 + \ldots & & + f_{jj-1} + & f_{jj}
\end{array}$$

Thus, at least for a uniform grid spacing Δx, the total quantity of f over the domain is unchanged by the dissipative interface: only its distribution is altered. Often only alternate grid points are smoothed in this way. Dissipative interfaces constitute a special class of *linear digital filters* (*see* Hamming, 1977). The amplification factor D of Equation (4.6.13), when it acts upon a Fourier component of wave number k, is easily calculated to be given by

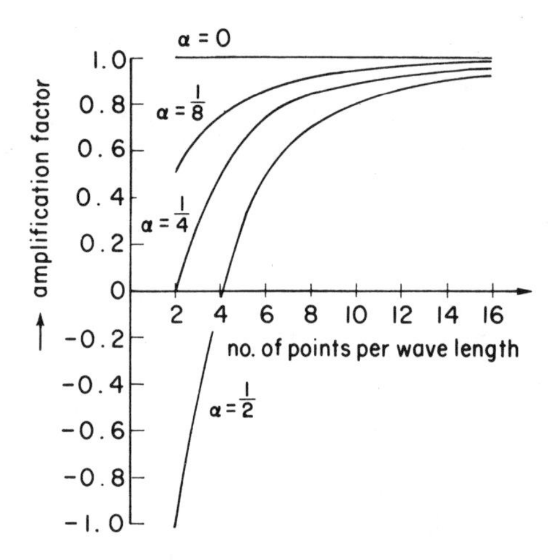

Fig. 4.29. Amplitude portrait of the centred dissipative interface (4.6.13)

$$D = 1 - 2\alpha\left(1 - \cos\left(\frac{k\pi\,\Delta x}{l}\right)\right)$$

so that it is entirely real with an amplification portrait as shown in Fig. 4.29.

Applying the centred dissipative interface (4.6.13) on the lower-time values of the time differences of the linearized form of Equation (4.6.2), namely (4.6.5), provides

$$\frac{f_j^{n+1} - (\alpha f_{j+1}^n + (1 - 2\alpha) f_j^n + \alpha f_{j-1}^n)}{\Delta t} + \frac{g_{j+1}^n - g_{j-1}^n}{2\Delta x} = 0 \tag{4.6.16}$$

and changes Equation (4.6.8) to

$$G = DI - \mathrm{i}\left(\frac{\Delta t}{\Delta x} \cdot A\right) \sin \gamma$$

so that, by the same reasoning as used earlier

$$g = D - \mathrm{i}\left(\frac{\Delta t}{\Delta x} \cdot \lambda\right) \sin \gamma$$

This is plotted in Fig. 4.30a from which it is seen that the eigenvalues form ellipses in the Argand diagram, with one axis of length 2α and the other of length $|\lambda|\,\Delta t/\Delta x$. It is seen from Fig. 4.30a that the most generous stability condition is that given by $\alpha = \frac{1}{2}$, namely $(\Delta t/\Delta x) \cdot |\lambda| \leqslant 1$. Figures 4.30b, c and d show amplitude portraits for $\alpha = 0$, $\alpha = \frac{1}{4}$ and $\alpha = \frac{1}{2}$ as obtained from Fig. 4.30a. The stability condition of the scheme with $\alpha = \frac{1}{2}$ can be written as

$$|u \pm (gh)^{\frac{1}{2}}| \cdot \frac{\Delta t}{\Delta x} \leqslant 1$$

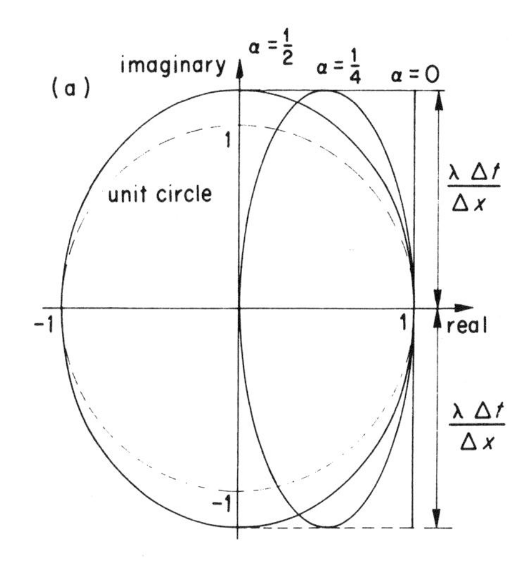

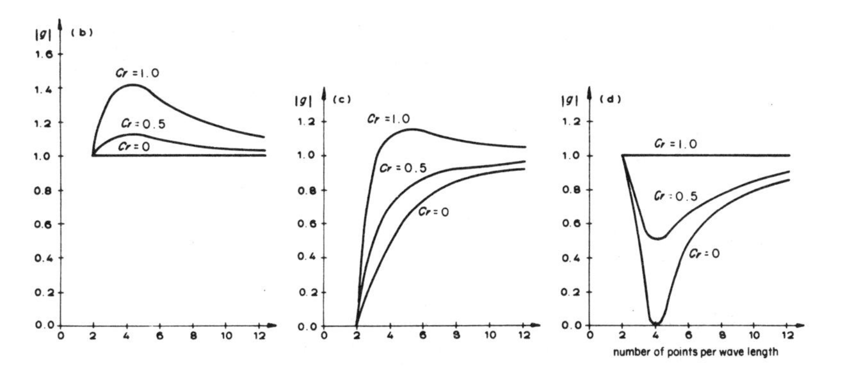

Fig. 4.30. (a) Argand diagram representation of the eigenvalues for Equation (4.6.2) when interfaced with a central dissipative interface (4.6.13) on its $\{f_j^n\}$ with dissipation degrees α of 0, $\frac{1}{4}$ and $\frac{1}{2}$. This last case of $\alpha = \frac{1}{2}$ provides the scheme of Lax (1954). (b), (c) and (d) The eigenvalue modulus for (b) $\alpha = 0$, (c) $\alpha = \frac{1}{4}$ and (d) $\alpha = \frac{1}{2}$

or

$$\Delta t \leqslant \frac{\Delta x}{|u \pm (gh)^{\frac{1}{2}}|}$$

The largest possible time step is seen in Fig. 4.31 to correspond to the Courant–Friedrichs–Lewy (CFL) condition introduced in the previous section. If a time step less than AC in Fig. 4.31, such as AB, is taken, then conditions at B can be defined by conditions given along the x-direction grid line corresponding to time $n\,\Delta t$, $(j-1)\,\Delta x$ to $(j+1)\,\Delta x$. If a time step greater than AC, such as AD in Fig.

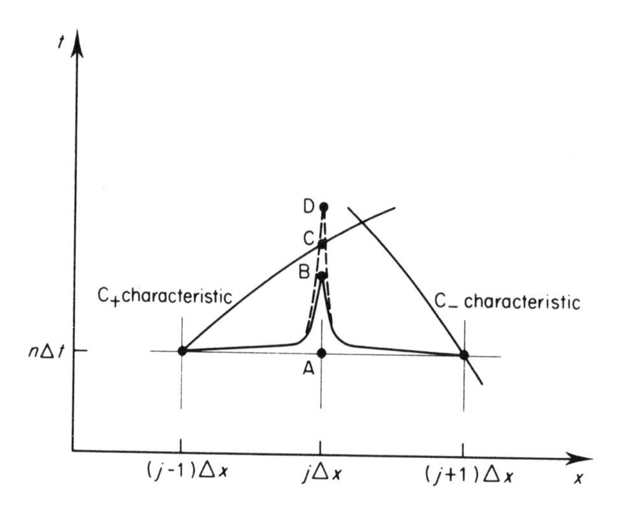

Fig. 4.31. Schematization of stable and unstable time steps for an explicit operator

4.31, is taken, then it is clear that conditions at D cannot be determined by conditions between $(j-1)\,\Delta x$ and $(j+1)\,\Delta x$ so that one is 'asking the operator to do the impossible'. A numerical 'nervous breakdown', manifest as instability, is the consequence.

Setting $\alpha = \frac{1}{2}$ in Equation (4.6.16) gives the scheme

$$\frac{f_j^{n+1} - (f_{j+1} + f_{j-1})^n/2}{\Delta t} + \frac{g_{j+1}^n - g_{j-1}^n}{2\,\Delta x} = 0 \tag{4.6.17}$$

first proposed by Lax (1954) and commonly called the 'Lax scheme'. It is seen from Fig. 4.30 that, whenever $|\lambda|\,\Delta t/\Delta x < 1$, this scheme is dissipative for all wave numbers. Now unless the bathymetry is at constant level and the Froude number is very low, it is impossible to maintain $|\lambda|\,\Delta t/\Delta x = 1$ throughout a computation with Equation (4.6.17), so that when one part is at the limit of stability, the rest is dissipating. The dissipation in such a first-order-accurate scheme can be very strong, as illustrated by the amplitude portrait of the Lax scheme in Fig. 4.30d and a practical case of the propagation of a related simple wave in Fig. 4.32.

It is seen that the numerical dissipation is quite unacceptable for most practical computations. On the other hand, it has been intimated in Section 4.5 that this dissipation could be useful for modelling a hydraulic jump, and indeed the Lax scheme has often been used for this purpose. An example is shown in Fig. 4.32. However, there are better methods even for modelling hydraulic jumps, as will appear further in this and the next chapter.

Consider now the other simple case, of the 'leapfrog' scheme (4.6.3) in its locally linearized form of Equation (4.6.6). This scheme relates values of the dependent variables at three time levels so that it constitutes what is often

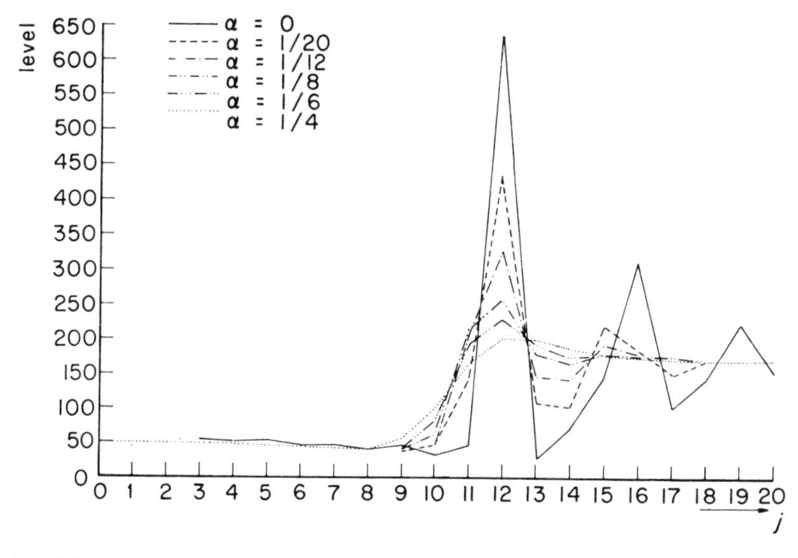

Fig. 4.32. A typical wave propagation solution showing the effect of dissipation

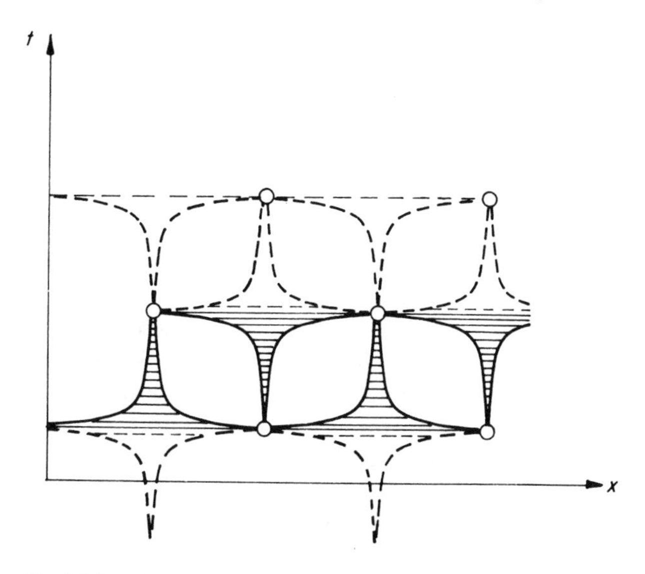

Fig. 4.33. Decomposition of the leapfrog operator

called a 'three-level' scheme. Although it is possible to reduce such a three-level scheme to a two-level one (e.g. Richtmyer and Morton, 1967, pp 260–263) the conditions governing the validity of this reduction are quite complicated in the general case. It is altogether simpler to use the method of Yanenko (1968) in which the Fourier coefficients ξ_k^n are expressed in the form

$$\xi_k^n = \xi_k \, e^{in\,\Delta t} = \xi_k \rho$$

so that ρ is itself the amplification factor over time Δt. Then (4.6.6) has the Fourier transform

$$\frac{(\rho - \rho^{-1})I}{\Delta t} + \frac{A}{\Delta x}\ \mathrm{i}2\sin\left(\frac{2\pi k\Delta x}{21}\right) = 0$$

or

$$\rho^2 I + \frac{\mathrm{i}A\Delta t}{\Delta x}\sin\left(\frac{2\pi k\Delta x}{21}\right)\rho - I = 0$$

If X is an eigenvector of A then

$$(\rho^2 + \mathrm{i}2a\rho - 1)X = 0$$

with

$a = \dfrac{\lambda\Delta t}{\Delta x} \cdot \sin\left(\dfrac{2\pi k\Delta x}{21}\right)$, so that for a non-zero X,

$$\rho = -\mathrm{i}a \pm \sqrt{-a^2 + 1}$$

When $|a| \leqslant 1$, then $|\rho|^2 = a^2 + (-a^2 + 1) = 1$ and the scheme is non-dissipative. For $|a| > 1$ it is unstable. In the event that $|a| = 1$ then (4.6.6) provides the exact solution. The stability condition can be written as

$$\Delta t \leqslant \frac{\Delta x}{|u \pm (gh)^{\frac{1}{2}}|}$$

In this case, then, the von Neumann condition coincides with the CFL necessary condition for stability.

It may be shown that this condition is simultaneously sufficient. Moreover, since the exact solution condition $|a| = 1$ necessitates that $u \pm (gh)^{\frac{1}{2}}$ reduce $\pm (gh)^{\frac{1}{2}}$, so to the linearized situation, it is easily seen that the difference scheme (4.6.6) then reduces to the 4-point method of characteristics.

The leapfrog scheme is centred in x and t so that it is at least of second-order accuracy. However, the centred dissipative interface (Equation (4.6.13)) is also of second-order accuracy, albeit in x only, so that the possibility then appears of combining these elements to make a dissipative scheme of second-order accuracy. Consider the case where the dissipative interface with $\alpha = \frac{1}{2}$ is applied to the lower-time variables of the time differences of the leapfrog scheme at each alternate time step, the scheme being locally decomposed in order to accommodate the interface so that it is centred between the leapfrog elements to maintain the accuracy. This process is schematized in Fig. 4.34. In Fig. 4.34a the decomposition is made at alternate time levels so that Equation (4.6.3) comes to read

$$\frac{f_j^{n+\frac{1}{2}} - f_j^n}{\Delta t} + \frac{g_{j+\frac{1}{2}}^{n+\frac{1}{2}} - g_{j-\frac{1}{2}}^{n+\frac{1}{2}}}{2\Delta x} = 0 \tag{4.6.18}$$

$$\frac{f_j^{n+1} - f_j^{n+\frac{1}{2}}}{\Delta t} + \frac{g_{j+\frac{1}{2}}^{n+\frac{1}{2}} - g_{j-\frac{1}{2}}^{n+\frac{1}{2}}}{2\Delta x} = 0 \tag{4.6.19}$$

The dissipative interface is now introduced on f with $\alpha = \frac{1}{2}$ (Fig. 4.34b) so that

Equation (4.6.19) becomes

$$\frac{f_j^{n+1} - (f_{j+1/2}^{n+\frac{1}{2}} + f_{j-1/2}^{n+\frac{1}{2}})/2}{\Delta t} + \frac{g_{j+\frac{1}{2}}^{n+\frac{1}{2}} - g_{j-\frac{1}{2}}^{n+\frac{1}{2}}}{2\Delta x} = 0 \tag{4.6.20}$$

which does not call upon any information at $(j\,\Delta x, n\,\Delta t)$, thus making Equation (4.6.18) redundant. The resulting scheme, schematized in Fig. 4.34c, has the appearance of a leapfrog scheme interposed with a Lax scheme.

Equation (4.6.3) can be expressed through (4.6.20), written locally for f_j^n, as

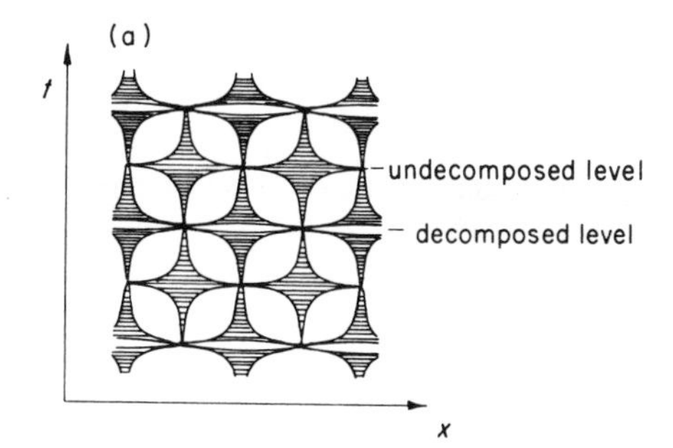

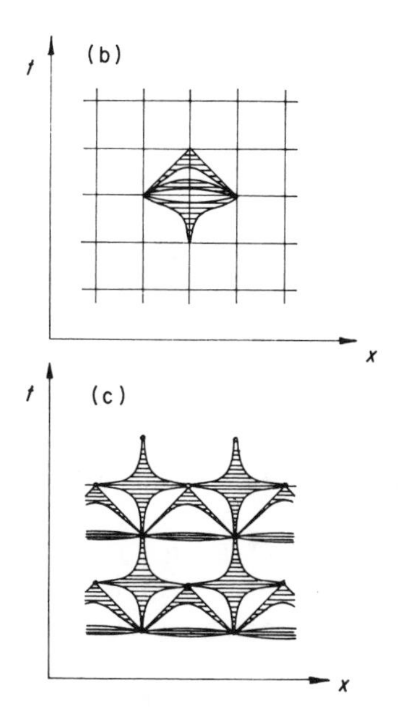

Fig. 4.34. (a) Decomposed operators interposed between undecomposed operators. (b) The introduction of centred dissipative interface with $\alpha = \frac{1}{2}$ makes the lower element redundant. (c) First-order errors subsequently introduced in the decomposed and dissipative operators will then be differenced to second-order errors in the undecomposed operators

$$\frac{f_j^{n+1} - f_j^n}{2\,\Delta t} + A\left[\frac{f_{j+1}^n - f_{j-1}^n}{4\,\Delta x} - \frac{\Delta t}{2\,\Delta x}\left(\frac{g_{j+1}^n - 2g_j^n + g_{j-1}^n}{2\,\Delta x}\right)\right] = 0$$

or, extending the localization,

$$f_j^{n+1} - f_j^n + \frac{A\,\Delta t}{2\,\Delta x}(f_{j+1}^n - f_{j-1}^n) - \frac{A^2\,\Delta t^2}{2\,\Delta x^2}(f_{j+1}^n - 2f_j^n + f_{j-1}^n) = 0 \tag{4.6.21}$$

The amplification matrix then becomes

$$G = I - (1 - \cos\gamma)\left(\frac{\Delta t}{\Delta x}\cdot A\right)^2 - \mathrm{i}\sin\gamma\left(\frac{\Delta t}{\Delta x}\cdot A\right)$$

where again $\gamma = 2\pi k\;\Delta x/2l$.

Since G is again a rational function of A, the eigenvalues of this scheme are given by

$$g = 1 - \left(\frac{\lambda\,\Delta t}{\Delta x}\right)^2(1 - \cos\gamma) - \mathrm{i}\left(\frac{\lambda\,\Delta t}{\Delta x}\right)\sin\gamma \tag{4.6.22}$$

This equation again provides an ellipse in the Argand diagram that lies within the unit circle so long as $|\lambda\,\Delta t/\Delta x| \leqslant 1$. It may be shown that the von Neumann necessary condition for stability thereby satisfied is simultaneously sufficient. The scheme (4.6.21) is, in effect, the *Lax–Wendroff* (1960) scheme, while Equations (4.6.3) and (4.6.20) constitute its equivalent two-stage form, as formulated by Richtmyer (1963). (*See* Richtmyer and Morton, 1967.)

These questions of amplitude error, including stability criteria, can be considered in a somewhat different way that, furthermore, provides a convenient method of analysis of *phase errors*. Following Leendertse (1967) and Vliegenthart (1968), a complex *propagation factor P* may be introduced, that is defined in terms of wave numbers by

$$P = P(x, \Delta t, k) = \frac{\mathrm{e}^{\mathrm{i}2\pi k(x - \delta t)/2l}}{\mathrm{e}^{\mathrm{i}2\pi k(x - \lambda t)/2l}} \tag{4.6.23}$$

Here δ is the (complex) celerity of the numerically computed wave, while λ is the celerity given by the differential equation.

The modulus of P for any period then provides a measure of the attenuation of the kth component over that period, while the argument of P is a measure of the corresponding phase error. It is convenient, for didactic reasons, to follow P over a time interval in which the component of wave number k propagates over its entire wave length, for in that case $t = 2l/\lambda k$ and Equation (4.6.23) gives

$$P = \mathrm{e}^{-\mathrm{i}2\pi(\delta/\lambda - 1)}$$

Recalling that δ is generally complex and $|\mathrm{e}^{\mathrm{i}\phi}| = 1$ for all real ϕ, it is found that

$$|P| = \mathrm{e}^{2\pi\cdot\mathrm{Im}(\delta)/\lambda} \tag{4.6.24a}$$

$$\text{Arg}\, P = 2\pi \left(1 - \frac{\text{Re}(\delta)}{\lambda}\right) = 2\pi \left(\frac{\lambda - \text{Re}(\delta)}{\lambda}\right) \tag{4.6.24b}$$

This last equation simply re-expresses the fact that the ratio Q of the computational celerity to the true celerity, called the *celerity ratio*, is given by $Q = \text{Re}(\delta)/\lambda$ while a measure of the phase error is provided by the departure of Q from unity, i.e. by the quantity $(\lambda - \text{Re}(\delta))/\lambda$. It should be noticed that, for any dissipative scheme, $\text{Im}(\delta) < 0$.

Now substituting $\xi_k \exp(i2\pi k(x - \delta t)/2l)$ in any two level scheme, explicit or implicit, consistent with Equation (2.6.6) provides

$$[G(x, \Delta t, k) - e^{-i2\pi k \delta \Delta t/2l} I]\, \xi_k = 0$$

where G is an amplification matrix of the difference scheme. Thence ξ_k is an eigenvector of G and $\exp(-i2\pi k\delta\, \Delta t/2l)$ is the corresponding eigenvalue. Since any G of this form can have at most two eigenvalues, it follows that

$$g(x, \Delta t, k) = \exp(-i2\pi k\delta\; \Delta t/2l)$$

Thence

$$g(x, \Delta t, k) = \exp(2\pi k \cdot \text{Im}(\delta) \cdot \Delta t/2l) \cdot \exp(-i2\pi k \cdot \text{Re}(\delta) \cdot \Delta t/2l) \tag{4.6.25}$$

or

$$|g| = \exp(2\pi k \cdot \text{Im}(\delta) \cdot \Delta t/2l)$$

This is the amplification over time Δt. It follows from the assumed linearity that the amplification over time $2l/\lambda k$ is

$$\exp(2\pi k \cdot \text{Im}(\delta) \cdot \Delta t/2l)^{2l/\lambda k \Delta t} = \exp(2\pi\, \text{Im}(\delta)/\lambda)$$

Comparing this result with Equation (4.6.24a) shows that the amplification factor provided by (4.6.23) corresponds to the usual definition of amplification factor, as the modulus of the eigenvalues of an amplification matrix.

Moreover, Equation (4.6.25) provides

$$\text{Re}(g) + i\, \text{Im}(g) = \exp(2\pi k \cdot \text{Im}(\delta) \cdot \Delta t/2l)[\cos(2\pi k \cdot \text{Re}(\delta) \cdot \Delta t/2l) - $$
$$- i \sin(2\pi k \cdot \text{Re}(\delta) \cdot \Delta t/2l)]$$

i.e.

$$\text{Re}(g) = \exp(2\pi k \cdot \text{Im}(\delta) \cdot \Delta t/2l)[\cos(2\pi k \cdot \text{Re}(\delta) \cdot \Delta t/2l)]$$

$$\text{Im}(g) = \exp(2\pi k \cdot \text{Im}(\delta) \cdot \Delta t/2l)[-\sin(2\pi k \cdot \text{Re}(\delta) \cdot \Delta t/2l)]$$

It is then seen that, since

$$Q = \frac{\text{Re}(\delta)}{\lambda} = \frac{2\pi k\, \Delta t/2l \cdot \text{Re}(\delta)}{2\pi k\, \Delta t/2l \cdot \lambda}$$

then

$$\tan(2\pi k\,\Delta t/2l\cdot\lambda\cdot Q) = \tan(2\pi k\,\Delta t/2l\cdot \mathrm{Re}(\delta))$$

$$= -\frac{\mathrm{Im}(g)}{\mathrm{Re}(g)}$$

Thence

$$Q = -\mathrm{arc\,tan}\left(\frac{\mathrm{Im}(g)}{\mathrm{Re}(g)}\right)\Big/\left(\frac{2\pi k\,\Delta t}{2l}\cdot\lambda\right) \tag{4.6.26}$$

This characteristic measure of phase shift is especially convenient computationally. Thus, it follows at once from Equation (4.6.26), that for the Lax scheme (4.6.16)

$$Q = (\mathrm{arc\,tan}\,(Cr\tan\gamma))/(\pi k\,\Delta t\cdot\lambda/l) \tag{4.6.27}$$

while for the two-step Lax Wendroff scheme (4.6.21)

$$Q = \left[\mathrm{arc\,tan}\left(\frac{Cr\sin\gamma}{1-(Cr)^2(1-\cos\gamma)}\right)\right]\Big/(\pi k\,\Delta t\cdot\lambda/l) \tag{4.6.28}$$

with Cr the *Courant number* $\lambda\,\Delta t/\Delta x$.

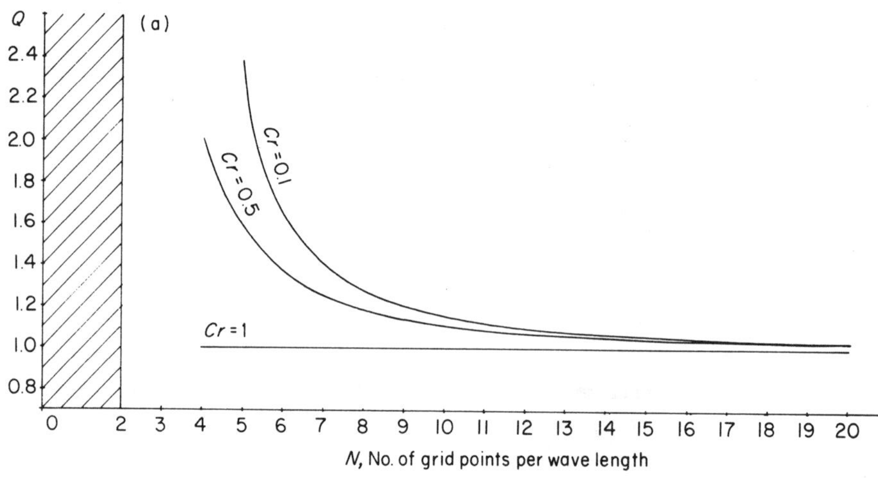

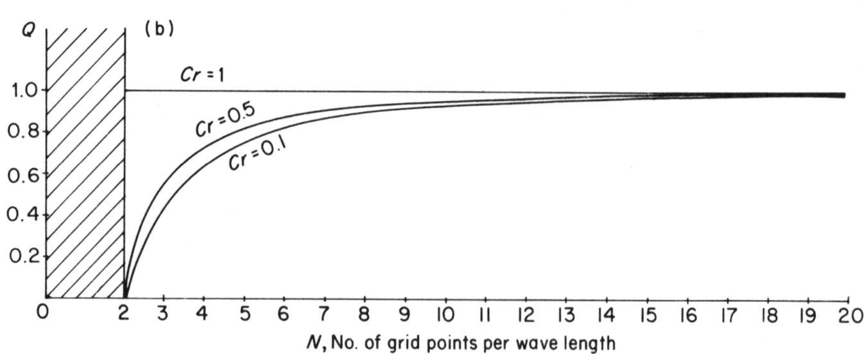

Fig. 4.35. Phase portraits of (a) the Lax (1954) and (b) the Lax–Wendroff (1960) schemes

It is seen that, when $Cr = 1$, both the Lax scheme and the Lax–Wendroff scheme provide $Q = 1$, so that there is no phase error in that case. Figures 4.35a and b show the phase portraits of the Lax and Lax–Wendroff schemes respectively, as determined from Equations (4.6.27) and (4.6.28).

Figure 4.36 schematizes the difference in amplitude and celerity of the true and computed components for the two cases of positive and negative characteristic directions. In general, of course, propagation occurs in both directions simultaneously.

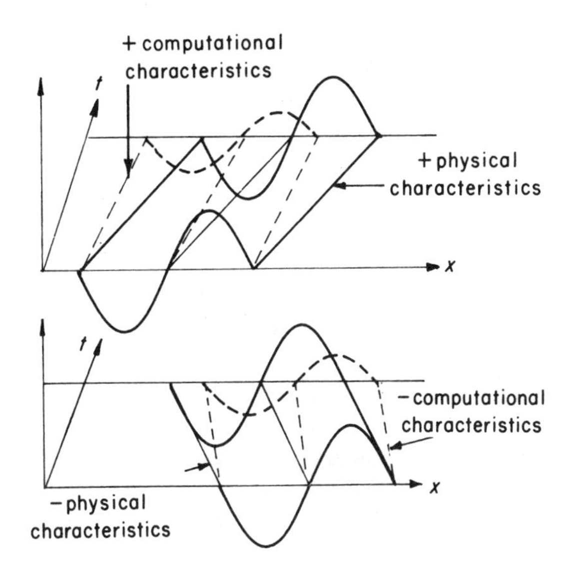

Fig. 4.36. Schematization of phase error in terms of differential and difference scheme directions of propagation in (h, x, t) space

The observation that, when $|\lambda \Delta t/\Delta x| = 1$, both the amplitude error and the phase error are zero in both Lax and Lax–Wendroff schemes implies that in this condition they are both exact. Thus both of these schemes must be equivalent to the method of characteristics when $|\lambda \Delta t/\Delta x| = 1$. Now this condition can only be approximated for *both* characteristics when $u \ll (gh)^{\frac{1}{2}}$, in which case the general form of the Lax scheme (4.6.16) simplifies. For example, the conservation law of Bernoulli

$$\frac{u_j^{n+1} - (u_{j+1}^n + u_{j-1}^n)/2}{\Delta t} + \frac{(u^2/2 + gh)_{j+1}^n - (u^2/2 + gh)_{j-1}^n}{2\,\Delta x} = 0$$

simplifies to

$$\frac{u_j^{n+1} - (u_{j+1}^n + u_{j-1}^n)/2}{\Delta t} + \frac{(c_{j+1}^n + c_{j-1}^n)(c_{j+1}^n - c_{j-1}^n)}{2\Delta x} = 0, \quad c = (gh)^{\frac{1}{2}}$$

so that, when $c\,\Delta t/\Delta x = 1$,

$$u_j^{n+1} = \tfrac{1}{2}(u_{j+1}^n + u_{j-1}^n) - (c_{j+1}^n - c_{j-1}^n)$$

which is the velocity statement of the three-point method of characteristics. Similarly, the mass law of the Lax scheme, viz.

$$\frac{h_j^{n+1} - (h_{j+1} + h_{j-1})^n/2}{\Delta t} + \frac{(uh)_{j+1}^n - (uh)_{j-1}^n}{2\,\Delta x} = 0$$

reduces to

$$c_j^{n+1} = \tfrac{1}{2}(c_{j+1}^n + c_{j-1}^n) - \tfrac{1}{4}(u_{j+1}^n - u_{j-1}^n)$$

which is the celerity statement of the three-point method. It follows at once that the linearized Lax–Wendroff scheme reduces to either the six-point or the three-point method of characteristics of Fig. 4.5 when $|\lambda\,\Delta t/\Delta x| = 1$.

There exists of course an infinity of other explicit difference schemes consistent with Equation (2.1.8) or its Eulerian, algorithmic or other forms, but the above suffices to introduce their main properties.

4.7 Implicit difference methods for one-dimensional nearly horizontal flows

In the previous section, it was seen how the time steps used by the explicit schemes are restricted by the Courant–Friedrichs–Lewy condition. This restriction will now be given a more precise formulation, as a preliminary to investigating the implicit schemes, where it can be made considerably less onerous.

It was seen in Chapter 3 that, in a conservative system, conditions at any point P are completely determined, through Riemann invariants J_+ and J_-, by conditions at points such as Q and R lying on C_+ and C_- characteristics through P (Fig. 4.37). In the more general case, where quasi-invariants appear, the way in which J_+

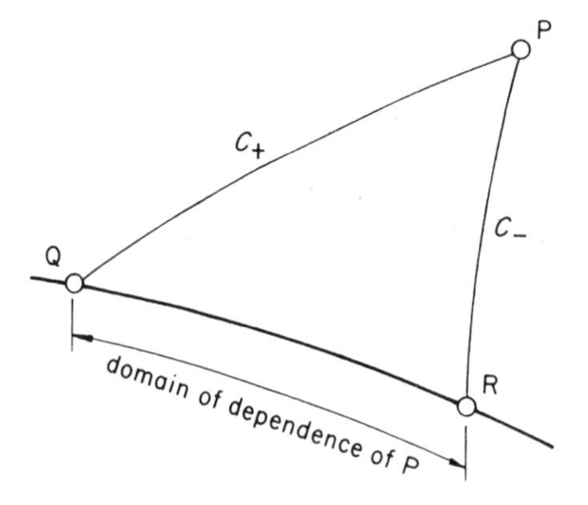

Fig. 4.37. Domain of dependence of a point P as defined by C_+ and C_- characteristics of a differential equation system

changes along any C_+ characteristic is influenced by the J_- values on the C_- characteristics intersecting that C_+ characteristic, and vice versa. Referring to Fig. 4.37, it will then be apparent that conditions at P are determined not only by points Q and R, but, more generally, by conditions along the arc QR. This determination then proceeds from QR to P through the intervention of all characteristic intervals inside the triangle PQR. The arc QR, intercepted by the C_+ and C_- characteristics through P, is called the 'domain of dependence' of P.

The significance of the domain of dependence as introduced above may be summarized in the following uniqueness theorem (Courant and Friedrichs, 1948, p. 51):

> *Consider a solution of the two-dependent-variable propagation problem within the region PQR bounded by the two characteristics through the point P and the domain of dependence QR cut out by them from some initial curve. Suppose another solution of the same problem is given in PQR which assumes on QR the same values as the first one. Then the second solution is identical with the first one in the region PQR.*

From a slightly different point of view, it may be seen that the characteristics PQ and PR essentially limit the domain, or region, which is controlled by any particular boundary conditions. The characteristics PQ and PR then behave as the limits of direct application of information on QR, and P is the furthest point to which that information can be independently utilized. In a Taylor-series representation, the lines PQ and PR define the furthest points to which a system of Taylor-series expansions from the domain of dependence can be extended. These are the information boundary interpretations of the characteristics, as related to the domain of dependence.

It should also be observed that, as the solution progresses across the integral surface, so any given point P influences a region such as Q_1R_1 at t_1, a region Q_2R_2 at t_2, and so on. It thus influences (but does not entirely determine) the solution in a region bounded by C_+ and C_- characteristics through P. This region,

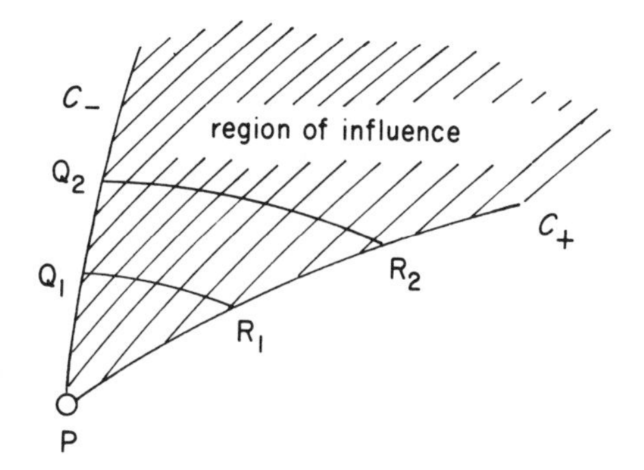

Fig. 4.38. Region of influence of a point P as defined by C_+ and C_- characteristics of a differential equation system

which is partly bounded by the diverging characteristics through P, is called the 'region of influence' of P.

Now if, in a mixed initial value/boundary value problem posed between $n\,\Delta t$ and $(n+1)\,\Delta t$, data are presented along the boundaries between $n\,\Delta t$ and $(n+1)\,\Delta t$ in such a way as to influence a part of the solution at $(n+1)\,\Delta t$, then the domain of dependence may encompass the corresponding domain of the solution. This solution is schematized in Fig. 4.39. Evidently, in order to realize such a presentation of data to a difference scheme consistent with a differential equation with real characteristics, data presented at $(n+1)\,\Delta t$ should influence the entirety of that part of the solution influenced by the boundary conditions through the characteristics of the differential equation, at least in some limit $\Delta t, \Delta x \rightarrow 0$. This is to say that, at least in some limit, the influence of information presented at $(n+1)\,\Delta t$ must propagate so far as to meet the characteristics entering the problem domain from the boundaries at $n\,\Delta t$. In principle, of course, further 'boundaries' may be introduced conceptually within the problem domain, so that each and every elemental numerical operator must be influenced by its neighbours and must in turn influence them as demanded by the characteristic conditions. This is only possible, however, if every elemental numerical operator at $j\,\Delta x$ contains items not just at $(j\,\Delta x; (n+1)\,\Delta t)$ but at $\{(j-l)\,\Delta x, \ldots, j\,\Delta x, \ldots, (j+m)\,\Delta x$, with $l+m \geqslant 1; (n+1)\,\Delta t\}$. Such a scheme is commonly called an *implicit* scheme.

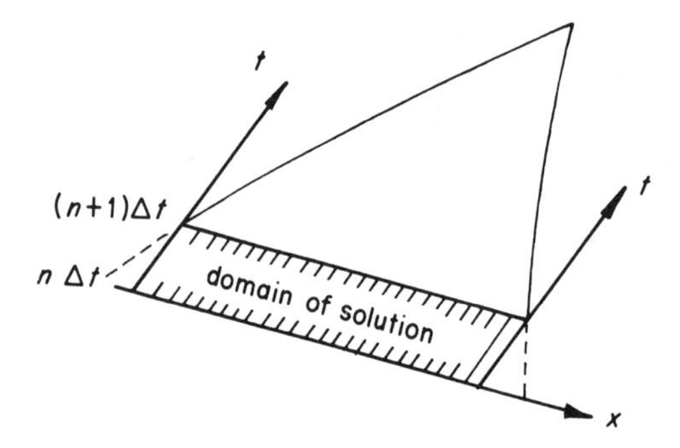

Fig. 4.39. Domain of a solution of an implicit difference equation system relative to the domain of dependence of the differential equation system with which it is consistent

The simplest examples of an implicit scheme can, once again, best be derived from the scalar wave equation

$$\frac{\partial z}{\partial t} + c\,\frac{\partial z}{\partial x} = 0$$

Consider, for example, the space and time centred scheme, or 'box' scheme, for this equation:

$$\left(\frac{z_{j+1}^{n+1} - z_{j+1}^{n}}{2\,\Delta t} + \frac{z_j^{n+1} - z_j^{n}}{2\,\Delta t}\right) + c\left(\frac{z_{j+1}^{n+1} - z_j^{n+1}}{2\,\Delta x} + \frac{z_{j+1}^{n} - z_j^{n}}{2\,\Delta x}\right) = 0 \tag{4.7.1}$$

Following any kth component of $z_j^n = \sum_k \xi_k^n \exp(i2\pi kj\,\Delta x/2l)$ gives the amplification factor

$$A_k = \frac{\cos\gamma_k - ir\sin\gamma_k}{\cos\gamma_k + ir\sin\gamma_k}$$

with $\gamma_k = \pi k\,\Delta x/2l$ and $r = c\,\Delta t/\Delta x$. Thus $|A_k| = 1$ for every k, r, so that the scheme is linearly unconditionally stable.

However, as mentioned in the definition of stability, a condition for the relevance of any stability condition to the real behaviour of the solution of a mixed initial value/boundary value problem is that the problem should be properly posed. Indeed, if the problem is not properly posed, it may be shown, for a wide class of problems, that no difference scheme can be both consistent and stable (e.g. Hadamard, 1923; Richtmyer and Morton, 1967). In the case of explicit schemes, this necessitates that, apart from those operators immediately adjacent to the boundaries, only the type and amount of *initial* data should be appropriate to the equation system that is being solved. In the case of the implicit schemes, however, it is necessary that both *initial* and *boundary* data should be appropriate for the equation system and, since boundaries can commonly be introduced conceptually anywhere in the domain, this necessary condition applies to every elemental operator, such as (4.7.1).

Consider now two solutions generated using (4.7.1), the one with $c > 0$ and the other with $c < 0$, as schematized in Fig. 4.40. It is seen that in the case of $c > 0$, (4.7.1) has to be used in a single sweep from the left to right,

$$z_{j+1}^{n+1} = \frac{(-1+r)z_j^{n+1} + (1-r)z_{j+1}^n + (1+r)z_j^n}{(1+r)} \tag{4.7.2A}$$

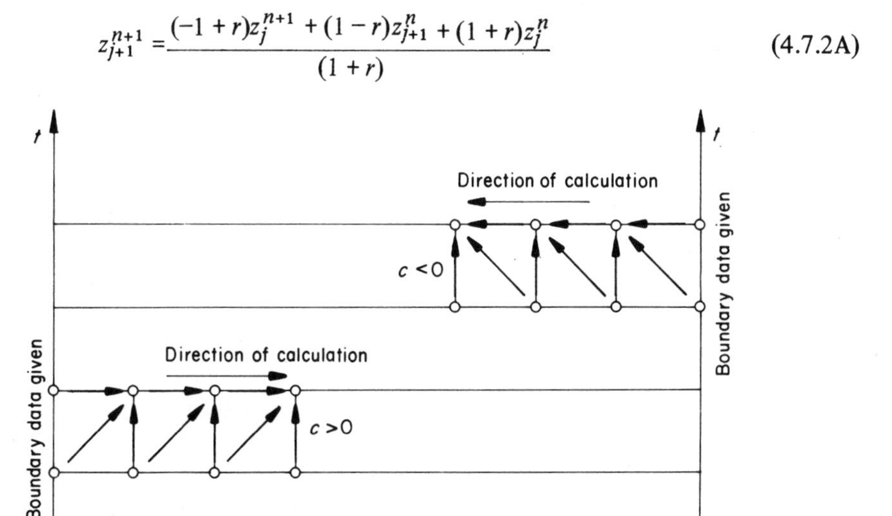

Fig. 4.40. The implicit scheme (4.7.1) for the scalar wave equation has two definite orders of computation, defining two 'algorithmic structures', the one appropriate when $c > 0$ and the other appropriate when $c < 0$. These structures are schematized here by arrows indicating the directions of information transfer

while when $c < 0$ it has to be used in a single sweep from right to left:

$$z_j^{n+1} = \frac{(-1-r)z_{j+1}^{n+1} + (1-r)z_{j+1}^{n} + (1+r)z_j^n}{(1-r)} \tag{4.7.2B}$$

The terminology of an *implicit scheme* will be applied here to any scheme that has a *definite order or direction of computation* in the space of the independent variables. Thus (4.7.1) is an implicit scheme. In such a scheme, two or more unknowns at the upper time level are related to one another in each elemental operator, so that unknowns are linked between operators and, through these, to the upper-time boundary conditions. The unknowns are thus defined implicitly and not explicitly relative to other values known at lower time levels. An *explicit scheme is then one that is indifferent to the order or direction of computation* in the space of the independent variables. An implicit scheme thus has cardinal and ordinal attributes, while on explicit scheme has only cardinal attributes.

The order of computation of an implicit scheme, as defined by the data structure of the differential equation with which the scheme is consistent (Section 3.3), will be called the *algorithmic structure* of the scheme. Thus the scheme (4.7.1) has two algorithmic structures, the one, a left to right sweep, being appropriate when, and only when, $c > 0$ and the other, a right to left sweep, being appropriate when, and only when, $c < 0$. It is easily verified in this case that, for any given non-trivial $c = c(x,t)$, the solution generated using one algorithmic structure is different from the solution generated using the other algorithmic structure and that, indeed, the use of the inappropriate structure, such as a left to right sweep with $c < 0$ in the above case, gives an unstable solution. The linear or 'local' stability analysis, then, describes the behaviour of the solution of the well-posed problem and not that of the ill-posed problem.

In the above example, both algorithmic structures of (4.7.1) provide unconditionally stable solutions when used appropriately, as predicted by the linear stability analysis. Transport schemes that are not space- and time-centred may have one algorithmic structure, used appropriately, providing stability up to one Δt, even so far as $\Delta t = \infty$, and another algorithmic structure, used appropriately, providing stability up to a quite other Δt, and even for $\Delta t = 0$. Such differences in behaviour between algorithmic structures used appropriately show up, of course, in the linear stability analysis.

It follows that implicit schemes, being sensitive to boundary data structures that are in turn transmitted through the implicit scheme to appear at all points within the domain of the solution, must be provided with the appropriate algorithmic structure if they are to survive in general use. In the sequel it will always be supposed that algorithmic structures are used appropriately. Later in this chapter, 'generalized algorithms', that combine two or more algorithmic structures in a compound structure, will be introduced. Explicit schemes are then schemes with no algorithmic structure and, of course, they can be defined in this way.

The type of construction of schemes that follows the data structure in the elementary manner of (4.7.2) is sometimes called 'upwind differencing' (Roache,

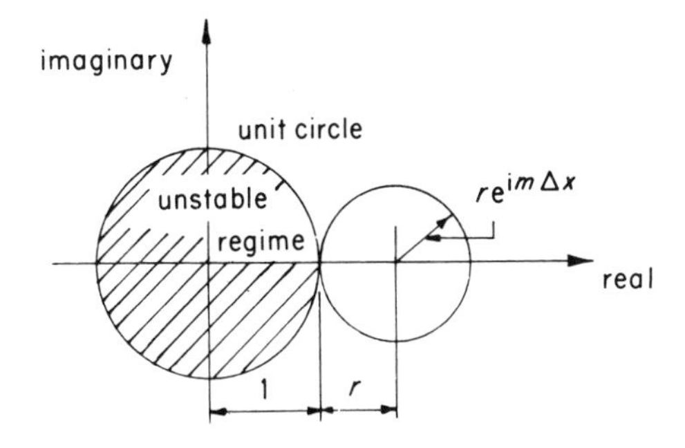

Fig. 4.41. Argand diagram representation of the inverse of the amplification factor of the implicit scheme (4.7.2)

1972). Cunge (1969) has shown that the well-known Muskingum method for flood routing is equivalent to upwind differencing with an implicit difference scheme consistent with the transport or scalar wave equation.

In order to extend these elementary notions to schemes for nearly horizontal flows, it is convenient to consider first the scheme of Abbott and Ionescu (1967). In its original form, this was built upon the algorithmic form (2.6.13) with (2.6.7), which may be written out as

$$\begin{aligned} h\frac{\partial u}{\partial t} - u\frac{\partial h}{\partial t} - (u^2 - gh)\frac{\partial h}{\partial x} &= 0 \\ g\frac{\partial h}{\partial t} - u\frac{\partial u}{\partial t} - (u^2 - gh)\frac{\partial u}{\partial x} &= 0 \end{aligned} \tag{4.7.3}$$

Centred space and time differences are introduced, to provide the scheme

$$\begin{aligned} &h\frac{(u_j^{n+1} - u_j^n)}{\Delta t} - \frac{u}{2}\left(\frac{h_{j+1}^{n+1} - h_{j+1}^n}{\Delta t} + \frac{h_{j-1}^{n+1} - h_{j-1}^n}{\Delta t}\right) - \frac{(u^2 - gh)}{4} \\ &\qquad \left(\frac{h_{j+1}^{n+1} - h_{j-1}^{n+1}}{\Delta x} + \frac{h_{j+1}^n - h_{j-1}^n}{\Delta x}\right) = 0 \\ &g\frac{(h_j^{n+1} - h_j^n)}{\Delta t} - \frac{u}{2}\left(\frac{u_{j+1}^{n+1} - u_{j+1}^n}{\Delta t} + \frac{u_{j-1}^{n+1} - u_{j-1}^n}{\Delta t}\right) - \frac{(u^2 - gh)}{4} \\ &\qquad \left(\frac{u_{j+1}^{n+1} - u_{j-1}^{n+1}}{\Delta x} + \frac{u_{j+1}^n - u_{j-1}^n}{\Delta x}\right) = 0 \end{aligned} \tag{4.7.4}$$

These equations can be written as

$$\begin{aligned} &\left(-\frac{u}{2\Delta t} - \frac{(u^2 - gh)}{4\Delta x}\right) h_{j+1}^{n+1} + \left(\frac{h}{\Delta t}\right) u_j^{n+1} + \left(-\frac{u}{2\Delta t} + \frac{(u^2 - gh)}{4\Delta x}\right) h_{j-1}^{n+1} = \\ &\quad = \left(-\frac{u}{2\Delta t} + \frac{(u^2 - gh)}{4\Delta x}\right) h_{j+1}^n + \left(\frac{h}{\Delta t}\right) u_j^n + \left(-\frac{u}{2\Delta t} - \frac{(u^2 - gh)}{4\Delta x}\right) h_{j-1}^n \end{aligned} \tag{4.7.5}$$

$$\left(-\frac{u}{2\Delta t}-\frac{(u^2-gh)}{4\Delta x}\right)u_{j+1}^{n+1}+\left(\frac{g}{\Delta t}\right)h_j^{n+1}+\left(-\frac{u}{2\Delta t}+\frac{(u^2-gh)}{4\Delta x}\right)u_{j-1}^{n+1}=$$
$$=\left(-\frac{u}{2\Delta t}+\frac{(u^2-gh)}{4\Delta x}\right)u_{j+1}^{n}+\left(\frac{g}{\Delta t}\right)h_j^{n}+\left(-\frac{u}{2\Delta t}-\frac{(u^2-gh)}{4\Delta x}\right)u_{j-1}^{n} \tag{4.7.6}$$

that is, in the form

$$A_j h_{j+1}^{n+1}+B_j u_j^{n+1}+C_j h_{j-1}^{n+1}=D_j \tag{4.7.7}$$

$$A_j u_{j+1}^{n+1}+B_j^* h_j^{n+1}+C_j u_{j-1}^{n+1}=D_j^* \tag{4.7.8}$$

Suppose, now, that u_j^{n+1} and h_j^{n+1} can be related such that

$$h_{j+1}^{n+1}=E_j u_j^{n+1}+F_j \tag{4.7.9}$$

$$u_{j+1}^{n+1}=E_j^* h_j^{n+1}+F_j^* \tag{4.7.10}$$

Then Equations (4.7.7) and (4.7.9) provide

$$A_j E_j u_j^{n+1}+B_j u_j^{n+1}+C_j h_{j-1}^{n+1}=D_j-A_jF_j$$

or

$$u_j^{n+1}=\frac{-C_j}{A_jE_j+B_j}h_{j-1}^{n+1}+\frac{D_j-A_jF_j}{A_jE_j+B_j} \tag{4.7.11}$$

while (4.7.8) and (4.7.10) provide

$$A_jE_j^* h_j^{n+1}+B_j^* h_j^{n+1}+C_j u_{j-1}^{n+1}=D_j^*-A_jF_j^*$$

or

$$h_j^{n+1}=\frac{-C_j}{A_jE_j^*+B_j^*}u_{j-1}^{n+1}+\frac{D_j^*-A_jF_j^*}{A_jE_j^*+B_j^*} \tag{4.7.12}$$

Comparison of Equations (4.7.9) and (4.7.10) with (4.7.11) and (4.7.12) then leads to the identities

$$E_{j-1}=\frac{-C_j}{A_jE_j^*+B_j^*}\qquad F_{j-1}=\frac{D_j^*-A_jF_j^*}{A_jE_j^*+B_j^*} \tag{4.7.13}$$

$$E_{j-1}^*=\frac{-C_j}{A_jE_j+B_j}\qquad F_{j-1}^*=\frac{D_j-A_jF_j}{A_jE_j+B_j} \tag{4.7.14}$$

The algorithm is initiated from one or the other of the boundary conditions. Thus, for example, if the velocity u_J is given at one end, say at $x=l\ (=J\,\Delta x$ for constant Δx), then the ultimate equation of the system, (4.7.7) and (4.7.8), provides

$$A_{J-1}u_J^{n+1}+B_{J-1}^*h_{J-1}^{n+1}+C_{J-1}u_{J-2}^{n+1}=D_{J-1}^*$$

Thence

$$h_{J-1}^{n+1} = -\frac{C_{J-1}}{B_{J-1}^*} u_{J-2}^{n+1} + \frac{D_{J-1}^* - A_{J-1} u_J^{n+1}}{B_{j-1}^*} \tag{4.7.15}$$

Comparing Equation (4.7.15) with (4.7.9) then suggests the identities

$$E_{J-2} = -\frac{C_{J-1}}{B_{J-1}^*}, \quad F_{J-2} = \frac{D_{J-1}^* - A_{J-1} u_J^{n+1}}{B_{J-1}^*} \tag{4.7.16}$$

Comparing Equation (4.7.16) with (4.7.13) indicates that

$$E_{J-1}^* = 0, \quad F_{J-1}^* = u_J^{n+1} \tag{4.7.17}$$

Similarly, if h_J^{n+1} is given as a boundary condition at $j = J$, it is seen that

$$E_{J-1} = 0, \quad F_{J-1} = h_J^{n+1} \tag{4.7.18}$$

Keeping to the case that u_J^{n+1} is given, it is seen that, from condition (4.7.17), E_{J-2} and F_{J-2} can be computed using Equations (4.7.13) and $A_{J-1}, B_{J-1}, C_{J-1}$ values at time $n\,\Delta t$. With E_{J-2} and F_{J-2} determined, E_{J-3}^* and F_{J-3}^* are found from (4.7.14), and so successively until arriving at either E_0, F_0 or $E_0^* F_0^*$, according to the nature of the other end conditions. Thus if h is given at the other end, the grid is chosen such that E_0^* and F_0^* are last determined. Then u_1^{n+1} is determined from Equation (4.7.10), h_2^{n+1} from (4.7.9) and so successively back to u_J^{n+1}. The grid used in this case is schematized in Figure 4.42.

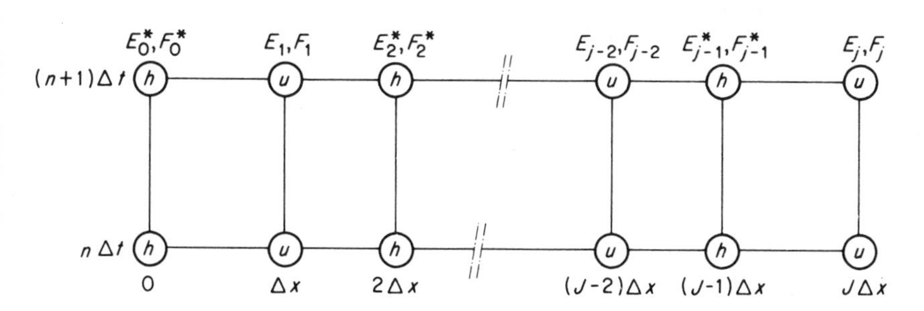

Fig. 4.42. Data layout for the implicit scheme of the Abbott–Ionescu type (Equation (4.7.4))

Referring to this grid and the corresponding boundary data structure it is seen that the *suppositions (4.7.9) and (4.7.10) correspond to the assumption of subcritical flow.* If the flow were *supercritical*, then two-point data would be given at the upstream boundary and Equations (4.7.7) and (4.7.8) could be solved directly through the recurrence relations:

$$\begin{aligned} h_{j+1}^{n+1} &= -\frac{1}{A_j}(B_j u_j^{n+1} + C_j h_{j-1}^{n+1} - D_j) \\ u_{j+1}^{n+1} &= -\frac{1}{A_j}(B_j^* h_j^{n+1} + C_j u_{j-1}^{n+1} - D_j^*) \end{aligned} \tag{4.7.19}$$

when flow is in the positive x-direction, or

$$h_{j-1}^{n+1} = -\frac{1}{C_j}(A_j h_{j+1}^{n+1} + B_j u_j^{n+1} - D_j)$$
$$u_{j-1}^{n+1} = -\frac{1}{C_j}(A_j u_{j+1}^{n+1} + B_j^* h_j^{n+1} - D_j^*) \tag{4.7.20}$$

when flow is in the negative x-direction.

The recurrence relations corresponding to supercritical flow, as defined by Equations (4.7.19) and (4.7.20), are examples of *single-sweep algorithms* while the process of first sweeping from one boundary to the other using Equations (4.7.13) and (4.7.14) from (4.7.17) or (4.7.18) and then sweeping back using (4.7.9) and (4.7.10), as corresponds to a subcritical flow, is an example of a *double-sweep algorithm*. Thus, the *single scheme (4.7.5), (4.7.6) or (4.7.7), (4.7.8) has three distinct algorithmic structures*. It then appears that the use of an inappropriate structure ought to lead to an improperly posed problem and some form of instability. Although this is true in certain limits, the pathology of algorithmic structures in *finite* schemes is rather more complicated than this, and outside the scope of this introduction.

It is, of course, possible that subcritical flow transforms to supercritical flow through a continuous nearly horizontal flow, so that two or more algorithmic structures have to be used in a single computation. The problem then arises of tracking the domains of each structure by following the point at which one of them transforms into the other. This point is approximately the critical-flow point, where $u = (gh)^{\frac{1}{2}}$, but it is useful for more general applications to define this point with the data-structure condition that one of the characteristics passes through zero:

$$(u - \dot{x})^2 - gh = 0$$

so that the function

$$H = u^2 - gh \tag{4.7.21}$$

passes through zero.

The change-over point for algorithmic structures then occurs when the function (4.7.21) passes through zero and, in general, when (4.7.21) changes sign. Thus a recursive sign control on (4.7.21) is sufficient to control the change-over of the algorithmic structure. Correspondingly, an expression of the form of (4.7.21) that is used in this way is called a *control function*.

It is possible also to construct a generalized double-sweep algorithm that will carry two pieces of information in both directions but such that one or both pieces can be effectively immobilized to suit the flow conditions. Such generalized algorithms have been outlined by Abbott and Grubert (1973).

The operation of the double-sweep algorithm upon the entire flow system can be seen more clearly by using the following argument (Bruce *et al.*, 1953; Jollife, 1968). Equations (4.7.7) and (4.7.8) can be written with constant boundary conditions, suppressing the $*$ notation at alternate points and opening out the $\{D_j\}$, as

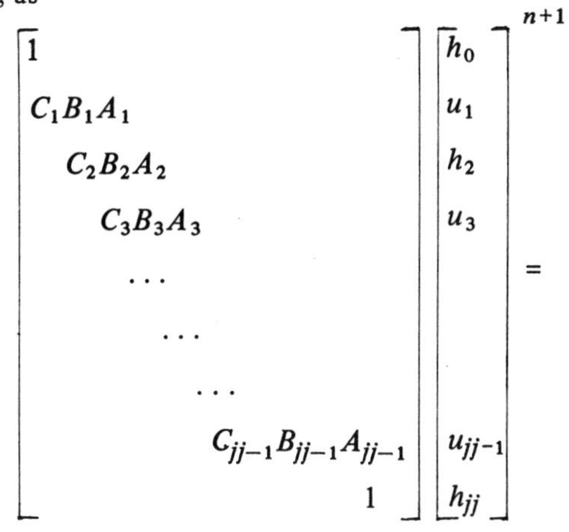

$$= \begin{bmatrix} 1 & & & & & & & \\ C_1'B_1'A_1' & & & & & & & \\ & C_2'B_2'A_2' & & & & & & \\ & & C_3'B_3'A_3' & & & & & \\ & & & \cdots & & & & \\ & & & & \cdots & & & \\ & & & & & \cdots & & \\ & & & & & & C_{jj-1}'B_{jj-1}'A_{jj-1}' & \\ & & & & & & & 1 \end{bmatrix} \begin{bmatrix} h_0 \\ u_1 \\ h_2 \\ u_3 \\ \\ \\ \\ u_{jj-1} \\ h_{jj} \end{bmatrix}^{n} = \begin{bmatrix} D_0 \\ D_1 \\ D_2 \\ D_3 \\ \\ \\ \\ D_{jj-1} \\ D_{jj} \end{bmatrix}$$

Formally:

$$[P]\,\bar{V} = [Q]\,\bar{U}$$

whence

$$\bar{V} = [P]^{-1}[Q]\,\bar{U}$$

Now let $[P] = [a]\,[b]$

where $[a] = \begin{bmatrix} a_0 & & & & & & & \\ C_1 a_1 & & & & & & & \\ & C_2 a_2 & & & & & & \\ & & C_3 a_3 & & & & & \\ & & & \cdots & & & & \\ & & & & \cdots & & & \\ & & & & & \cdots & & \\ & & & & & & & a_{jj} \end{bmatrix}$, $[b] = \begin{bmatrix} 1\ b_0 & & & & & & & \\ & 1\ b_1 & & & & & & \\ & & 1\ b_2 & & & & & \\ & & & 1\ b_3 & & & & \\ & & & & \cdots & & & \\ & & & & & \cdots & & \\ & & & & & & \cdots & \\ & & & & & & & 1\ b_{jj-1} \\ & & & & & & & 1 \end{bmatrix}$

Thence $[P] =$

$$\begin{bmatrix} a_0 & a_0 b_0 & 0 & & \\ C_1 & C_1 b_0 + a_1 & a_1 b_1 & 0 & \\ 0 & C_2 & C_2 b_1 + a_2 & a_2 b_2 & \\ 0 & 0 & C_3 & & \text{etc.} \end{bmatrix} \equiv \begin{bmatrix} 1 & & & & \\ C_1 & B_1 & A_1 & & \\ & C_2 & B_2 & A_2 & \\ & & & & \text{etc.} \end{bmatrix} \quad (4.7.22)$$

Now introduce an auxiliary vector $\bar{g}$ such that

$$[b]\bar{V} = \bar{g}$$

$$[a]\bar{g} = \bar{D}$$

that is

$$\begin{bmatrix} 1\ b_0 & & & \\ & 1\ b_1 & & \\ & & 1\ b_2 & \\ & & & 1 b_3 \end{bmatrix} \begin{bmatrix} v_0 \\ v_1 \\ v_2 \\ v_3 \end{bmatrix} = \begin{bmatrix} g_0 \\ g_1 \\ g_2 \\ g_3 \end{bmatrix} = \begin{bmatrix} v_0 + b_0 v_1 \\ v_1 + b_1 v_2 \\ v_2 + b_2 v_3 \\ \end{bmatrix}$$

$$\begin{bmatrix} a_0 & & & \\ C_1 a_1 & & & \\ & C_2 a_2 & & \\ & & C_3 a_3 & \end{bmatrix} \begin{bmatrix} g_0 \\ g_1 \\ g_2 \\ g_3 \end{bmatrix} = \begin{bmatrix} D_0 \\ D_1 \\ D_2 \\ D_3 \end{bmatrix} = \begin{bmatrix} a_0 g_0 \\ C_1 g_0 + a_1 g_1 \\ C_2 g_1 + a_2 g_2 \\ C_3 g_2 + a_3 g_3 \end{bmatrix}$$

If the latter vector equality is multiplied out, the first term is seen to be

$$g_0 = \frac{D_0}{a_0}$$

Then, recursively,

$$g_j = \frac{D_j - C_j g_{j-1}}{a_j}$$

so that $\bar{g}$ is found provided [a] is known.

Similarly with the former vector equality, the first term is

$$v_{jj} = g_{jj}$$

so that

$$v_{jj-1} = g_{jj-1} - b_{jj-1}\, v_{jj}$$

and then, recursively,

$$v_{j-1} = g_{j-1} - b_{j-1}\, v_j$$

and $\bar{V}$ is found provided [b] is known.

However, [a] and [b] are related in Equation (4.7.22) such that

$$B_j = C_j b_{j-1} + a_j$$
$$A_j = a_j b_j$$

so that

$$B_j = C_j b_{j-1} + \frac{A_j}{b_j}$$

or

$$b_j = \frac{A_j}{B_j - C_j b_{j-1}}$$

$$a_j = \frac{A_j}{b_j}$$

Hence $\{a_j, b_j\}$ can be found provided b_0 is known. But again, from Equation (4.7.22),

$$a_0 = 1$$

$$a_0 b_0 = 0$$

whence $b_0 = 0$.

The double-sweep operation then appears in matrix terms as:

(a) find [a] [b], with correspondence $a_j = B_j + C_j E_{j-1}$ and $b_j = -E_j$
(b) compute $\bar{g}$, with correspondence $g_j = F_i$
(c) compute $\bar{V}$, with correspondence $v_{j-1} = E_{j-1}\, v_j + F_{j-1}$

In the case of supercritical flows, the inversion follows directly, so that for these flows the matrices (4.7.22) reduce either to upper-triangular or to lower triangular form, corresponding to right-to-left supercritical flow and left-to-right supercritical flow, respectively. The data structure and the algorithmic structure thus reappear in the matrix formulation.

Because of the tri-diagonal form of the matrices in Equation (4.7.22), the double-sweep algorithm described above is sometimes referred to specifically as the 'tri-diagonal algorithm'. It will now be clear that it is precisely the differential forms (2.6.13), written out here as Equations (4.7.3), that provide a tri-diagonal form in Equation (4.7.22) or its lower and upper triangular variations and it is therefore this differential form that provides both the single-sweep algorithm and the double-sweep algorithm so appropriate to the solution of the equations of nearly horizontal flow. That is why forms such as (2.6.13), (4.7.3) are called 'algorithmic forms'.

The tri-diagonal algorithm becomes more complicated in the presence of bifurcated canal systems, and especially in multiply connected systems. The solution in this case belongs more to a specialized work on one-dimensional flows.

As developed above the scheme has time and space symmetry in first derivatives, and thus has truncation error $0(\Delta x^2, \Delta t^2)$ in these parts, but uses coefficients A_j, B_j and C_j taken entirely at $n\,\Delta t$, and thus introducing truncation errors as great as $0(\Delta t)$. This effect can be much reduced, to give an error of approximately $0(\Delta x^2, \Delta t^2)$ by repeating the above process using averages of $\{u_j^n, h_j^n\}$ and the newly acquired $\{u_j^{n+1}, h_j^{n+1}\}$. For all reasonably smooth data and representation Δx, one such repetition usually appears to be quite sufficient. More complex procedures exist that centre the most important coefficients as well as derivatives, but these are outside the scope of the present introduction.

In the event that unequal distances are required, the terms in the larger size of brackets in Equation (4.7.4) must be replaced by

$$\left\{\frac{2}{\Delta t}\,[(h_{j+1}^{n+1} - h_{j+1}^{n})q_j + (h_{j-1}^{n+1} - h_{j-1}^{n})(1 - q_j)]\right\}\ldots$$

$$\ldots\left\{\frac{2}{\Delta x_j + \Delta x_{j+1}}\,[(h_{j+1}^{n+1} - h_{j-1}^{n+1}) + (h_{j+1}^{n} - h_{j-1}^{n})]\right\}$$

and

$$\left\{\frac{2}{\Delta t}\,[u_{j+1}^{n+1} - u_{j+1}^{n})q_j + (u_{j-1}^{n+1} - u_{j-1}^{n})(1 - q_j)]\right\}\ldots$$

$$\ldots\left\{\frac{2}{\Delta x_j + \Delta x_{j+1}}\,[(u_{j+1}^{n+1} - u_{j-1}^{n+1}) + (u_{j+1}^{n} - u_{j-1}^{n})]\right\}$$

where

$$\Delta x_j = x_j - x_{j-1}$$

$$q_j = \frac{\Delta x_j}{\Delta x_j + \Delta x_{j+1}}$$

in order that the truncation error should remain $\approx 0(\Delta x^2, \Delta t^2)$. It should be remarked that the weighting terms are constant for all n so long as the $\{\Delta x_j\}$ are independent of time.

For the analysis of stability, it is again supposed that equations linear in their difference coefficients are stable if and only if all their localized forms, in which coefficients are taken as constants, are stable. It is consistent with this supposition that it is possible to transform the dependent variables, by multiplying them by functions of the locally constant coefficients to obtain new variables. It is particularly convenient, for reasons that will appear in the next chapter, to transform to homogeneous coordinates $r_j^n = g^{\frac{1}{2}}h_j^n$ and $s_j^n = h^{\frac{1}{2}}u_j^n$. For the moment it suffices to remark that the sum of the squares of the $\{r_j^n\}$ provides a measure of the total potential energy at $n\,\Delta t$ and the sum of the squares of the $\{s_j^n\}$ provides a measure of the total kinetic energy at $n\,\Delta t$. Introducing r and s into Equations (4.7.4) written in the form (4.7.5), (4.7.6) provides

$$\left(-\frac{u}{2\Delta t}-\frac{u^2-gh}{4\Delta x}\right)r_{j+1}^{n+1}+\left(\frac{(gh)^{\frac{1}{2}}}{\Delta t}\right)s_j^{n+1}+\left(-\frac{u}{2\Delta t}+\frac{u^2-gh}{4\Delta x}\right)r_{j-1}^{n+1}$$
$$=\left(-\frac{u}{2\Delta t}+\frac{u^2-gh}{4\,\Delta x}\right)r_{j+1}^{n}+\left(\frac{(gh)^{\frac{1}{2}}}{\Delta t}\right)s_j^{n}+\left(-\frac{u}{2\Delta t}-\frac{u^2-gh}{4\Delta x}\right)r_{j-1}^{n}$$

and

$$\left(-\frac{u}{2\Delta t}-\frac{u^2-gh}{4\Delta x}\right)s_{j+1}^{n+1}+\left(\frac{(gh)^{\frac{1}{2}}}{\Delta t}\right)r_j^{n-1}+\left(-\frac{u}{2\Delta t}+\frac{u^2-gh}{4\Delta x}\right)s_{j-1}^{n+1}$$
$$=\left(-\frac{u}{2\Delta t}+\frac{u^2-gh}{4\Delta x}\right)s_{j+1}^{n}+\left(\frac{(gh)^{\frac{1}{2}}}{\Delta t}\right)r_j^{n}+\left(-\frac{u}{2\Delta t}-\frac{u^2-gh}{4\Delta x}\right)s_{j-1}^{n}$$

These equations are now seen to be unaltered under interchange of r and s.

Taking the Fourier transform and simplifying in the same way as described in the last section, provides the relation

$$\begin{bmatrix} a+ib & c \\ c & a+ib \end{bmatrix}\begin{bmatrix} \xi_{1m}^{n+1} \\ \xi_{2m}^{n+1} \end{bmatrix}=\begin{bmatrix} a-ib & c \\ c & a-ib \end{bmatrix}\begin{bmatrix} \xi_{1m}^{n} \\ \xi_{2m}^{n} \end{bmatrix} \tag{4.7.23}$$

between the mth Fourier coefficients ξ_{1m}^n and ξ_{2m}^n of r_j^n and s_j^n respectively. In Equation (4.7.23) the coefficients are seen to be

$$a = -\frac{u}{\Delta t}\cdot\cos(m\,\Delta x),\quad b = -\frac{(u^2-gh)}{2\Delta x}\cdot\sin(m\,\Delta x),\quad c = \frac{(gh)^{\frac{1}{2}}}{\Delta t}$$

so that the coefficient b becomes zero when a is a maximum, and vice versa. As in all such locally linear wave equations, there exists an equipartition of energy between the kinetic and the potential, over the periods of all Fourier components, so that the system (4.7.23) is also invariant under interchanges of potential and kinetic terms, i.e. under interchanges of the corresponding Fourier coefficients ξ_{1m} and ξ_{2m}. Moreover, the left-hand matrix operator of Equation (4.7.23) appears as the complex conjugate of the right-hand matrix operator, so that the modulus of each vector $\{\xi_{1m}^{n+1}, \xi_{2m}^{n+1}\}$ is amplified to the same extent as is the vector $\{\xi_{1m}^{n}, \xi_{2m}^{n}\}$.

It follows at once that the corresponding amplification matrix:

$$\frac{1}{(a-\mathrm{i}b)^2 - c^2}\begin{bmatrix} a^2+b^2-c^2 & \mathrm{i}2bc \\ \mathrm{i}2bc & a^2+b^2-c^2 \end{bmatrix}$$

must have eigenvalues with a modulus of unity, or possess a 'unit spectral radius'. However, it is still informative to consider the eigenvalues of this matrix:

$$g = \frac{a^2 - (\mathrm{i}b \pm c)^2}{(a+\mathrm{i}b)^2 - c^2} \tag{4.7.24}$$

It is seen that

$$g = \frac{(a + (\mathrm{i}b \pm c))\,(a - (\mathrm{i}b \pm c))}{((a+\mathrm{i}b)+c)\,((a+\mathrm{i}b)-c)} = \frac{(a \mp c) - \mathrm{i}b}{(a \mp c) + \mathrm{i}b}$$

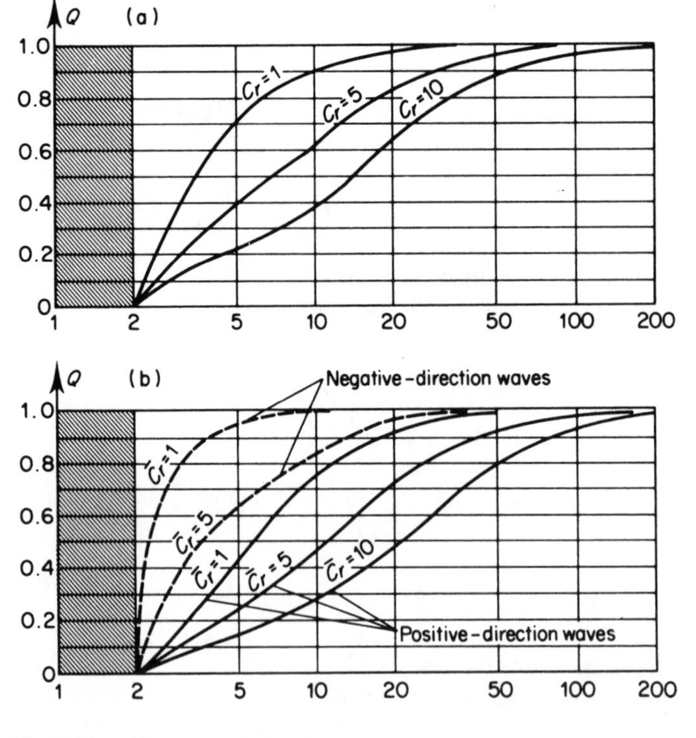

Fig. 4.43. Phase portrait for the implicit scheme (4.7.4). (a) $Fr = 0$; (b) $Fr = 0.5$ (flow in the positive direction, $\overline{Cr} = \sqrt{gh}\,\Delta t/\Delta x$)

so that $|\lambda| = 1$. The phase portrait, determined from (4.7.24), is shown in Fig. 4.43. It is seen that the phase error is never zero in any practical computation.

Now it has been shown that the Abbott-Ionescu scheme is linearly unconditionally stable so that it can be run at arbitrary Courant numbers, and so allows the use of much larger time steps than can any accurate explicit scheme. However, although the Abbott-Ionescu scheme has no amplification error, it is now seen to have a phase error that increases rapidly with increasing Courant number over a wide range of operating flow regimes (characterized in Fig. 4.43 by Froude numbers). It may then at first sight appear that the advantages of using large time steps can only be purchased at the price of a rapidly increasing inaccuracy. However, it should be realized that the interpretation of these phase-errors as real computational errors is strictly justified only in the case of isolated unidirectional waves, while these errors may not be realized to anything like the same extent in many other flow situations. An important example of these last situations is that of forced oscillations, where it is easily seen from Fig. 3.48 that considerable errors in the wave celerities, or characteristic directions, will have minor effects upon the solution for all situations well removed from the resonant situation. Another important example is that of kinematic wave formation, where Fig. 3.36 illustrates that considerable errors in the dynamic characteristic directions will have a quite negligible influence upon magnitudes in the kinematic wave solution over a considerable and practically important range of Froude numbers. Moreover, of course, there are usually a large number of grid points available per wave length in flood-routing problems.

This insensitivity of special solutions to errors in celerities can, of course, as well be ascertained by setting corresponding constraints on the differential equations, or from the corresponding closed-form solutions. It thus follows that the non-dissipative implicit schemes, represented here by the Abbott-Ionescu scheme, can frequently be run at much larger Courant numbers than their phase portraits might at first suggest. For example, well-constructed tidal oscillation models are commonly run at $Cr \simeq 10$ or, with fine space discretisations, at $Cr \simeq 20$ or more, while river flows can sometimes be simulated using $Cr \simeq 100$. Since the kinematic, transport or scalar wave equations provide only one direction of propagation, or one family of characteristics, they cannot exhibit standing waves or further degenerations, so that they are effectively restricted to $Cr \simeq 1$. In the dynamic two-characteristic case, simulations of unidirectional wave trains and near-resonant oscillations (e.g. seiches) are similarly restricted to low Courant number working.

The above analysis is restricted to the primitive equations and is, in that sense also, incomplete. Other terms in the general equations, and especially lateral inflows and resistances, may influence the stability of a more complete scheme. Consider, for example, the case of a resistance following Chézy's law. The law of motion may be degenerated to

$$\frac{u_j^{n+1} - u_j^n}{\Delta t} + \frac{g|u_j^n|u_j^{n+\theta}}{C^2 h} = 0, \quad 0 < \theta < 1$$

Taking $\theta = 0$ gives

$$u_j^{n+1} = \left(1 - \frac{g|u_j^n|/\Delta t}{C^2 h}\right) u_j^n$$

so that instability occurs from this formulation alone when

$$\frac{g|u_j^n|\Delta t}{C^2 h} > 2$$

or

$$\Delta t > \frac{2C^2 h}{g|u_j^n|}$$

This is, in fact, an onerous condition in practical computations.

Setting $\theta = 1$, on the other hand, gives

$$\frac{u_j^{n+1}}{u_j^n} = \frac{1}{1 + \dfrac{g|u_j^n|\Delta t}{C^2 h}} < 1 \text{ for all } \Delta t$$

Thus, this last formulation, in itself, always tends to stabilize the computation. As it is also time-centred in the manner of scheme (4.7.4), it is the most commonly employed, being refined further only under certain very special flow conditions.

On the basis of the above discussion, an appreciation can now be made of the much earlier scheme of Preissmann (1961; also Preissman and Cunge, 1961). This is of a type that is sometimes called a 'box scheme', as schematized in Fig. 4.44a. It can be most easily constructed on the Eulerian forms

$$\frac{\partial u}{\partial t} + u\,\frac{\partial u}{\partial x} + g\,\frac{\partial h}{\partial x} = 0$$

$$\frac{\partial h}{\partial t} + u\,\frac{\partial h}{\partial x} + h\,\frac{\partial u}{\partial x} = 0$$

when it reads

$$(1-\psi)\left(\frac{u_j^{n+1} - u_j^n}{\Delta t}\right) + \psi\left(\frac{u_{j+1}^{n+1} - u_{j+1}^n}{\Delta t}\right) + u\left[(1-\theta)\left(\frac{u_{j+1}^n - u_j^n}{\Delta x}\right) + \right.$$
$$\left. + \theta\left(\frac{u_{j+1}^{n+1} - u_j^{n+1}}{\Delta x}\right)\right] + g\left[(1-\theta)\left(\frac{h_{j+1}^n - h_j^n}{\Delta x}\right) + \theta\left(\frac{h_{j+1}^{n+1} - h_j^{n+1}}{\Delta x}\right)\right] = 0$$

$$(1-\psi)\left(\frac{h_j^{n+1} - h_j^n}{\Delta t}\right) + \psi\left(\frac{h_{j+1}^{n+1} - h_{j+1}^n}{\Delta t}\right) + u\left[(1-\theta)\left(\frac{h_{j+1}^n - h_j^n}{\Delta x}\right) + \right.$$
$$\left. + \theta\left(\frac{h_{j+1}^{n+1} - h_j^{n+1}}{\Delta x}\right)\right] + h\left[(1-\theta)\left(\frac{u_{j+1}^n - u_j^n}{\Delta x}\right) + \theta\left(\frac{u_{j+1}^{n+1} - u_j^{n+1}}{\Delta x}\right)\right] = 0$$

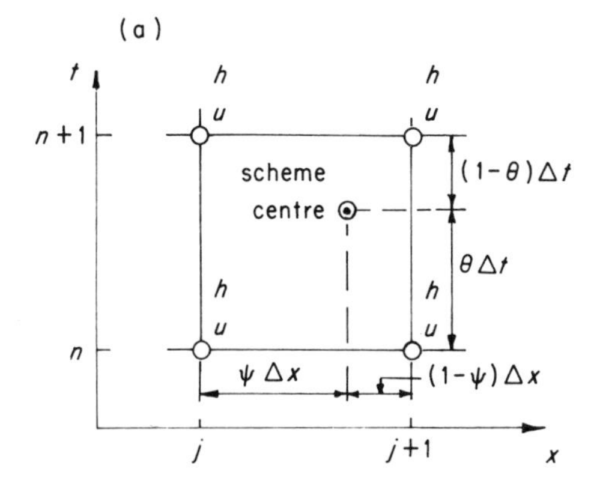

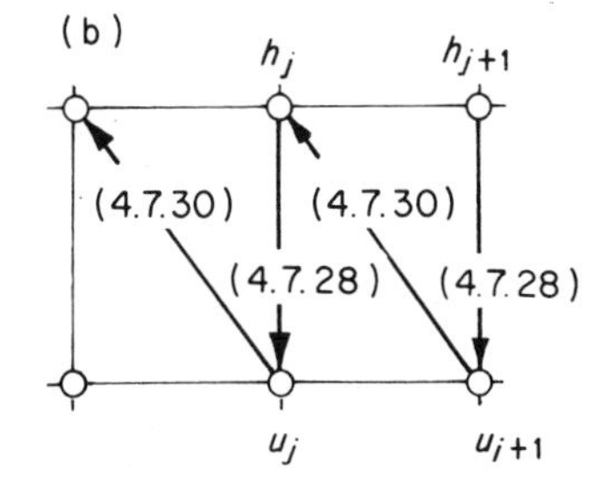

Fig. 4.44. (a) Data layout for the implicit scheme of the Preissmann type (Equations (4.7.25), (4.7.26)). (b) Schematization of order of computation in a scheme of the Preissman type

with ψ a space weighting and θ a time weighting.

Collecting terms and adopting the convention that all dependent variables appearing are defined at $(n + 1)\,\Delta t$, the following scheme emerges.

$$\begin{aligned} A_j u_{j+1} + B_j u_j + C_j h_{j+1} + D_j h_j &= E_j \\ A_j^* u_{j+1} + B_j^* u_j + C_j^* h_{j+1} + D_j^* h_j &= E_j^* \end{aligned} \tag{4.7.27}$$

Following the method described by Cunge and Wegner (1964; *see also* Cunge 1975, and Preissman and Cunge, 1961) and introducing the return sweep recurrence

$$u_j = E_j h_j + F_j \tag{4.7.28}$$

into Equations (4.7.27) provides

$$\begin{aligned} A_j u_{j+1} + (B_j E_j + D_j) h_j + C_j h_{j+1} &= E_j - B_j F_j \\ A_j^* u_{j+1} + (B_j^* E_j + D_j^*) h_j + C_j^* h_{j+1} &= E_j - B_j^* F_j \end{aligned} \tag{4.7.29}$$

The first of these gives

$$h_j = -\left(\frac{C_j}{B_jE_j + D_j}\right) h_{j+1} - \left(\frac{A_j}{B_jE_j + D_j}\right) u_j + \left(\frac{E_j - B_jF_j}{B_jE_j + D_j}\right)$$

which has the form

$$h_j = L_j h_{j+1} + M_j u_{j+1} + N_j \tag{4.7.30}$$

On the other hand, h_j can be eliminated from Equations (4.7.29) to give

$$u_{j+1} = \frac{[(B_j^*E_j + D_j^*)C_j - (B_jE_j + D_j)C_j^*]}{[(B_j^*E_j + D_j^*)A_j - (B_jE_j + D_j)A_j^*]} h_{j+1} + \\ + \frac{[(B_j^*E_j + D_j^*)(E_j - B_jF_j) - (B_jE_j + D_j)(E_j - B_jF_j)]}{[(B_j^*E_j + D_j^*)A_j + (B_jE_j + D_j)A_j^*)]}$$

which, in turn, provides the initiating sweep recurrences for E_j and F_j in Equation (4.7.28).

The algorithmic structure of the scheme in the above formulation then has the form schematized in Fig. 4.44b for its return sweep: Equation (4.7.28) is used to obtain u_{j+1} from a given h_{j+1}, then Equation (4.7.30) is used to obtain h_j from the h_{j+1} and the u_{j+1} then available, and so on.

The initiating sweep has itself to be initiated at the boundaries. Suppose, for example, that $h_0 = h_0(t)$ is given at $j = 0$. Then Equation (4.7.28) provides

$$h_0 = \frac{u_0}{E_0} - \frac{F_0}{E_0}$$

However, in a subcritical flow, h_0 should be independent of u_0. This condition can be introduced by setting

$$E_0 = \alpha$$

$$F_0 = -\alpha h_0$$

where $\alpha \gg u_0$. Then, formally, $h_0 \approx 0 - (-h_0) = h_0$. Cunge proposes the use of an α value of the order of 10^4–10^6.

If $u = u_0(t)$ is given, then the condition that u_0 should be independent of h_0 provides $E_0 = 0$ and $F_0 = u$ directly.

It should be remarked that the original scheme used a better scaling of variables through the use of variables $\Delta u_j^{n+1} = u_j^{n+1} - u_j^n$ (and, in fact, even $\Delta Q_j^{n+1} = Q_j^{n+1} - Q_j^n$) and $\Delta h_j^{n+1} = h_j^{n+1} - h_j^n$. The whole question of scaling and dimensioning variables for double-sweep algorithms is, however, so complicated as to be again outside the scope of this work.

This approach to the Preissmann scheme indicates that the double-sweep algorithm is by no means restricted to tri-diagonal schemes. In this case, however, it is easy to see that an equivalent tri-diagonal scheme can be constructed. In fact, following an approach of Verwey (unpublished), the coefficient matrix of the unknowns has the form

$$\begin{bmatrix} A_0 & B_0 & C_0 & D_0 & 0 & 0 & 0 & . & . & . \\ A_0^* & B_0^* & C_0^* & D_0^* & 0 & 0 & 0 & . & . & . \\ 0 & 0 & A_1 & B_1 & C_1 & D_1 & 0 & . & . & . \\ 0 & 0 & A_1^* & B_1^* & C_1^* & D_1^* & 0 & . & . & . \\ & & & & & & & & & \text{etc.} \end{bmatrix}$$

Clearly, D_0 can be eliminated in the first row using the first two equations and then A_0^* can be eliminated from the resulting three-element expression and the second equation, and so on recursively. Following this, the double sweep proceeds just as for the Abbott–Ionescu scheme. Moreover, the application of the Preissmann scheme to supercritical flows is also simplified through this procedure.

The Preissmann scheme is a convenient one for introducing the effect of off-centring the space differences, so using θ values other than $\frac{1}{2}$. By analogy with the results obtained in Section 4.5, one expects that values of $\theta < \frac{1}{2}$ would provide either conditionally stable or unstable schemes while values of $\theta > \frac{1}{2}$ would provide dissipative schemes. In the latter case, if the schemes were themselves built upon mass and momentum conservation laws, the energy

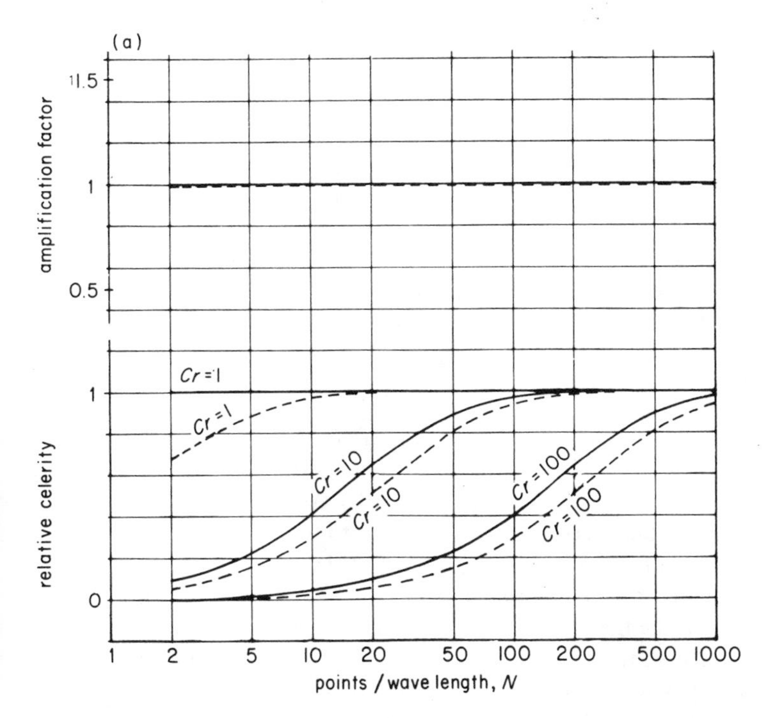

Fig. 4.45. Properties of the Preissman scheme. (a) Amplitude and phase portrait: (——) $Fr = 0$, $\theta = 0.5$, $C = \infty$, (– – –) $Fr = 0.5$, $\theta = 0.5$, $C = \infty$; (b) amplitude and phase portrait – dissipative interface; (c) amplitude and phase portraits, with friction: $Cr = 1$, $C = 50$; (d) amplitude and phase portraits: $Fr = 0.5$, $\theta = 0.7$, $C = \infty$. (*See pages 185–188*)

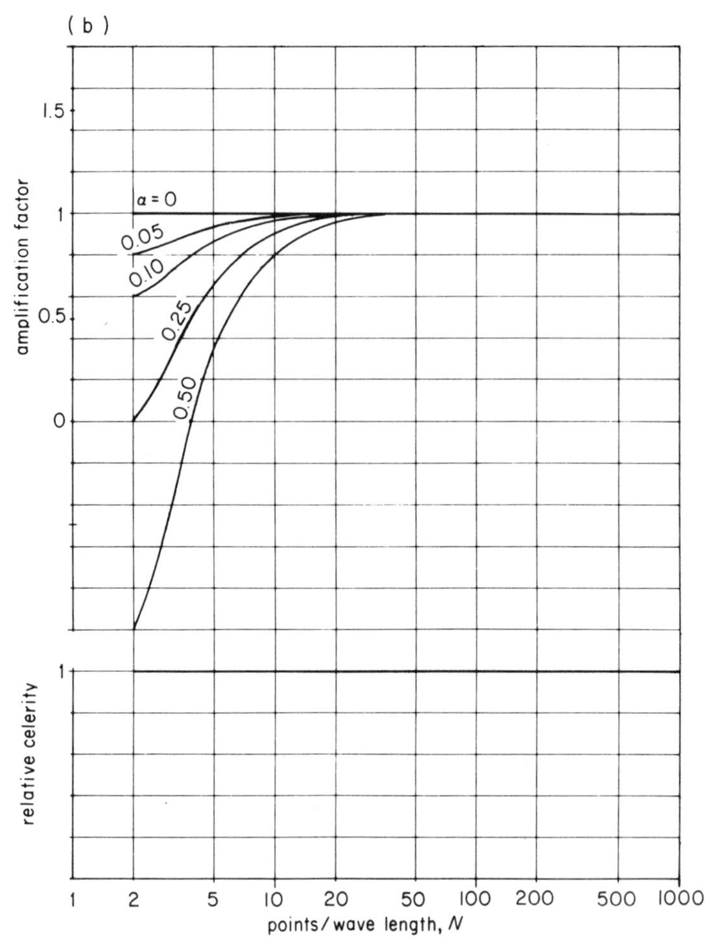

Fig. 4.45 (cont.)

would be dissipated. In order to illustrate the effects of using other θ values, Fig. 4.45a shows amplitude and phase portraits for the Preissmann scheme built upon Equations (4.7.3), taken from Evans (1977). At a Froude number $Fr = 0$, corresponding to a linearized form of (4.7.3), so to the equation system (2.6.15), (2.6.16), the use of $\theta = 0.5$ and $\psi = 0.5$ with $Cr = 1$ provides an exact solution. In this limit, therefore, the Preissmann scheme is equivalent to the method of characteristics. While the scheme is kept centred, the amplification factor remains constant at unity, so that the scheme is non-dissipative, but the phase error increases rapidly with Courant number. Thus, at a Courant number of 100, some 350 points per wave length are required to maintain a celerity ratio of 0.8 at $Fr = 0$ (linear conditions) and some 500 points per wave length are required to maintain this celerity ratio for $Fr = 0.5$.

As the scheme is taken forward from time centring so the anticipations arising from the scalar wave equation analysis are seen to be fulfilled. In particular, a

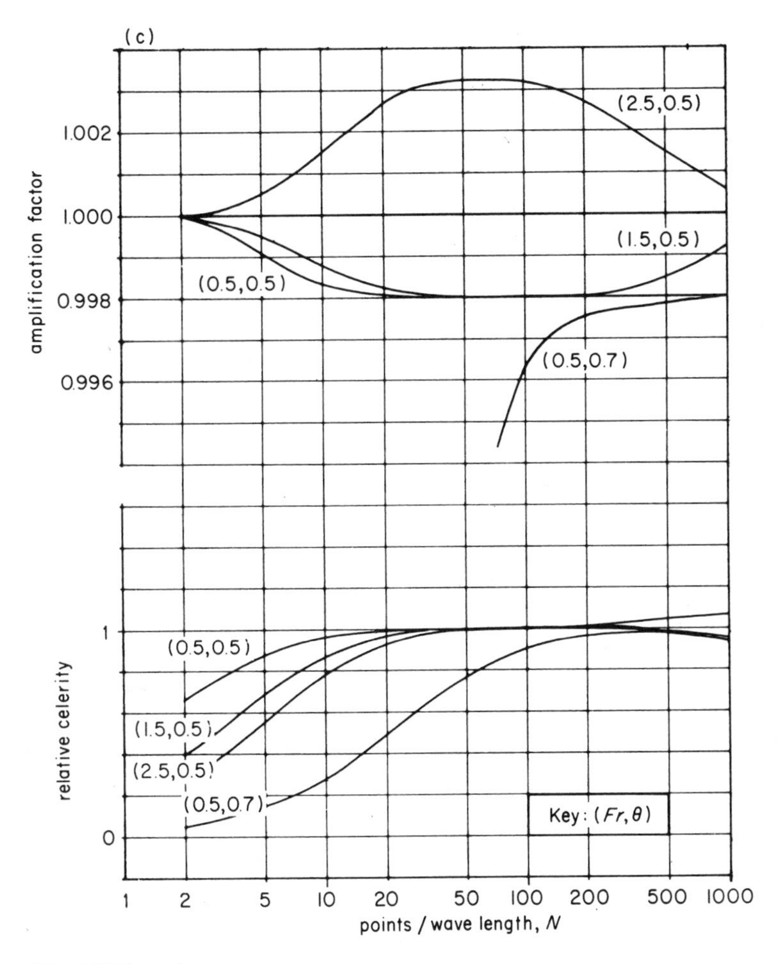

Fig. 4.45 (cont.)

heavy numerical damping is introduced, that spreads up into very large numbers of points per wave length as the Courant number is increased. By way of comparison, the amplitude portrait of the centred dissipative interface (Equation (4.6.13)) is plotted in Fig. 4.45b. It will be clear that the effect of off-centring the differences is not simply equivalent to the introduction of a dissipative interface. (The nature of the equivalence is more subtle, and it can be conveniently related to the introduction of the physical effects of turbulent stresses at the grid scale. It will have to form part of a more specialized treatise.)

The analysis of the full Preissmann scheme was extended by Evans to account for the effects of resistance. Figure 4.45c illustrates some of his results. Of particular interest is the result that, at $Fr = 2.5$, the scheme is unstable, a result that Evans has related to the work of Barnett (1975) on the computational-hydraulic representation of Jeffreys–Vedernikov instabilities.

In principle, the Preissmann scheme is generally more convenient than the

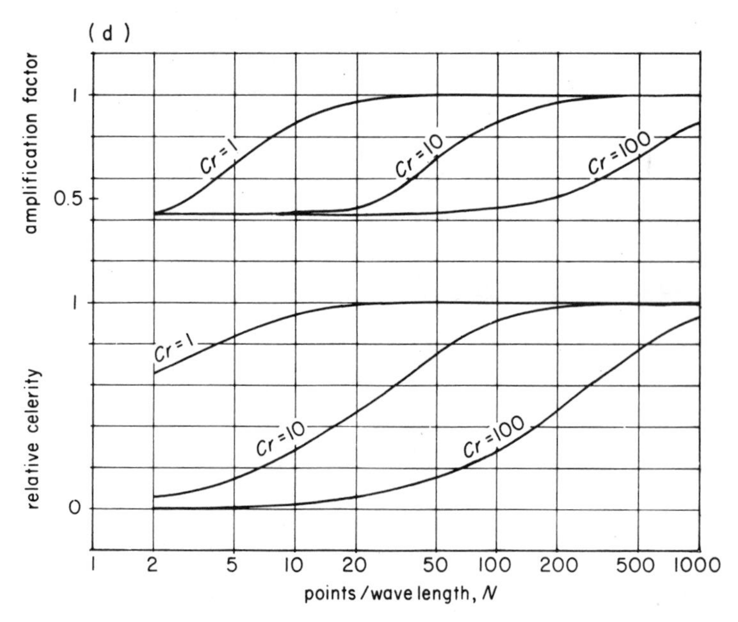

Fig. 4.45 (cont.)

Abbott–Ionescu scheme for one-dimensional flow simulations. No weighting of differences is required for local centring to second-order accuracy with non-equidistant grid points while special conditions that relate the dependent variables at discrete points, such as a weir condition, are more conveniently introduced. The Preissmann scheme is particularly convenient for supercritical flow computations, since the single sweep is then initiated from two-point boundary data defined at one and the same point.

In order to show the influence of forward weighting on the dissipativeness of the scheme, some results with $\theta = 0.7$, again taken from the report of Evans, are shown in Fig. 4.45d. The use of $\theta = 0.7$ is seen to have a severe dissipative effect in this example, and one that is stronger than that obtained using a relatively heavy dissipative interface.

This is not the place to discuss the relative merits of the two implicit schemes discussed here, which, together with the scheme of Vasiliev, *et al.* (1965) (*see also* Liggett and Cunge, 1975), are the schemes most widely used in engineering practice. It will be clear, however, simply by comparing the phase portraits of Figs. 4.43 and 4.45a that the Abbott-Ionescu scheme and the Preissmann scheme are not equivalent.

4.8 Inverse difference methods for one-dimensional nearly horizontal flows

In most descriptions of continuum mechanics, a distinction is made between independent variables and dependent variables. It is usual in hydraulics to assign position coordinates and time to the roles of independent variables, while such measures of fluid behaviour as pressure, velocity, discharge, energy and momentum are assigned to the roles of dependent variables. Implicit in this distinction is a posing of problems in terms of determining what happens at a particular place and time; that is, of determining the values of the dependent variables at given values of the independent variables.

In the same way, the books on numerical analysis (e.g. Hartree, 1952; Kopal, 1955; Richtmyer and Morton, 1967) first introduce the notion of a table of values of the dependent variables corresponding to given values of the independent variables and they then introduce difference coefficients as differences between tabulated values of the dependent variables divided by the corresponding given increments, or arguments, in the independent variables. Difference schemes constructed from this traditional point of view may be called 'dependent variable difference schemes', abbreviated to 'DV schemes', or they may simply be called 'direct' difference schemes.

Now although this point of view is adopted almost universally and appears unquestioned in most works on hydraulics and numerical analysis, it is clearly only one of a whole continuum of possible descriptive systems. Another, and equally special point of view, is that in which one asks where a particular event occurs at a particular time, i.e. one poses the problem in terms of finding values of certain independent variables at given values of the dependent variables and time. In the latter case, it is certain of the independent variables, essentially the space coordinates, that are differenced over arguments in the dependent variables and time. Difference schemes constructed from this point of view will be called either 'independent variable difference schemes', abbreviated to 'IV schemes', or simply 'inverse' difference schemes. (See also Lamb, 1932, pp. 72–75.)

The inverse schemes are defined upon a net of points or 'grid' of the type schematized for one dependent variable f and one independent space variable x in Fig. 4.46. Constant increments Δf have been introduced in f, for simplicity of exposition. With each $f_j = j\,\Delta f$ is associated an $x^n_{j(1)}$ at which this f_j occurs on the first occasion in x at time $n\,\Delta t$, reading in the positive x-direction, an $x^n_{j(2)}$ at which it occurs on the second occasion, and so on up to any $x^n_{j(mm)}$. It follows from the compatibility of the notation that

$$l \leqslant m \text{ implies } x^n_{j(l)} \leqslant x^n_{j(m)} \text{ for every } j, n \tag{4.8.1}$$

The set $\{x^n_{j(m)}\}$ goes into the set $\{x^{n+1}_{j(m)}\}$ under the inverse scheme.

In the simplest (f, x, t) space, lines of constant f satisfy

$$\mathrm{d}f = \frac{\partial f}{\partial t}\,\mathrm{d}t + \frac{\partial f}{\partial x}\,\mathrm{d}x = 0 \tag{4.8.2}$$

so that their celerity is given by

$$\frac{\mathrm{d}x}{\mathrm{d}t} = \dot{x} = -\left(\frac{\partial f}{\partial t}\right)_x \Big/ \left(\frac{\partial f}{\partial x}\right)_t = \left(\frac{\partial x}{\partial t}\right)_f \tag{4.8.3}$$

When this is applied to the scalar wave equation

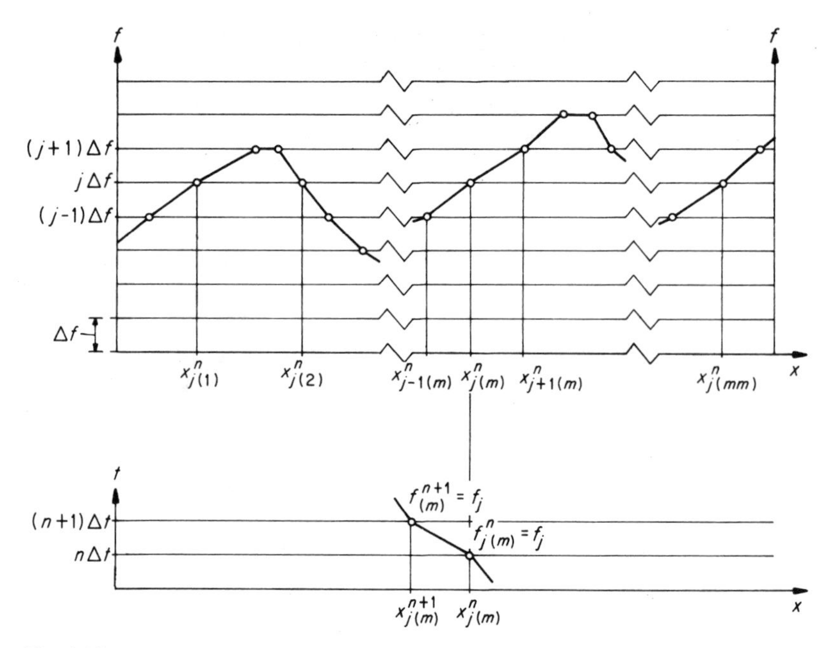

Fig. 4.46. The grid for an inverse scheme on fixed increments in one dependent variable

$$\frac{\partial f}{\partial t} + c\,\frac{\partial f}{\partial x} = 0$$

there results simply

$$\left(\frac{\partial x}{\partial t}\right)_f = c$$

which is just the characteristic solution (3.7.10). In the large, therefore, the exact solution is defined by

$$f(x, t_2) = f\left(x - \int_{t_1}^{t_2} c \, \mathrm{d}t, t_1\right)$$

The characteristics $\dot{x} = c$ can be considered as defining a new frame of reference (x', t) related to the original (x, t) frame by the stream geometry and flow. The quantity

$$\left(\frac{\partial x'}{\partial x}\right)_t$$

is then the *coefficient of distortion* of the transformation (Aramanovich *et al.*, 1965; O'Kane, 1971), being the ratio of the length of a small element measured in x' units to the length measured in x units.

Applying Equation (4.8.3) to the simplest conservation laws (Equation (2.1.8))

$$\frac{\partial f}{\partial t} + \frac{\partial g(f)}{\partial x} = 0$$

gives

$$\dot{x} = \left(\frac{\partial g(f)}{\partial x}\right) \bigg/ \left(\frac{\partial f}{\partial x}\right) \tag{4.8.4}$$

Then, introducing the approximations

$$\begin{aligned} \frac{\partial g(f)}{\partial x} &\approx \frac{g_{j+1} - g_{j-1}}{x_{j+1} - x_{j-1}} \\ \frac{\partial f}{\partial x} &\approx \frac{f_{j+1} - f_{j-1}}{x_{j+1} - x_{j-1}} \end{aligned} \tag{4.8.5}$$

reduces Equation (4.8.4) to

$$\dot{x} \approx \frac{g_{j+1} - g_{j-1}}{f_{j+1} - f_{j-1}} \tag{4.8.6}$$

or, using the notation of Chapter 1

$$\dot{x}[f]_{j-1}^{j+1} = [g]_{j-1}^{j+1}$$

which is simply the discrete form of the conservation law as given in Equation (1.4.1).

Equation (4.8.4)–(4.8.6) is solved in the form

$$(x_f)_j^{r+1} - (x_f)_j^n = \Delta t\, R_j^r \left[\theta \left(\frac{(x_f)_{j+1}^{r+1} - (x_f)_{j-1}^{r+1}}{f_{j+1} - f_{j-1}}\right) + (1-\theta)\left(\frac{(x_f)_{j+1}^n - (x_f)_{j-1}^n}{f_{j+1} - f_{j-1}}\right)\right] \tag{4.8.7}$$

where

$$R_j^r = \frac{\theta\left(\dfrac{(g(f))_{j+1}^r - (g(f))_{j-1}^r}{(x_f)_{j+1}^r - (x_f)_{j-1}^r}\right) + (1-\theta)\left(\dfrac{(g(f))_{j+1}^n - (g(f))_{j-1}^n}{(x_f)_{j+1}^n - (x_f)_{j-1}^n}\right) + R_j^{r-1}}{2} \tag{4.8.8}$$

and r is an iteration for $(n+1)$, for each component of f in turn. The coefficients of the $\{(x_f)_j^{r+1}\}$ form a tri-diagonal system so that the double-sweep procedure, along the lines of Equations (4.7.9)–(4.7.14), may again be used at each iteration.

The iteration is continued until

$$|x_j^{r+1} - x_j^r| < \epsilon, \text{ for all } j$$

Evidently any such solution of Equation (4.8.4) as (4.8.7), (4.8.8) necessitates knowing f_1 at the points $\{(x_{(f_2)})_j\}$ and f_2 at the points $\{(x_{(f_2)})_j\}$ in order to compose the appropriate $g(f)$, and these must be obtained by interpolation. A parabolic (second-degree) interpolation has been found to be sufficiently accurate.

It is seen that this implicit inverse scheme preserves, of necessity, the algorithmic structure of the direct implicit schemes.

In the inverse schemes, points must be introduced and removed from the computation. The removal is effected using the condition (4.8.1) and the introduction by curve fitting at extrema, as described by Abbott and Vium (1977).

The conservation of the component quantities $f_{(1)}$ and $f_{(2)}$ of f under the operations of the scheme (4.8.7) with $\theta = 0$ is discussed by Harlow (1971).

In order to simulate hydraulic jumps, fronts, etc. a dissipative interface can be introduced. This now takes the simplest forms

$$x_j^*: = \alpha x_{j-1} + (1 - 2\alpha)x_j + \alpha x_{j+1} \text{ when } f_{j-1} \neq f_j \text{ and } f_j \neq f_{j+1}$$

$$x_j^*: = (1 - \alpha)x_j + \alpha x_{j+1} \text{ when } f_{j-1} = f_j \text{ and } f_j \neq f_{j+1}$$

$$x_j^*: = \alpha x_{j-1} + (1 + \alpha)x_j \text{ when } f_{j-1} \neq f_j \text{ and } f_j = f_{j+1}$$

as is usual for direct schemes over a restricted region. The index j refers here to the numbering sequence in the positive x-direction. At the cost of some dissipation in f, a weighted scheme can also be conveniently used, of the form

$$x_j = \alpha_1 x_{j-1} + (1 - \alpha_1 - \alpha_2)x_j + \alpha_2 x_{j+1}$$

where

$$\alpha_1 = \frac{4(f_{j+1} - f_j)(f_j - f_{j-1})}{(f_{j+1} - f_{j-1})^2} \cdot \frac{(x_{j+1} - x_j)}{(x_{j+1} - x_{j-1})} \cdot 2\alpha$$

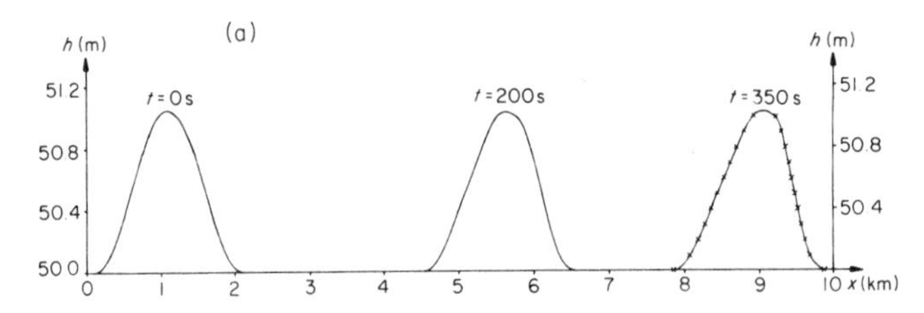

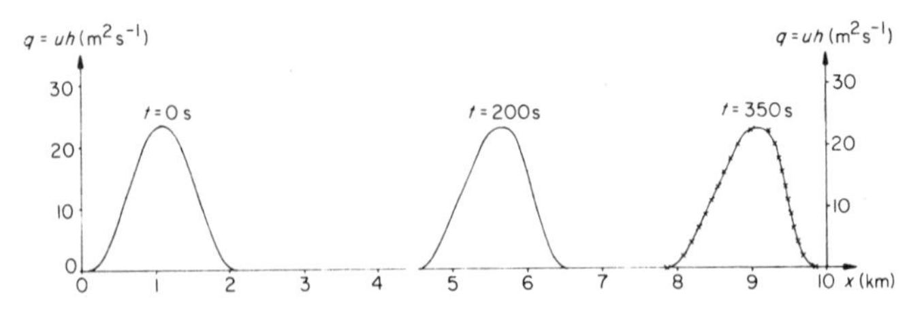

Fig. 4.47. Some results obtained for nearly horizontal flow in a uniform channel using an inverse scheme. (a) Propagation of an isolated wave: $\Delta t = 1$ s, $\Delta h = 0.1$ m, $\Delta q = 2\text{m}^2 \text{ s}^{-1}$, $\theta = 1$, $\alpha = 0$. (b) Propagation of a negative simple wave: $\Delta t = 2$ s, $\Delta h = 0.05$ m, $\Delta q = 1.5 \text{ m}^2 \text{ s}^{-1}$, $\theta = \frac{1}{2}$, $\alpha = 0.05$. (c) A positive simple wave forming up into a hydraulic jump: $\Delta t = 2$ s, $\Delta q \approx 1.5 \text{ m}^2 \text{ s}^{-1}$, $\theta = \frac{1}{2}$, $\alpha = 0.02$. In all three cases the crosses indicate results obtained with the graphical method of characteristics (*Continued on page 193*)

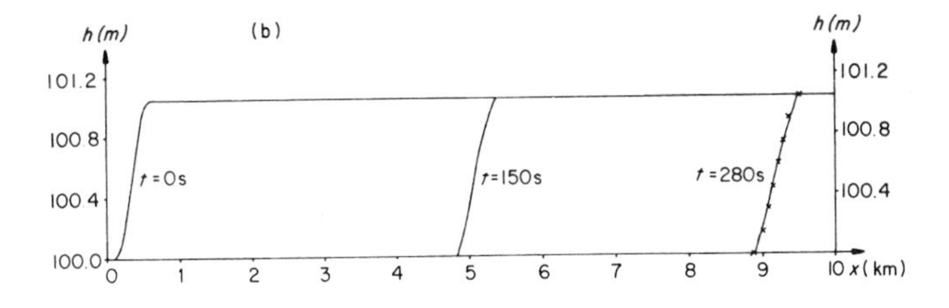

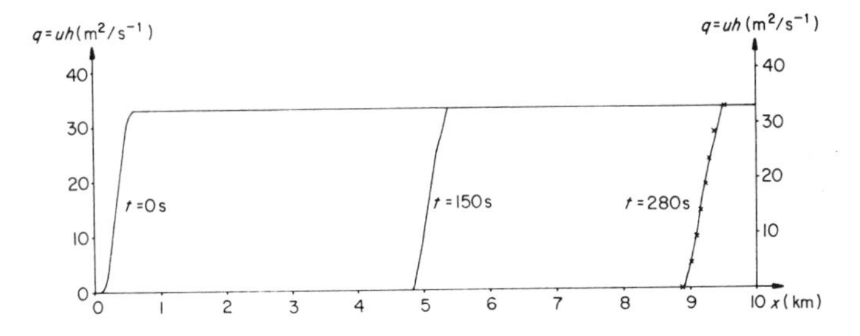

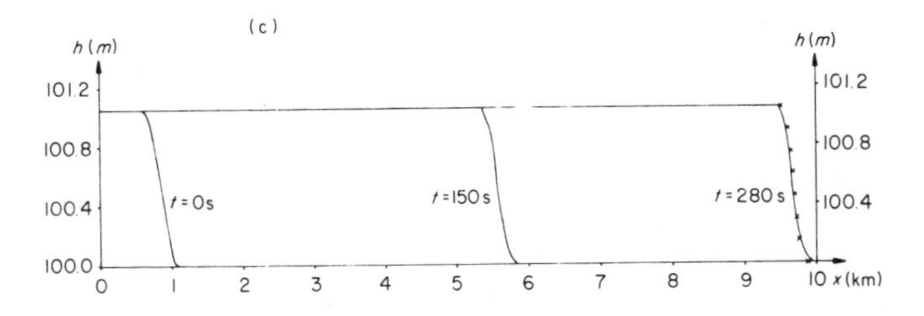

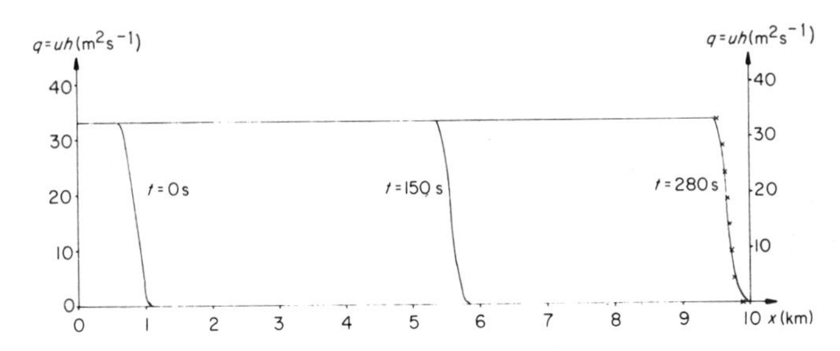

Fig. 4.47 (cont.)

$$\alpha_2 = \frac{4(f_{j+1} - f_j)(f_j - f_{j-1})}{(f_{j+1} - f_{j-1})^2} \cdot \frac{(x_j - x_{j-1})}{(x_{j+1} - x_{j-1})} \cdot 2\alpha$$

Some test results of these methods are shown in Fig. 4.47. Figure 4.47a shows an isolated wave propagating along a channel, Fig. 4.47b a negative simple wave and Fig. 4.47c a positive simple wave forming up into a hydraulic jump or bore. The results obtained using the method of characteristics are shown for comparison.

Alternatively, the transformation (3.2.6) can be regarded as one that generates a pair of transport equations

$$\begin{aligned} \frac{\partial J_+}{\partial t} + c_+ \frac{\partial J_+}{\partial x} = 0 \\ \frac{\partial J_-}{\partial t} + c_- \frac{\partial J_-}{\partial x} = 0 \end{aligned} \qquad (4.8.9)$$

where $J_\pm = u \pm 2(gh)^{\frac{1}{2}}$ and $c_\pm = u \pm (gh)^{\frac{1}{2}}$.

The inverse scheme of Equation (4.8.9) then provides

$$\begin{aligned} J_+(x, t_2) = J_+ \left(x - \int_{t_1}^{t_2} c_+ \, \mathrm{d}t, t_1 \right) \\ J_-(x, t_2) = J_- \left(x - \int_{t_1}^{t_2} c_- \, \mathrm{d}t, t_1 \right) \end{aligned} \qquad (4.8.10)$$

which is simply a reformulation of the method of characteristics, as an inverse difference scheme.

The inverse schemes are of particular interest for relativizing the direct schemes that are considered elsewhere in this book. The direct schemes work best when space variations in the quantities of interest are of the same order of magnitude over most of the computational domain, while inverse schemes are particularly well suited to situations where the areas of most interest are concentrated in small areas of the domain. These 'best areas of application' are clearly extreme situations corresponding to the extreme nature of the two types of scheme concerned. This notion can be formulated in terms of the problem of determining which function is best described by a given type of scheme. Each scheme might then be characterized by the (generalized) function (or set of functions) that it best describes, in some sense. From this point of view, the direct schemes on uniform increments are characterized for finite f by the ϵ-distribution of Equation (1.5.1) while the inverse schemes on uniform increments are then characterized by the Dirac δ-distribution, defined by

$$\int_{-\infty}^{\infty} \delta(x) \, w(x) \, \mathrm{d}x = w(0)$$

where $w(x)$ is a suitable test function (e.g. Marchand, 1962). Since all functions of physical reality lie between these two extremes, it may be supposed that all best descriptions lie between uniform-increment direct scheme descriptions and uniform-increment inverse scheme descriptions.

The generation of 'best descriptions' of objects and laws, as they change in time during a computation, currently appears as one of the central problems of computational hydraulics (e.g. Abbott, Castro and Tas, 1967; Malvern, 1969 and Prenter, 1975).

4.9 Difference methods for one-dimensional flows of non-homogeneous fluids

In practice, the density of a fluid may change both in the horizontal and in the vertical directions. This variation is due, in the first place, to differences in salinity, as occur in estuaries or in straits between water bodies of different constitution. In such situations, the dynamic equations must be complemented by behavioural equations for the density variations. The simplest and therefore the most commonly used of these behavioural equations is the transport – diffusion equation.*

Now, as remarked on several occasions above, the numerical solutions of the transport–diffusion equation are themselves numerous and often very sophisticated (e.g. Molenkamp, 1968; Anderson and Fattahi, 1974). It is quite outside the range of this work even to enter into a discussion of these solutions, the only objective here being to indicate how they can be combined with the solutions of the nearly horizontal flow equations and how the latter can be modified in order to resolve density variations.

Consider first a variation of density confined to the horizontal, so that the density still remains constant in any vertical. This is the typical situation of the 'well-mixed estuary'. The simplest one-dimensional transport–diffusion equation is then

$$\frac{\partial c}{\partial t} + u\frac{\partial c}{\partial x} - D_F\frac{\partial^2 c}{\partial x^2} = 0 \tag{4.9.1}$$

This may be regarded as combining the influences of two processes: a pure transport process, corresponding to $D_F = 0$ and so described by

$$\frac{\partial c}{\partial t} + u\frac{\partial c}{\partial x} = 0 \tag{4.9.2}$$

and a pure diffusion process, corresponding to $u = 0$ and so described by

$$\frac{\partial c}{\partial t} - D\frac{\partial^2 c}{\partial x^2} = 0, D \text{ real positive} \tag{4.9.3}$$

As shown in Section 3.7, the transport process described by Equation (4.9.2) introduces one-point boundary data upstream and zero-point boundary data downstream while it is easily seen from the characteristics of Equation (4.9.3) that the dispersion process (4.9.3) introduced one-point data at both boundaries. This data structure suggests that, in general, the influences of the two processes might best be computed separately, as two 'stages' in a two-staged difference scheme. For example, the explicit scheme (4.9.2) could be used for the transport stage

* 'Dispersion' and 'diffusion' are often used interchangeably in hydraulics. In some cases, where such distinctions are possible, 'dispersion' is used to denote a consequence of spatial averaging processes while diffusion denotes a consequence of temporal averaging processes. In numerical terms, as in wave theory generally, 'dispersion' is used to denote the spreading out in space of Fourier solution components of different celerity, while 'diffusion' is used to denote the damping in time of magnitudes (e.g. Whitham, 1974, Holley and Cunge, 1975).

$$\left(\frac{c_j^{n+1} - c_j^n}{\Delta t}\right)_{\text{trans}} + u\left(\frac{c_j^n - c_{j-1}^n}{\Delta x}\right) = 0, \quad u > 0 \tag{4.9.4}$$

while the diffusion stage (4.9.3) might be approximated with the explicit scheme

$$\left(\frac{c_j^{n+1} - c_j^n}{\Delta t}\right)_{\text{diff}} + D\left(\frac{c_{j+1}^n - 2c_j^n + c_{j-1}^n}{\Delta x^2}\right) = 0 \tag{4.9.5}$$

Then the processes could be combined through

$$\frac{c_j^{n+1} - c_j^n}{\Delta t} = \left(\frac{c_j^{n+1} - c_j^n}{\Delta t}\right)_{\text{trans}} + \left(\frac{c_j^{n+1} - c_j^n}{\Delta t}\right)_{\text{diff}}$$

or

$$\frac{c_j^{n+1} - c_j^n}{\Delta t} + u\left(\frac{c_j^n - c_{j-1}^n}{\Delta x}\right) - D\left(\frac{c_{j+1} - 2c_j + c_{j-1}}{\Delta x^2}\right) = 0 \tag{4.9.6}$$

Equation (4.9.6) is seen to be term by term consistent with Equation (4.9.1), corresponding in this example, to the individual term by term consistency of Equations (4.9.4) and (4.9.5) with (4.9.2) and (4.9.3) respectively.

For the purpose of conceiving of such a 'multi-staging', it is sometimes helpful to think of one process being 'frozen' or 'locked' while the other is implemented. It is then also possible to visualize a situation where one of the processes is locked for 'some part of the time' while the other process is locked for 'the rest of the time'. Now it has become common to introduce the notation $\{c_j^{n+\frac{1}{2}}\}$ for the values of c that are obtained from the first process and which enter the computation of the second process. Then Equations (4.9.4) and (4.9.5) are written simply as

$$\frac{c_j^{n+\frac{1}{2}} - c_j^n}{\Delta t} + u\left(\frac{c_j^n - c_{j-1}^n}{\Delta x}\right) = 0, \quad u > 0 \tag{4.9.7}$$

$$\frac{c_j^{n+1} - c_j^{n+\frac{1}{2}}}{\Delta t} - D\left(\frac{c_{j+1}^n - 2c_j^n + c_{j-1}^n}{\Delta x^2}\right) = 0 \tag{4.9.8}$$

which again sum to Equation (4.9.6). It should be emphasized that writing '$c_j^{n+\frac{1}{2}}$' is only a convention, merely corresponding to a convenient notation, the $\{c_j^{n+\frac{1}{2}}\}$ always cancelling out in the representation of the total operator. It can be quite misleading to interpret these values as 'the values of c at $(n + \frac{1}{2})t$' and correspondingly incorrect, in many cases, to use these values for computing coefficients for any other ancillary purpose. In the same vein, Equations (4.9.7) and (4.9.8) are individually consistent with

$$\text{'}\frac{1}{2}\frac{\partial c}{\partial t} + u\frac{\partial c}{\partial x} = 0\text{'}$$

$$\text{'}\frac{1}{2}\frac{\partial c}{\partial t} - D\frac{\partial^2 c}{\partial x^2} = 0\text{'}$$

which 'add up' to Equation (4.9.1). However, the process of taking individual consistencies over to component differential forms and adding these differential component forms to obtain a resultant differential form is generally meaningless. In fact it is easy to construct examples where this practice leads to serious errors. The method of splitting equations into components is often translated from the Russian as 'the method of splitting-up' and its very extensive methodology is called the 'fractioned-step method' (e.g. Yanenko, 1968).

The above concept and nomenclature were introduced through the example of component explicit schemes providing a resultant explicit scheme, so that all considerations of algorithmic structure could be temporarily put aside. Thus Equations (4.7.9) and (4.9.8) could be solved in any order, as could (4.9.6) also. However, multi-staging really comes into its own when one is using implicit schemes. In order to see its desirability in this case, consider the following scheme described by Richtmyer and Morton (1967), that is often recommended for the solution of Equation (4.9.1):

$$\frac{1}{6}\left(\frac{c_{j+1}^{n+1}-c_{j-1}^{n}}{\Delta t}\right)+\frac{2}{3}\left(\frac{c_{j}^{n+1}-c_{j}^{n}}{\Delta t}\right)+\frac{1}{6}\left(\frac{c_{j-1}^{n+1}-c_{j-1}^{n}}{\Delta t}\right)+$$
$$+\frac{u}{2}\left(\frac{c_{j+1}^{n+1}-c_{j-1}^{n+1}}{2\Delta x}+\frac{c_{j+1}^{n}-c_{j-1}^{n}}{2\Delta x}\right)- \tag{4.9.9}$$
$$-\frac{D_{\mathrm{F}}}{2}\left[\left(\frac{c_{j+1}^{n+1}-2c_{j}^{n+1}+c_{j-1}^{n+1}}{\Delta x^{2}}\right)+\left(\frac{c_{j+1}^{n}-2c_{j}^{n}+c_{j-1}^{n}}{\Delta x^{2}}\right)\right]=0$$

The scheme is linearly unconditionally stable while the truncation error of the transport part is an attractive $0(\Delta x^4, \Delta t^2)$ and of the dispersion part an entirely acceptable $0(\Delta x^2, \Delta t^2)$. It is seen that Equation (4.9.9) provides the tri-diagonal form

$$A_j c_{j+1}^{n+1} - B_j c_j^{n+1} + C_j c_{j-1}^{n+1} = D_j$$

so that it necessitates the use of a double-sweep algorithmic structure. This structure is appropriate to the diffusion process, but entirely inappropriate to the transport process. As can then be anticipated, the scheme can easily come to present problems at downstream boundaries while these problems are inherent precisely in the ($\frac{1}{6}$, $\frac{2}{3}$, $\frac{1}{6}$) weightings of the time derivatives that provide its impressive truncation error.

The same accuracy can be obtained by using an explicit fourth-order interpolation scheme for a transport stage (Everett interpolation away from boundaries with Stirling interpolation or a lower-order interpolation at the boundaries) and then the diffusion part of Equation (4.9.9) for a separate diffusion stage. The latter stage retains the correct algorithmic structure while the scheme of the transport stage is explicit, so not demanding any algorithmic structure at all.

Now, as shown in Section 2.1, the specific mass, or density variations will themselves influence the dynamics of the fluid system while the velocities and flow areas arising from the dynamics will influence in turn the transports and diffusion of the densities. The dynamic and transport–diffusion computations must then be coupled into one computation. This is usually attained through a form of multi-staging that is called 'parallel running', whereby the grid of the dynamic stages is supposed to be staggered by $\frac{1}{2}\Delta t$ relative to the grid of the transport–diffusion stages. The information exchange between the two main stages is then centred according to the schematization of Fig. 4.48.

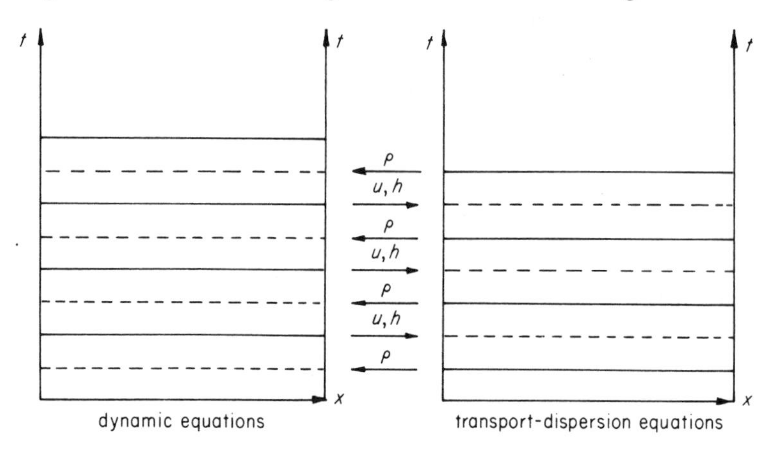

Fig. 4.48. Information exchange between the dynamic and transport – dispersion stages

Transport–diffusion equations can also be solved using inverse schemes, as described by Abbott and Vium (1977), and a procedure for parallel running with the dynamic equations can as well be implemented in this case also.

The density may vary not only in the horizontal but also in the vertical, and in some situations it may be necessary to resolve the internal motions of the fluid, so, of necessity, taking account of the vertical density variations. The simplest approach to the flow description is that introduced in Sections 2.6 and 3.5, where a schematization into two conceptually distinct layers was considered in some detail. It was seen that the five characteristic structures of Fig. 3.25 gave nine data structures, which can now be set out as in Fig. 4.49. The two lines in each 2 x 2 matrix element represent the two layers, while the integer on the left gives the number of data points at the left-hand boundary with the integer on the right given the number at the right-hand boundary. Of these only the five unshaded examples are normally encountered in hydraulics.

The simplest algorithmic forms for the two-layered fluid, Equations (2.5.1) and (2.5.2) in the form (2.6.13) with ρ_0 and ρ_1 constant, can now be written as

$$L\frac{\partial H}{\partial t} + M\frac{\partial U}{\partial t} + N\frac{\partial U}{\partial x} = 0 \tag{4.9.10}$$

with

$$H = \begin{bmatrix} h_1 \\ h_0 \end{bmatrix}, \qquad U = \begin{bmatrix} u_1 \\ u_0 \end{bmatrix}$$

$$L = \begin{bmatrix} g(1-\sigma) & 0 \\ 0 & g(1-\sigma) \end{bmatrix}, \qquad M = \begin{bmatrix} -u_1 & +u_1 \\ +\sigma u_0 & -u_0 \end{bmatrix}$$

$$N = \begin{bmatrix} (g(1-\sigma)h_1 - (u_1)^2) & u_1 u_0 \\ \sigma u_1 u_0 & (g(1-\sigma)h_0 - (u_0)^2) \end{bmatrix}$$

while Equations (2.5.5) and (2.5.6) can be expressed in the form of (2.6.13) as

$$R\frac{\partial U}{\partial t} + S\frac{\partial H}{\partial t} + T\frac{\partial H}{\partial x} = 0 \tag{4.9.11}$$

with

$$R = \begin{bmatrix} h_1 & 0 \\ 0 & h_0 \end{bmatrix}, \qquad S = \begin{bmatrix} -u_1 & 0 \\ 0 & -u_0 \end{bmatrix}$$

$$T = \begin{bmatrix} (gh_1 - (u_1)^2) & gh_1 \\ \sigma g h_0 & (gh_0 - (u_0)^2) \end{bmatrix}$$

These provide time- and space-centred difference equations, as follows

$$L\left(\frac{H_j^{n+1} - H_j^n}{\Delta t}\right) + \frac{M}{2}\left(\frac{U_{j+1}^{n+1} - U_{j+1}^n}{\Delta t} + \frac{U_{j-1}^{n+1} - U_{j-1}^n}{\Delta t}\right) + \frac{N}{2}\left(\frac{U_{j+1}^{n+1} - U_{j-1}^{n+1}}{2\Delta x} + \frac{U_{j+1}^n - U_{j-1}^n}{2\Delta x}\right) = 0 \tag{4.9.12}$$

$$R\left(\frac{U_j^{n+1} - U_j^n}{\Delta t}\right) + \frac{S}{2}\left(\frac{H_{j+1}^{n+1} - H_{j+1}^n}{\Delta t} + \frac{H_{j+1}^{n+1} - H_{j-1}^n}{\Delta t}\right) + \frac{T}{2}\left(\frac{H_{j+1}^{n+1} - H_{j-1}^{n+1}}{2\Delta x} + \frac{H_{j+1}^n - H_{j-1}^n}{2\Delta x}\right) = 0 \tag{4.9.13}$$

These equations have nine different data structures corresponding to nine algorithmic structures: there are nine different ways of solving them. Once again, in principle, the only satisfactory algorithmic structure at any point is that appropiate to the characteristics defined by the instantaneous values of $\{h_0, h_1, u_0, u_1\}$ at that point. Obviously, a detailed account of all these structures, and their corresponding algorithms and generalized algorithms cannot be entertained in this introduction, so only the one case, corresponding to the top left-hand matrix element of Fig. 4.49, will be considered. By analogy with homogeneous flows, this case is often called the 'subcritical–subcritical' case. Equations (4.9.12), (4.9.13) are expressed as

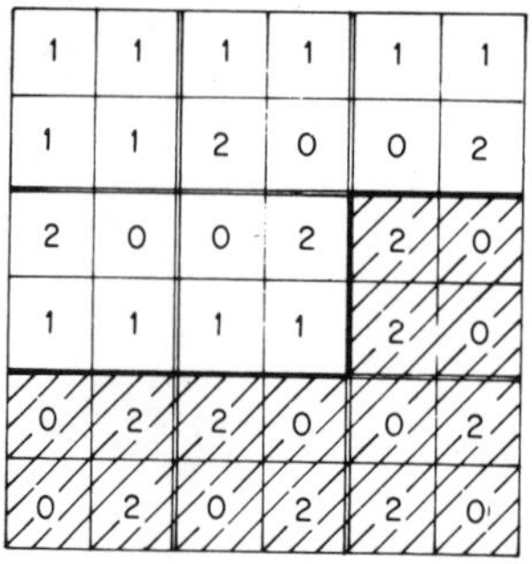

1	1	1	1	1	1
1	1	2	0	0	2
2	0	0	2	2	0
1	1	1	1	2	0
0	2	2	0	0	2
0	2	0	2	2	0

Fig. 4.49. Schematization of the nine data structures of two-layer, nearly horizontal flow, showing the number of points of data to be given at each boundary for each layer

$$A_j^* U_{j+1}^{n+1} + B_j^* H_j^{n+1} + C_j^* U_{j-1}^{n+1} = D_j^*$$

$$A_j H_{j+1}^{n+1} + B_j U_j^{n+1} + C_j H_{j-1}^{n+1} = D_j$$

Introducing E_j, F_j defined by

$$H_{j+1}^{n+1} = E_j U_j^{n+1} + F_j$$

$$U_{j+1}^{n+1} = E_j^* H_j^{n+1} + F_j^*$$

then provides the recurrence relations for the initiating sweep:

$$E_{j-1} = (A_j^* E_j^* + B_j^*)^{-1}(-C_j^*)$$

$$F_{j-1} = (A_j^* E_j^* + B_j^*)^{-1}(D_j^* - A_j^* F_j^*)$$

$$E_{j-1}^* = (A_j E_j + B_j)^{-1}(-C_j)$$

$$E_{j-1}^* = (A_j E_j + B_j)^{-1}(D_j - A_j F_j)$$

The recurrence is used matrix-wise just as in the case of the scalar DS algorithm, while the use of the boundary data is also entirely analogous.

More generally, the two-layer schematization will also allow horizontal density variations in each of its component layers, and indeed this is necessary if the influence of inter-layer exchanges is to be introduced. In this case the transport–diffusion schemes can be run in parallel with the dynamic scheme.

The above scheme used a matrix generalization of the tri-diagonal double-sweep algorithm, but there are advantages, especially for generalizations to other structures of Fig. 4.49, in making computations of the two layers separately, or of the total fluid and the internal processes separately. The computation will then be multi-staged into two-component computations, with associated difficulties in coupling these components properly.

The problem of changing algorithmic structure within a sweep structure again arises and in this case a suitable control function H follows from Equation (3.5.9) as

$$H = (u_1{}^2 - gh_1)(u_0{}^2 - gh_0) - \lambda g^2 h_1 h_0$$

Changes of sign of this function then trigger changes in the algorithmic structure. Generalized double-sweep algorithms may also be introduced, as outlined by Abbott and Grubert (1973).

4.10 Difference methods for linearized two-dimensional nearly horizontal flows

Once again, only the simplest elements of the theory can be introduced. The simplest equations for a two-dimensional nearly horizontal flow are those appertaining to the linearized situation, derived in Equation (2.6.22) as

$$\frac{\partial^2 h}{\partial t^2} - c^2 \left(\frac{\partial^2 h}{\partial x^2} + \frac{\partial^2 h}{\partial y^2}\right) = 0 \tag{4.10.1}$$

where h can as well be taken as the variation in water elevation. This can be approximated by the time- and space-centred scheme

$$\frac{(h^{n+1} - 2h^n + h^{n-1})_{jk}}{\Delta t^2} - c^2 \left[\frac{(h_{j+1} - 2h_j + h_{j-1})_k^n}{\Delta x^2} + \frac{(h_{k+1} - 2h_k + h_{k-1})_j^n}{\Delta y^2}\right] = 0 \tag{4.10.2}$$

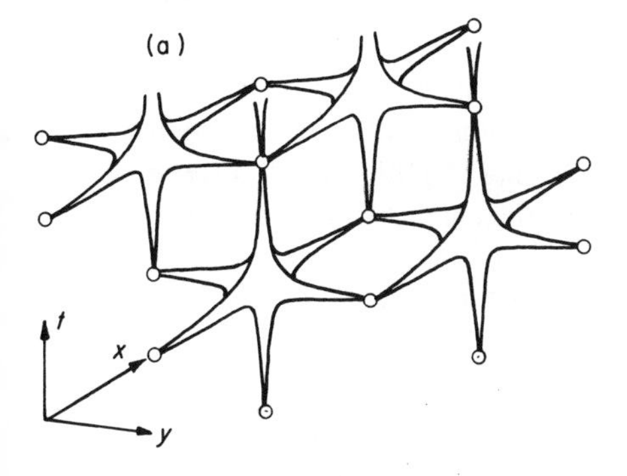

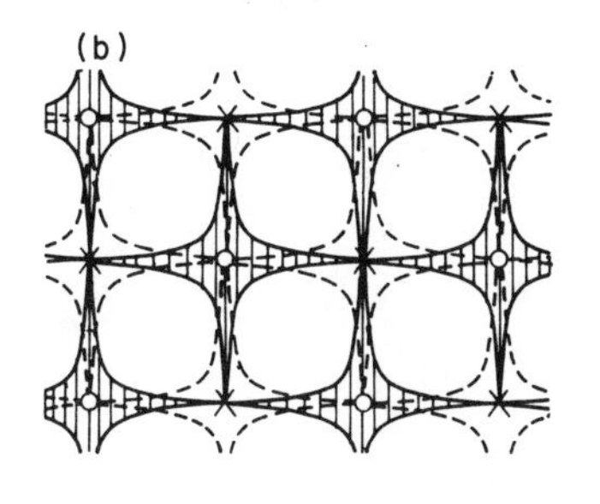

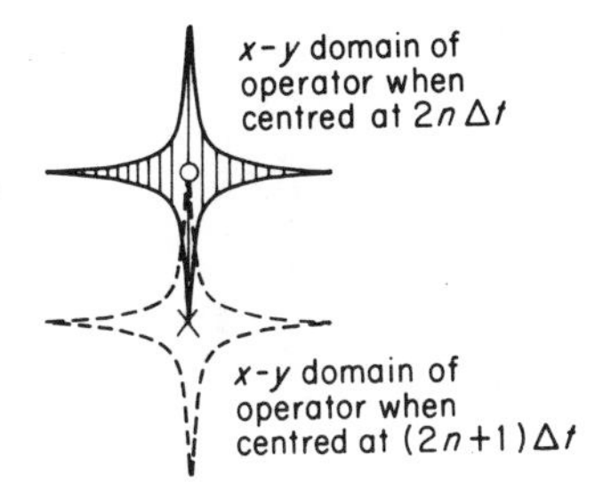

Fig. 4.50. (a) Three-dimensional representation of the layout or tiling of the leap-frog operator for two-dimensional flow. (b) Plan view on Fig. 4.50a, so taken at some t = const.

The truncation error is then $0(\Delta t^2, \Delta x^2, \Delta y^2)$. The scheme is shown in the (x, y, t) space in Fig. 4.50a, where it is seen to provide a 'leapfrog' structure. Looking down on Fig. 4.50a into the (x, y) plane gives the structure shown in Fig. 4.50b. It is seen that one system of operators transforms values of h from one time level to another, or it *accumulates* on h values, at grid points indicated by circles, O, while a second system of operators accumulates on h values at grid points indicated by crosses, x.

Simplifying further to $\Delta x = \Delta y = \Delta s$ in Equation (4.10.2) gives

$$h_{j,k}^{n+1} = \frac{c^2 \Delta t^2}{\Delta s^2} (h_{j+1,k}^n + h_{j-1,k}^n + h_{j,k+1}^n + h_{j,k-1}^n) + 2\left[1 - 2\left(\frac{c^2 \Delta t^2}{\Delta s^2}\right)\right] h_{j,k}^n - h_{j,k}^{n-1} \quad (4.10.3)$$

It is seen that when

$$1 - 2\left(\frac{c^2 \Delta t^2}{\Delta s^2}\right) = 0 \quad \text{or} \quad \Delta s = 2^{\frac{1}{2}} c\, \Delta t \quad (4.10.4)$$

then Equation (4.10.3) reduces to

$$h_{j,k}^{n+1} = \tfrac{1}{2}(h_{j+1,k} + h_{j-1,k} + h_{j,k+1} + h_{j,k-1})^n - h_{j,k}^{n-1} \quad (4.10.5)$$

When $\Delta s = c\Delta t$, (4.10.3) gives the scheme obtained in Section 4.3 through an approximation on the four-point method of characteristics. The condition (4.10.4) is schematized in Fig. 4.51, where it is seen to correspond to the condition that the characteristic cone centred upon the operator's $(n + 1)\Delta t$ element must lie within the region of interpolation of the values that determine conditions at that element. Thus this simplifying condition again corresponds to a coincidence of the operator structure with the characteristic structure.

In order to investigate stability in such a three-time-level scheme, it is usual, after Yanenko (1968), to set $h_j^n = e^{\omega n \Delta t} e^{imj\Delta x}$ in Equation (4.10.3) to obtain

$$\rho^2 - [2 - \alpha]\rho + 1 = 0$$

where

$$\alpha = 8\left(\frac{c^2 \Delta t^2}{\Delta s^2}\right) \sin^2\left(\frac{m\,\Delta s}{2}\right)$$ and $\rho, = e^{\omega\,\Delta t}$, is the amplification factor.

Thence

$$\rho = \frac{(2 - \alpha) \pm \sqrt{-4\alpha + \alpha^2}}{2}$$

When $(-4\alpha + \alpha^2) \leqslant 0$, or $\alpha \leqslant 4$, this provides

$$\rho^2 = \frac{(2 - \alpha)^2 + (4\alpha - \alpha^2)}{4} = 1$$

and a non-dissipative scheme. When $(-4\alpha + \alpha^2) > 0$, $\rho > 1$ and the scheme is unstable. The condition $\alpha \leqslant 4$ implies that

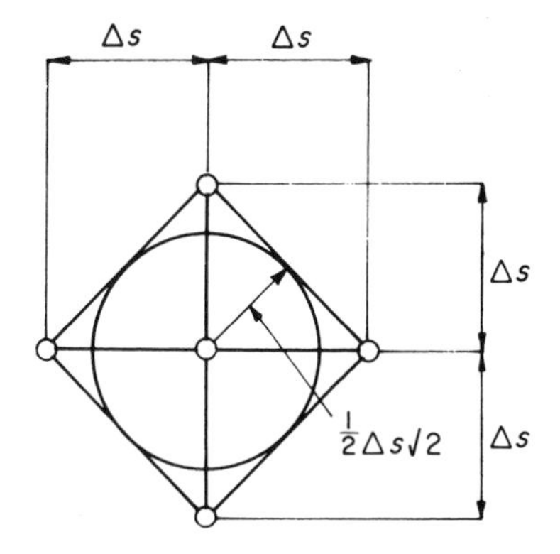

Fig. 4.51. The stability condition of the linearized two-dimension leap-frog operator corresponds to the CFL condition that the characteristic cone should lie on or within the region of interpolation of surrounding grid points

$$8\left(\frac{c^2\Delta t^2}{\Delta s^2}\right) \leqslant 4 \text{ or } \Delta t \leqslant \frac{\Delta s}{2^{\frac{1}{2}}c}$$

Referring again to Fig. 4.51, it is seen that this corresponds to a two-dimensional form of the Courant–Friedrichs–Lewy condition.

Relating Equation (4.10.5) to the double system of accumulation points in Fig. 4.50b, it is seen that values in one system are related to values in another system only through *differences* between values in the other system. This has the consequence that two solutions, both smooth and slowly varying, may accumulate in time in the two systems in such a way that they diverge, the one from the other. This occurs, in fact, in any such scheme and necessitates a periodic realignment of the two schemes through the use of a dissipative interface. Figure 4.52 shows a part of a typical double solution and its elimination through the use of the interface.

Evidently the scheme (4.10.1) can be easily modified into a Lax–Wendroff form by introducing a dissipative interface

$$f_{j,k}^* = (1-4\alpha)f_{j,k} + \alpha(f_{j+1,k} + f_{j-1,k} + f_{j,k+1} + f_{j,k-1}),\ \alpha < \tfrac{1}{4}$$

at alternate time levels. The resulting second-order scheme can then be made so dissipative that it can be used for wave breaking problems, dam break simulations and other applications, as outlined in the next chapter.

The implicit schemes can also be introduced through the case of the linearized equations, in this case taken in linearized Eulerian form

$$\frac{\partial h}{\partial t} + h_0\left(\frac{\partial u}{\partial x} + \frac{\partial v}{\partial y}\right) = 0 \tag{4.10.6}$$

$$\frac{\partial u}{\partial t} + g\frac{\partial h}{\partial x} = 0 \tag{4.10.7}$$

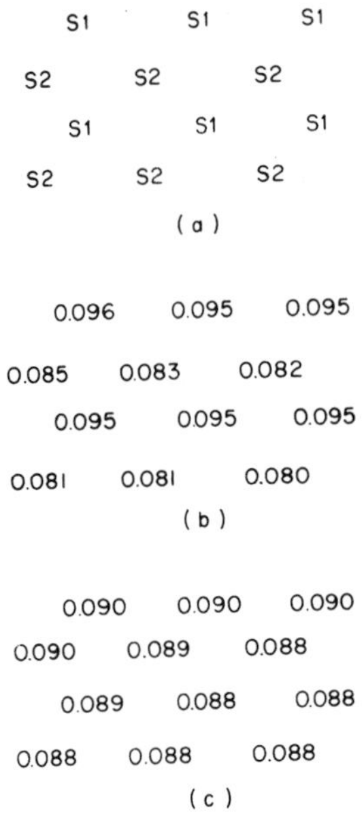

S1 S1 S1

S2 S2 S2

S1 S1 S1

S2 S2 S2

(a)

0.096 0.095 0.095

0.085 0.083 0.082

0.095 0.095 0.095

0.081 0.081 0.080

(b)

0.090 0.090 0.090

0.090 0.089 0.088

0.089 0.088 0.088

0.088 0.088 0.088

(c)

Fig. 4.52. A typical double solution generated by a second order (leapfrog) operator and its elimination through an application of a centred dissipative interface

$$\frac{\partial v}{\partial t} + g\frac{\partial h}{\partial y} = 0 \tag{4.10.8}$$

Then consider the corresponding fractioned step scheme of Leendertse (1967):

$$\frac{u^{n+\frac{1}{2}}_{j,k} - u^{n}_{j,k}}{\frac{1}{2}\Delta t} + g\frac{h^{n+\frac{1}{2}}_{j+1,k} - h^{n+\frac{1}{2}}_{j-1,k}}{2\Delta s} = 0 \tag{4.10.9}$$

$$\frac{v^{n+\frac{1}{2}}_{j,k} - v^{n}_{j,k}}{\frac{1}{2}\Delta t} + g\frac{h^{n}_{j,k+1} - h^{n}_{j,k-1}}{2\Delta s} = 0 \tag{4.10.10}$$

$$\frac{h^{n+\frac{1}{2}}_{j,k} - h^{n}_{j,k}}{\frac{1}{2}\Delta t} + h_0\left(\frac{u^{n+\frac{1}{2}}_{j+1,k} - u^{n+\frac{1}{2}}_{j-1,k}}{2\Delta s} + \frac{v^{n}_{j,k+1} - v^{n}_{j,k-1}}{2\Delta s}\right) = 0 \tag{4.10.11}$$

$$\frac{u^{n+1}_{j,k} - u^{n+\frac{1}{2}}_{j,k}}{\frac{1}{2}\Delta t} + g\frac{h^{n+\frac{1}{2}}_{j+1,k} - h^{n+\frac{1}{2}}_{j-1,k}}{2\Delta s} = 0 \tag{4.10.12}$$

$$\frac{v^{n+1}_{j,k} - v^{n+\frac{1}{2}}_{j,k}}{\frac{1}{2}\Delta t} + g\frac{h^{n+1}_{j,k+1} - h^{n+1}_{j,k-1}}{2\Delta s} = 0 \tag{4.10.13}$$

$$\frac{h^{n+1}_{j,k} - h^{n+\frac{1}{2}}_{j,k}}{\frac{1}{2}\Delta t} + h_0\left(\frac{u^{n+\frac{1}{2}}_{j-1,k} - u^{n+\frac{1}{2}}_{j-1,k}}{2\Delta s} + \frac{v^{n+1}_{j,k+1} - v^{n+1}_{j,k-1}}{2\Delta s}\right) \tag{4.10.14}$$

on the grid shown in the x–y plane in Fig. 4.53.

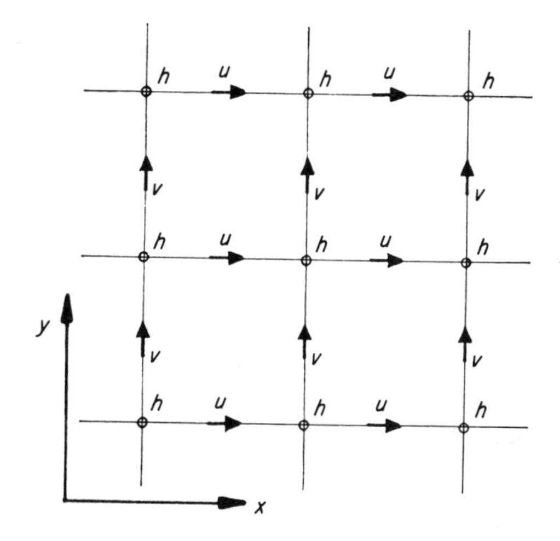

Fig. 4.53. Data layout for two-dimension implicit schemes of the Leendertse (1967) type and Abbott (1968) type

Equations (4.10.9) and (4.10.11) are solved by double-sweep algorithms in the x-direction while Equation (4.10.10) is solved explicitly, for all k. This process takes h, u, v from their values at $n\,\Delta t$ to some intermediate values, placed in time at $(n+\frac{1}{2})\Delta t$ by convention. These values at '$(n+\frac{1}{2})\,\Delta t$' are ill-defined in that the scheme so far is dissipative in the x-direction for $Cr \neq 0$ and unstable in the y-direction, as can be deduced from Section 4.6. Equations (4.10.13) and (4.10.14) are then solved by double sweeps in the y-direction while Equation (4.10.12) is solved explicitly, for all j. This process takes the intermediate values of h, u, v to their values at $(n+1)\,\Delta t$, to complete one *cycle* of the complete algorithm. The procedure used in this case, of first making all the sweeps in the one direction with the other direction variables locked and then making all the sweeps in this other direction with all the original direction variables locked, is an example of an *alternating-direction* algorithm, often abbreviated to 'AD algorithm'.

Leendertse has shown that his scheme can be defined between well-defined levels $n\,\Delta t$ and $(n+1)\,\Delta t$, through eliminating ill-defined values at '$(n+\frac{1}{2})\,\Delta t$' so that a finite Fourier transform on the basis of Equation (4.6.7) can be written as

$$D^{-1}C\xi^{n+1} = A^{-1}B\xi^{n}$$

with

$$\xi^{n} = \begin{bmatrix} \xi_1 \\ \xi_2 \\ \xi_3 \end{bmatrix}^{n}, \text{ corresponding to } \begin{bmatrix} u \\ v \\ h \end{bmatrix}^{n}$$

$$A = \begin{bmatrix} 1 & 0 & ig\alpha \\ 0 & 1 & 0 \\ ih\alpha & 0 & 1 \end{bmatrix}, \qquad B = \begin{bmatrix} 1 & 0 & 0 \\ 0 & 1 & -ig\psi \\ 0 & -ih\psi & 1 \end{bmatrix}$$

$$C = \begin{bmatrix} 1 & 0 & 0 \\ 0 & 1 & ig\psi \\ 0 & ih\psi & 1 \end{bmatrix}, \qquad D = \begin{bmatrix} 1 & 0 & -ig\alpha \\ 0 & 1 & 0 \\ -ih\alpha & 0 & 1 \end{bmatrix}$$

with

$$\alpha = \frac{\Delta t}{\Delta s} \sin \frac{(\sigma_1 \Delta s)}{2} \text{ and } \psi = \frac{\Delta t}{\Delta s} \sin \frac{(\sigma_2 \Delta s)}{2}, \; \sigma_1 = \frac{2\pi K_j \Delta x}{2l}$$

and

$$\sigma_2 = \frac{2\pi Kk\,\Delta y}{2l}, \text{ where } K \text{ is the dimensionless wave number.}$$

It may be shown that the spectral radius of the resulting amplification matrix is unity for all Courant numbers and regardless of how waves propagate across the grid. The scheme is thus unconditionally stable. The celerity ratios are not unity, however, and, as in all two-dimensional schemes, they are now also functions of the direction of propagation relative to the grid. Figure 4.54, based upon Sobey (1970), shows phase portraits in terms of phase shifts as functions of numbers of points per wave length when propagation occurs at angles γ to the grid lines, x or

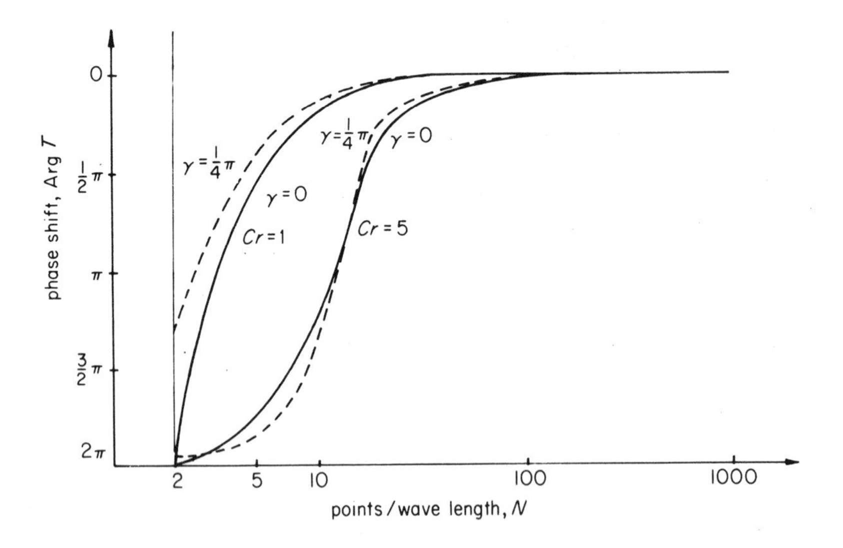

Fig. 4.54. Phase portraits (after Sobey, 1970) of the linearized Leendertse (1967) and linearized Abbott (1968) type schemes

y. As will appear presently, $\gamma = 0$ corresponds simply to a linearized form of Equation (4.7.4). Results are shown for $Cr = 1$ and $Cr = 5$. The relative insensitivity of the phase error to the orientation of the wave propagation relative to the grid is a valuable asset of such a scheme. It is seen, however, that, as the Courant number increases, there is a corresponding increase in the number of grid points required per wave length for a given phase error. In order to use the capabilities of the scheme for working at the high Courant number, something in the range of 20–100 points is required for the main wave components, while when this number of points per wave length is prohibitive in terms of grid size, then either a lower Courant number has to be used or a more accurate difference scheme has to be introduced, as outlined in Section 4.12.

Consider now another implicit scheme, of Abbott (described by Sobey, 1970):

$$\frac{u_{j,k}^{n+1} - u_{j,k}^{n}}{\Delta t} + g\,\frac{h_{j+1,k}^{n+\frac{1}{2}} - h_{j-1,k}^{n+\frac{1}{2}}}{2\Delta s} = 0 \tag{4.10.15}$$

$$\frac{h_{j,k}^{n+\frac{1}{2}} - h_{j,k}^{n}}{\frac{1}{2}\Delta t} + h_0\left(\frac{1}{2}\,\frac{u_{j+1,k}^{n+1} - u_{j-1,k}^{n+1}}{2\Delta s} + \frac{1}{2}\,\frac{u_{j+1,k}^{n} - u_{j-1,k}^{n}}{2\Delta s} + \frac{v_{j,k+1}^{n} - v_{j,k-1}^{n}}{2\Delta s}\right) = 0 \tag{4.10.16}$$

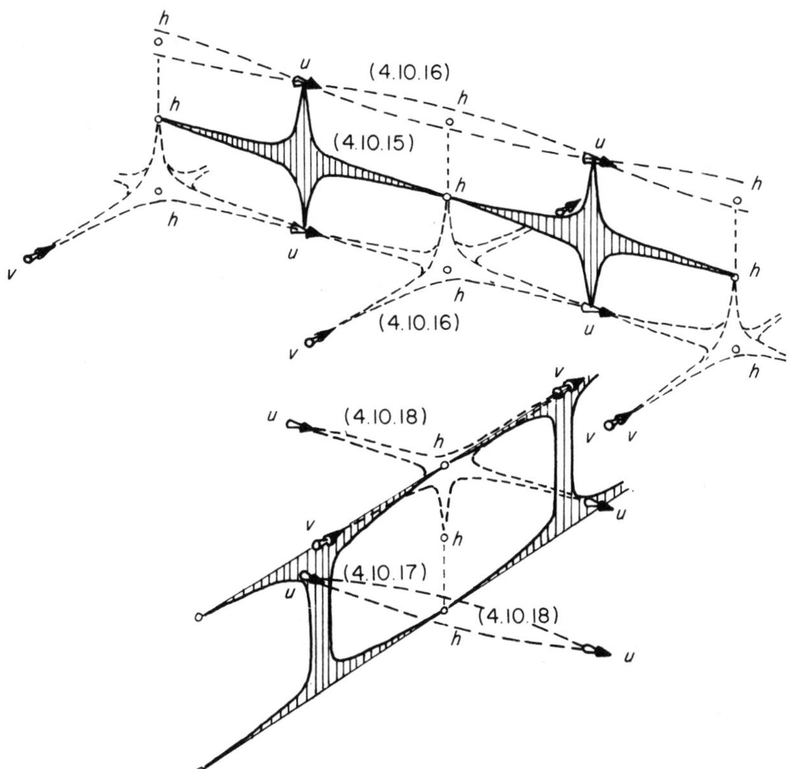

Fig. 4.55. Operator layout or tiling for the Abbott (1968) scheme

$$\frac{v_{j,k}^{n+1} - v_{j,k}^{n}}{\Delta t} + g\left(\frac{1}{2}\,\frac{h_{j,k+1}^{n+1} - h_{j,k-1}^{n+1}}{2\Delta s} + \frac{1}{2}\,\frac{h_{j,k+1}^{n} - h_{j,k-1}^{n}}{2\Delta s}\right) = 0 \tag{4.10.17}$$

$$\frac{h_{j,k}^{n+1} - h_{j,k}^{n+\frac{1}{2}}}{\frac{1}{2}\Delta t} + h_0\left(\frac{1}{2}\,\frac{u_{j+1,k}^{n+1} - u_{j-1,k}^{n+1}}{2\Delta s} + \frac{1}{2}\,\frac{u_{j+1,k}^{n} - u_{j-1,k}^{n}}{2\Delta s} + \frac{v_{j,k+1}^{n+1} - v_{j,k-1}^{n+1}}{2\Delta s}\right) = 0 \tag{4.10.18}$$

This scheme is shown in Fig. 4.55. It is seen that the first two equations, viz. (4.10.15), (4.10.16), are *algorithmically connected* in a first stage in such a way as to advance u directly from $n\,\Delta t$ to $(n+1)\,\Delta t$ while advancing h only from $n\,\Delta t$ to $(n+\frac{1}{2})\,\Delta t$. The second stage connects Equations (4.10.17) and (4.10.18) algorithmically to advance v directly from $n\,\Delta t$ to $(n+1)\,\Delta t$ while advancing h from $(n+\frac{1}{2})\,\Delta t$ to $(n+1)\,\Delta t$. In this scheme, information is accumulated, during the cycle, on the h values only. Correspondingly, the elements of the *algorithmically connected chains* (e.g. $(h_0, u_0, h_1, u_1, \ldots h_j, u_j, \ldots h_{jj})_k$) do not appear to be at the same time level in one of the operations.

It is often convenient to represent the information-accumulating properties of such schemes by a 'cycle diagram' or 'clock diagram', as exemplified in Fig. 4.56.

The scheme (4.10.15)–(4.10.18) can be brought to an equivalent two-level form

$$\begin{aligned} u_{j,k}^{n+1} - u_{j,k}^{n} &+ \frac{g\,\Delta t}{4\,\Delta s}\left[(h_{j+1,k}^{n+1} - h_{j-1,k}^{n+1}) + (h_{j+1,k}^{n} - h_{j-1,k}^{n})\right] + \\ &+ \frac{g h_0 \Delta t^2}{16\Delta s^2}\left[(v_{j+1,k+1}^{n+1} - v_{j+1,k-1}^{n+1}) - (v_{j+1,k+1}^{n} - v_{j+1,k-1}^{n}) - \right. \\ &\left. - (v_{j-1,k+1}^{n+1} - v_{j-1,k-1}^{n+1}) + (v_{j-1,k+1}^{n} - v_{j-1,k-1}^{n})\right] = 0 \end{aligned} \tag{4.10.19}$$

$$v_{j,k}^{n+1} - v_{j,k}^{n} + \frac{g\Delta t}{4\Delta s}\left[(h_{j,k+1}^{n+1} - h_{j,k-1}^{n+1}) + (h_{j,k+1}^{n} - h_{j,k-1}^{n})\right] = 0 \tag{4.10.20}$$

$$\begin{aligned} h_{j,k}^{n+1} - h_{j,k}^{n} &+ \frac{h_0\,\Delta t}{4\Delta s}\left[(u_{j+1,k}^{n+1} - u_{j-1,k}^{n+1}) + (u_{j+1,k}^{n} - u_{j-1,k}^{n}) + \right. \\ &\left. + (v_{j,k+1}^{n+1} - v_{j,k-1}^{n+1}) + (v_{j,k+1}^{n} - v_{j,k-1}^{n})\right] = 0 \end{aligned} \tag{4.10.21}$$

The additional terms of the order of the truncation error in (4.10.19) are typical of those introduced when decomposing an essentially two-dimensional problem into two one-dimensional problems through the use of the algorithmic form (2.6.14) and its associated alternating direction algorithm. For the scheme (4.10.19) to (4.10.21), Sobey (1970) derived the following expression for the complex propagation factor P:

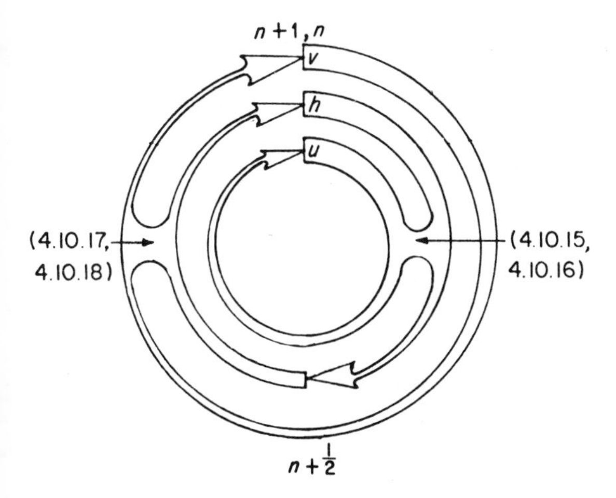

Fig. 4.56. 'Cycle' or 'clock' diagram for the Abbott (1968) scheme

$$\text{Det}\begin{bmatrix} P-1 & -(P-1)\dfrac{gh_0\,\Delta t^2}{4\,\Delta s^2}\sin\sigma_1\,\Delta s\cdot\sin\sigma_2\,\Delta s & \mathrm{i}(P+1)\dfrac{g\,\Delta t}{2\,\Delta s}\sin\sigma_1\,\Delta s \\ 0 & P-1 & \mathrm{i}(P+1)\dfrac{g\,\Delta t}{2\,\Delta s}\sin\sigma_2\,\Delta s \\ \mathrm{i}(P+1)\dfrac{h_0\,\Delta t}{2\,\Delta s}\sin\sigma_1\,\Delta s & \mathrm{i}(P+1)\dfrac{h_0\,\Delta t}{2\,\Delta s}\sin\sigma_2\,\Delta s & P-1 \end{bmatrix} = 0$$

He showed that this, again, provided amplification factors of modulus unity while its phase errors were identical to those of the Leendertse scheme. Thus Sobey showed formally that the scheme (4.10.15)–(4.10.18) was entirely equivalent to the scheme (4.10.9)–(4.10.14).

In many cases, it suffices to consider only the stability of the one-dimension form of a two-dimensional scheme in order to show that it is untenable. For example, the scheme (4.10.19)–(4.10.21) has a one-dimensional form, or one-dimensional *descent*:

$$\frac{u_{j,k}^{n+1}-u_{j,k}^{n}}{\Delta t}+\frac{g}{2}\left(\frac{(h_{j+1}-h_{j-1})_k^{n+1}}{2\Delta x}-\frac{(h_{j+1}-h_{j-1})_k^{n}}{2\Delta x}\right)=0$$

$$\frac{h_{j,k}^{n+1}-h_{j,k}^{n}}{\Delta t}+\frac{h_0}{2}\left(\frac{(u_{j+1}-u_{j-1})_k^{n+1}}{2\Delta x}-\frac{(u_{j+1}-u_{j-1})_k^{n}}{2\Delta x}\right)=0$$

with the same form of expression in the y-direction. This is simply the linearized form of the Abbott–Ionescu (1967) scheme, so it is certainly stable. If its descent was found to be unstable, of course, there would be no hope at all for the success of a two-dimensional scheme. The stability of its descents is thus necessary but not sufficient for stability of a multi-dimensional scheme.

4.11 Difference methods for general two-dimensional nearly horizontal flows

As this is one of the most important areas of computational-hydraulic applications, the work that has gone into the corresponding difference schemes is very great. It is obviously far outside the scope of this introduction to describe such schemes in any detail, and only a few principles can be introduced.

It is a sound rule to start the investigation of any proposed two-dimensional scheme with a study of its descent. For example, the first version of the System 21 'Jupiter' (Abbott, Damsgaard and Rodenhuis, 1973) used a scheme with the following algorithmic structure:

(1) Sweep in y-direction, laid down with decreasing x
(2) Sweep in x-direction, laid down with decreasing y
(3) Sweep in x-direction, laid down with increasing y
(4) Sweep in y-direction, laid down with increasing x

The corresponding clock diagram is shown in Fig. 4.57. As set out in that figure the scheme is 'framed' by its y-sweeps, so that when the x-sweeps are locked with $u = 0$ the y-sweeps should add up to a suitable one-dimensional scheme, in this

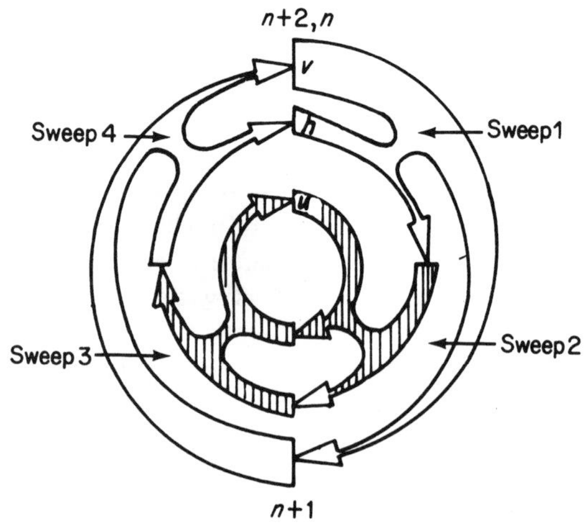

Fig. 4.57. Clock diagram for a System 21-type scheme allowing side-feeding for convective momentum terms

case the Abbott–Ionescu scheme. One can just as well lock the y-sweeps to obtain the condition that the x-sweeps should add up, albeit now at $(n + 1)\Delta t$, to provide the identical scheme. This is indeed necessary, in order that the primitive scheme should be invariant under interchange of x and y. Thus, in the y-sweeps, nothing need appear in descent with any other time address than n and $n + 2$, operator elements at $(n + 1)\Delta t$ cancelling out in the addition, and correspondingly in the x-sweeps over the equivalent range. This can be attained in several ways for the convective terms that remain following the descent but then problems are introduced through the need to centre the cross derivatives for

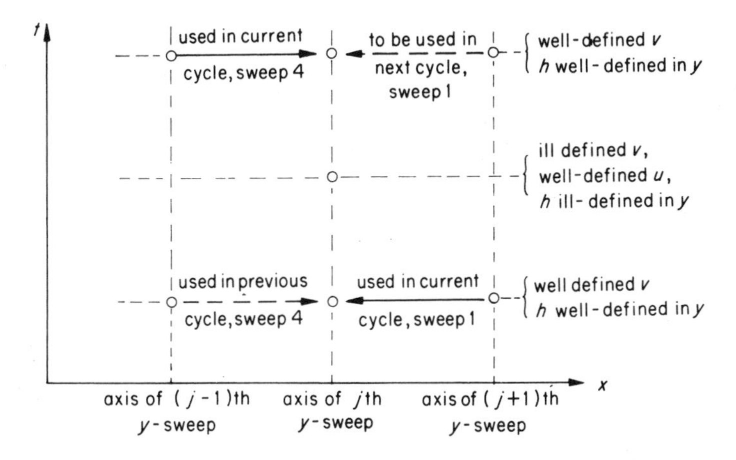

Fig. 4.58. Schematization of side feeding-process

a fully non-dissipative operation. The algorithmic structure as shown in Fig. 4.58 provides one possibility for solving this problem. It is seen how, in relation to the y-sweep so singled out, the convective terms can be centred in their h and v components by 'side feeding' from the immediately preceding sweep. The u components are already centred in this view. Taking averages of the u, v and h appearing in this scheme may provide a certain measure of dissipation, further to stabilize the cross-momentum term. Evidently, the choice of coefficients is very restricted in such a formulation, at first sight appearing to be restricted to lower time levels only. Once again, however, side-feeding processes can be used to improve centring.

The introduction of convective momenta through such algorithmic devices at first appears as 'mere technique', but in fact it introduces certain quite fundamental notions. Consider, for example the primitive form of the x-momenta conservation law with ρ constant:

$$\frac{\partial p}{\partial t} + \frac{\partial}{\partial x}\left(\frac{p^2}{h\rho}\right) + gh\rho\,\frac{\partial h}{\partial x} + \frac{\partial}{\partial y}\left(\frac{pq}{h\rho}\right) = 0 \tag{4.11.1}$$

This can be expanded to read

$$\frac{\partial p}{\partial t} + \frac{2p}{\rho h}\frac{\partial p}{\partial x} + \left(gh\rho - \frac{p^2}{\rho h^2}\right)\frac{\partial h}{\partial x} + \left(\frac{q}{\rho h}\frac{\partial p}{\partial y} + \frac{p}{\rho h}\frac{\partial q}{\partial y} - \frac{pq}{\rho h^2}\frac{\partial h}{\partial y}\right) = 0 \tag{4.11.2}$$

and then the mass equation can be used to 'algorithmize' the equation to

$$\frac{\partial p}{\partial t} - \frac{2p}{\rho h}\frac{\partial h}{\partial t} + \left(gh\rho - \frac{p^2}{\rho h^2}\right)\frac{\partial h}{\partial x} + \left(\frac{q}{\rho h}\frac{\partial p}{\partial y} + \frac{p}{\rho h}\frac{\partial q}{\partial y} - \frac{pq}{\rho h^2}\frac{\partial h}{\partial y} - \frac{2p}{\rho h}\frac{\partial q}{\partial y}\right) = 0 \tag{4.11.3}$$

The first, unbracketed part of this equation leads to a tri-diagonal algorithm, while the second, bracketed part contains terms in derivatives of y that are often called the 'cross terms'. It is these cross terms that then have to be treated by side-feeding and other algorithmic devices.

Now however simple the development from Equation (4.11.1) to Equation (4.11.3) is in terms of differentials, it poses exceedingly difficult problems in terms of differences. In particular, there will always be some combinations of bathymetric configurations and flows that will cause the various convective terms to become large but opposing in sign, so that fractionally small errors in the individual terms will give rise to large and even accumulating errors in the solution.

This source of error can be traced back to the need to expand derivatives of products, ideally:

$$d(uv) = u\ dv + v\ du \tag{4.11.4}$$

Now out of the infinity of possible (consistent) difference approximations to Equation (4.11.4), there is, essentially, one and only one difference *equality*:

$$(uv)_{j+1} - (uv)_{j-1} = \left(\frac{u_{j+1} + u_{j-1}}{2}\right)(v_{j+1} - v_{j-1}) + \left(\frac{v_{j+1} + v_{j-1}}{2}\right)(u_{j+1} - u_{j-1}) \tag{4.11.5}$$

It is unusual, however, that this particular expansion fits the algorithmic possibilities provided by the scheme, so that expansions other than this have to be used, and in these the terms consistent with the right-hand side of Equation (4.11.4) can come to differ by any amount from the required '$d(uv)$' under one or another combination of flow patterns. In effect, the schemes become 'ill-conditioned', and this ill-conditioning manifests itself as local singularities of coefficient matrices, as ill-conditioning of the double-sweep algorithms and even as singularities in the amplification matrices in the Fourier transforms of the schemes. Some relief is found through increasing the accuracy of the partial differencing procedures, bringing them closer to Equation (4.11.5), through the use of higher-precision ('double-precision') working and a very careful scaling of coefficients, but the problem appears to be intrinsic to discrete, finite operations, and its effects can only be reduced and dissipated, but never entirely eliminated.

(That this is no new experience may perhaps be inferred from the histories of mathematics, that repeat the anecdote of Leonhard Euler's dismay at the atheism that permeated the Tsarist court at St Petersburg, and of how he undertook to provide a mathematical proof that 'God exists'. Various versions of the anecdote exist, but it appears that Euler wrote one or another algebraic equality, the simplest of which would have been

$$(x + y)^2 = x^2 + 2xy + y^2 \tag{4.11.6}$$

He added, '*Ergo*, God exists', and then left his thoroughly bewildered audience to contemplate this marvel of compactness. His audience apparently made nothing out of it, and neither have the historians since that time, who have dismissed this happening as a mere prank. But, of course, if there is any truth in the story at all (Struik, 1948), Euler must have been his usual ethical self, for although an algebraic identity, exemplified by Equation (4.11.6), insists algebraically that, in this case,

$$\frac{(x+y)^2}{x^2+2xy+y^2}=1 \tag{4.11.7}$$

there is always some $\epsilon_N > 0$ in $x = -y + \epsilon_N$ such that the left-hand side of Equation (4.11.7) will differ by any amount from unity for any finite number of significant figures N used in its calculation. Only a computer with an infinite word length could realize Equation (4.11.7) for every x and y, and neither man nor any of his productions can have this capacity.)

It should again be emphasized that operational schemes differ considerably from the above sketch: the distinction between longitudinal convective momentum and cross-convective momentum is no longer maintained, the equivalent terms themselves are offset to represent various shear-stress effects while the whole balancing of the scheme is taken to a higher level than the second order suggested here (e.g. to third order) through the use of memory arrays and an operational basis of more complex computational molecules. For homogeneous fluids, it is thus possible to produce schemes that will run stably without any iteration whatsoever, thus maintaining running speeds that are commercially viable.

Associated with problems of describing the effects of convective momenta are the problems of posing of boundary data, as introduced in Section 3.10. Consider, for example, the data layout at a boundary with inflow, as shown in the x–y plane in Fig. 4.59. In order to determine the effect of the convective momenta emanating from point P on conditions at (j, k), the values of $q_{j-1,k+1}$

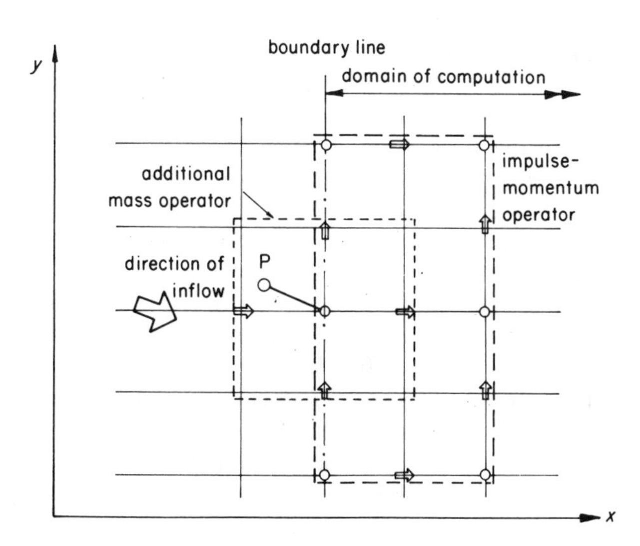

Fig. 4.59. Data layout at an open boundary with inflow showing the effect of introducing an additional mass balance operator

and $q_{j-1,k-1}$ are required along the boundary, for only then can this convected momentum be determined by a well defined interpolation. If these q values are available, then the computation can proceed from the boundary line without further difficulty, but, of course, they are rarely if ever available in practice, owing to the great expense involved in measuring the required average velocities. The q values can be conditioned, however, by an additional mass equation, consistent with

$$\frac{\partial h}{\partial t} + \frac{\partial p}{\partial x} + \frac{\partial q}{\partial y} = 0 \qquad (4.11.8, = 2.1.1)$$

Thus, if any relation between p and q is given that is independent of Equation (4.11.8) then the effects of the convective momenta can be determined. The simplest such relation is the algebraic one of direction, so that p is simply proportional to q, and in many problems the otherwise one-point data on a boundary with inflow are most conveniently supplemented in this way.

It will be seen from Fig. 4.59 that on a boundary with outflow, the determination of the momentum convected into (j, k) is determined by an interpolation that does not involve more than a 0.5 weighting of values of velocities at the boundary and so can be obtained stably by interpolation. Similarly, the problem does not arise in problems of linearized two-dimensional flow, where the effects of the convected momenta are ignored. It also disappears automatically from one-dimensional flows, where the cross velocities are by definition zero, so that the flow directions are then always known (Section 3.11).

The boundary-value problem in two dimensions is, of course, greatly complicated by these and other influences, principally Coriolis accelerations and resistances, and the formulation of general boundary conditions and their numerical realization as boundary modules necessitates a very considerable effort.

Most of the above notions can be applied to the case of the two-layer fluid and even to full three-dimensional modelling. However, in these cases, the sensitivity of the solution is much greater owing to the much greater variations in internal properties associated with a given change in potential energy distribution, as compared with the total fluid properties. The solution is correspondingly more sensitive to noise from the difference scheme and even rapidly converging iterations may sometimes have to be used for long-term stable operation.

Two-dimensional schemes often require controlling or steering routines, especially when they are built as 'modelling systems' or 'design systems'. These routines constitute the 'frame' of the design system, with such functions as allocating the appropriate composition of coefficients – depending upon whether the grid point concerned has water on all four sides or is constrained by adjacent land points – and flooding and drying land areas. When the models generated by the system may have different grid sizes, solved simultaneously through each sweep, then the frame must also order the dissociation and reassociation of coefficients in the alternating-direction algorithms.

Among the more important elements of the frame are the so-called 'grid code' routines. According to whether a grid point has water on all four sides, or water on three sides but land in the negative y-direction or whatever other configuration, so it receives an integer that identifies its situation. In a modelling system such as System 21, the grid codes are generated from the given topography/bathymetry within the system. This may be done in the simplest case by sweeping along each grid line in turn and setting the Boolean 'topograph/bathymetry below given surface water level?' at each grid point (j, k) in turn along this line. In the case of a positive reply, an integer α_1 is added to the existing grid code number of the adjacent grid point of the one parallel grid line, an integer α_2 is added to the existing grid code of the point (j, k) itself and an integer α_3 is added to the existing grid code number of the adjacent grid point of the other parallel grid line. If the reply to the Boolean is negative, then numbers β_1, β_2 and β_3 are placed instead of the α_1, α_2 and α_3. A typical grid point (j, k) on an x-sweep may then be identified by the grid codes in the following table.

$j, k-1$	j, k	$j, k+1$	Grid-code-generated		Code required
sea	sea	sea	$\alpha_1 + \alpha_2 + \alpha_3$	=	γ_1
land	sea	sea	$\beta_3 + \alpha_2 + \alpha_3$	=	γ_2
sea	sea	land	$\alpha_1 + \alpha_2 + \beta_1$	=	γ_3
land	sea	land	$\beta_3 + \alpha_2 + \beta_1$	=	γ_4
sea	land	sea			
(and others of no practical interest in this example)					

There are in this case four equations for the six unknowns $(\alpha_1, \alpha_2, \alpha_3; \beta_1, \beta_2, \beta_3)$ so that it may at first appear that one could set $\gamma_j = j$ and still have a certain measure of freedom – or redundancy. However, the equations relating the α_j, β_j and γ_j are *equations in integers*, and these provide much less freedom and, in effect, little or no redundancy. The general theory of equations in integers has been outlined by Gelfond (1961). It should be remarked that the grid code numbers have no cardinal or even ordinal significance, being used essentially topologically. Thus a grid code, together with the Boolean filters that are described next, constitutes a 'topological interface'.

Once the grid code has been constructed, it can be subsequently used to steer the computation through a 'Boolean filter'. The principle of such a device is illustrated by the simplest of such filters in Fig. 4.60. As regions are flooded or dried, so the grid code is updated correspondingly, while the Boolean filter continues to ensure that the correct composition of coefficients is being maintained.

Other grid codes and filters can be used, as for example to identify the number of characteristics of a given sense at each grid point of a flow with both subcritical and supercritical regions. The resulting identifying integer is then used to select the appropriate algorithmic structure. Such a device is particularly useful for stratified flow computations.

The design of grid codes and their associated filters is a very specialized and very important part of the modelling operation. If it is not done very thoroughly,

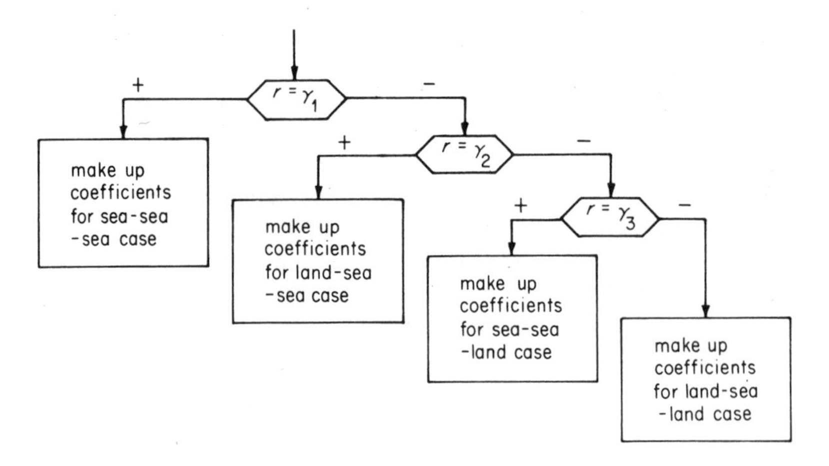

Fig. 4.60. The simplest of Boolean filters of the type used in the control frames of modelling systems

the number of grid code elements can proliferate rapidly and the filters can become excessively cumbersome. A general theory, drawing upon related elements of coding theory and graph theory, does not yet appear to have been put forward.

Completely different topological methods can be taken over from the analysis of open-channel flow using the now traditional methods of non-linear mechanics (e.g. Iwasa, 1966). Other methods again can be used to identify the number of degrees of connectedness of multiply connected one-dimensional flows in terms of topological invariants of the flow network, very much as the number of degrees of redundancy of a structure can be related to the first Betti number (Henderson 1960; Patterson, 1959). A further use of topological methods, for the design of a difference scheme that will become numerically unstable under the same conditions as the corresponding physical system becomes unstable, has been described by Abbott and Rodenhuis (1972).

4.12 An implicit difference method for a Boussinesq equation

The Boussinesq equation extends the impulse–momentum equation of nearly horizontal flow by a term in third derivatives, while the truncation error of the usual implicit approximations to the nearly horizontal flow equations is also of second order, so dominated by terms in third derivatives. Now it has long been known (Lighthill and Whitham, 1955) that small variations in the energy or momentum flux incident upon a wave system can have a very great effect upon the wave lengths and celerities and a lesser but still considerable effect upon the wave amplitude. Accordingly the variations in energy or momentum flux generated by the truncation errors of the nearly horizontal flow equations will be very significant whenever third derivatives become significant, as in short waves (Abbott and Rodenhuis, 1972). Considerable effort must then be expended to reduce the truncation error generated by the difference approximations to the nearly horizontal flow equations if unacceptable errors are to be avoided in short-wave modelling.

The truncation error can, of course, be reduced by using a very large number of grid points per wave length in space and time. For example, Street, Chan and Fromm (1970) used a distance step corresponding to about 4 m and a time step corresponding to about 0.2 s for a 12-s period wave in 8 m deep water, corresponding to a wave length of 100 m. This approach gives rise to quite unacceptable computation costs, however, since a 1 km^2 harbour then requires the computation of 2×10^8 grid-point values for a single 10-prototype-minute simulation, and this still (1977) corresponds to many hours of machine time. An alternative approach is to reduce the truncation error of the scheme to such a level that a relatively coarse grid can still resolve the flow with acceptable accuracy.

Consider, for example, the fully centred Preissmann scheme:

$$\rho\left(\frac{h_{j+1}^{n+1}-h_{j+1}^{n}}{2\Delta t}+\frac{h_{j}^{n+1}-h_{j}^{n}}{2\Delta t}\right)+\frac{p_{j+1}^{n+1}-p_{j}^{n+1}}{2\Delta x}+\frac{p_{j+1}^{n}-p_{j}^{n}}{2\Delta x}=0 \qquad (4.12.1)$$

$$\frac{p_{j+1}^{n+1}-p_{j+1}^{n}}{2\Delta t}+\frac{p_{j}^{n+1}-p_{j}^{n}}{2\Delta t}+\frac{2p}{\rho h}\left(\frac{p_{j+1}^{n+1}-p_{j}^{n+1}}{2\Delta x}+\frac{p_{j+1}^{n}-p_{j}^{n}}{2\Delta x}\right)+$$

$$+\left(\rho g h-\frac{p^2}{\rho h^2}\right)\left(\frac{h_{j+1}^{n+1}-h_{j}^{n+1}}{2\Delta x}+\frac{h_{j+1}^{n}-h_{j}^{n}}{2\Delta x}\right)=0 \qquad (4.12.2)$$

consistent with the conservation laws, differentiated by parts,

$$\rho\,\frac{\partial h}{\partial t}+\frac{\partial p}{\partial x}=0$$

$$\frac{\partial p}{\partial t}+\frac{2p}{\rho h}\frac{\partial p}{\partial x}+\left(\rho g h-\frac{p^2}{\rho h^2}\right)\frac{\partial h}{\partial x}=0$$

The problem of centring the coefficients is put aside in order to concentrate upon the difference terms. The truncation error can then be expressed through Taylor-series expansions such as

$$\left(\frac{\partial h}{\partial t}\right)_{j+\frac{1}{2}}^{n+\frac{1}{2}}=\frac{h_{j+1}^{n+1}-h_{j+1}^{n}}{2\Delta t}+\frac{h_{j}^{n+1}-h_{j}^{n}}{2\Delta t}-$$

$$-\frac{1}{3!}\left[\frac{3}{4}\left(\frac{\partial^3 h}{\partial x^2\,\partial t}\right)_{j+\frac{1}{2}}^{n+\frac{1}{2}}\Delta x^2+\frac{1}{4}\left(\frac{\partial^3 h}{\partial t^3}\right)_{j+\frac{1}{2}}^{n+\frac{1}{2}}\Delta t^2\right]+\ldots$$

$$\left(\frac{\partial p}{\partial x}\right)_{j+\frac{1}{2}}^{n+\frac{1}{2}}=\frac{p_{j+1}^{n+1}-p_{j}^{n+1}}{2\Delta x}+\frac{p_{j+1}^{n}-p_{j}^{n}}{2\Delta x}-$$

$$-\frac{1}{3!}\left[\frac{1}{4}\left(\frac{\partial^3 p}{\partial x^3}\right)_{j+\frac{1}{2}}^{n+\frac{1}{2}}\Delta x^2+\frac{3}{4}\left(\frac{\partial^3 p}{\partial x\,\partial t^2}\right)_{j+\frac{1}{2}}^{n+\frac{1}{2}}\Delta t^2\right]+\ldots$$

Then the truncation error of the volume equation (4.12.1), for example, is

$$\frac{1}{3!}\left(\frac{3}{4}\frac{\partial^3 h}{\partial x^2\,\partial t}\Delta x^2+\frac{1}{4}\frac{\partial^3 h}{\partial t^3}\Delta t^2+\frac{1}{4}\frac{\partial^3 p}{\partial x^3}\Delta x^2+\frac{3}{4}\frac{\partial^3 p}{\partial x\,\partial t^2}\Delta t^2\right)+\ldots$$

Now use is made of the property of the Boussinesq equation noted and exemplified in Section 2.7, that one form of the higher order terms in the equation can be transformed into another form simply by use of the linearized nearly horizontal flow equations. Then the truncation error, which is of the same order as these terms, can be transformed in the same way, so as to reduce to

$$-\frac{1}{3!}\cdot\frac{1}{gh}\frac{\partial^3 p}{\partial x\partial t^2}\left(\frac{\Delta x^2}{2}-\frac{gh\,\Delta t^2}{2}\right)+\ldots \tag{4.12.3}$$

This is seen to disappear for a Courant number $((gh)^{\frac{1}{2}}\,\Delta t/\Delta x)$ of unity, as already presaged by the phase analysis of the Preissmann scheme given in Section 4.7. It may similarly be shown that the truncation error disappears for all successive orders of derivatives when $Cr = 1$ and, furthermore, that this property carries over to the momentum equation so long as the convective momenta are ignored, or so long as a linearized system is considered, again in agreement with the phase portrait results of Section 4.7.

In order to eliminate a truncation error of the form of Equation (4.12.3), it is most convenient to retain p values at $(n-1)\,\Delta t$ and to correct Equation (4.12.1) with a term of the form

$$\frac{1}{3!}\frac{1}{gh\Delta x}\left(\frac{p_{j+1}^{n+1}-2p_{j+1}^{n}+p_{j+1}^{n-1}}{\Delta t^2}-\frac{p_j^{n+1}-2p_j^{n}+p_j^{n-1}}{\Delta t^2}\right)\left(\frac{\Delta x^2-gh\,\Delta t^2}{2}\right) \tag{4.12.4}$$

However, this is centred at $n\,\Delta t$ while the truncation error to be corrected is centred at $(n+\frac{1}{2})\Delta t$. When a memory element is to be maintained in any case for Equation (4.12.4), it is possible to modify the Preissmann scheme to span also from $(n-1)\,\Delta t$ to $(n+1)\,\Delta t$ to maintain a fourth-order accuracy, so having only fifth-order derivatives. For example, the required centring can be provided by

$$\left(\frac{\partial h}{\partial t}\right)_{j+\frac{1}{2}}^{n}\approx\frac{h_{j+1}^{n+1}-h_{j-1}^{n-1}}{4\Delta t}+\frac{h_j^{n+1}-h_j^{n-1}}{4\Delta t} \tag{4.12.5}$$

$$\left(\frac{\partial p}{\partial x}\right)_{j+\frac{1}{2}}^{n}\approx\theta\left(\frac{p_{j+1}^{n+1}-p_j^{n+1}}{\Delta x}\right)+(1-2\theta)\left(\frac{p_{j+1}^{n}-p_j^{n}}{\Delta x}\right)+$$

$$+\,\theta\left(\frac{p_{j+1}^{n-1}-p_j^{n-1}}{\Delta x}\right),\ 0<\theta<\tfrac{1}{2} \tag{4.12.6}$$

where θ is chosen to hold an acceptable stability while preventing the development of a double solution structure (Section 4.10). Equations (4.12.5) and (4.12.6) again define a truncation error that can be given the form of Equation (4.12.3) and eliminated with a difference term of the form of Equation (4.12.4).

The extension and correction of the third derivatives engendered by the vertical accelerations are also introduced.

The remarkable property of Boussinesq equations, that they can be transformed from one form to another through use of the linearized nearly horizontal flow equations, is not only of great utility in the numerical treatment of these equations. It also indicates that even the single advantage of differential formulations over difference formulations – that the former are unique – may not in fact hold when these equations contain terms of different orders of magnitude. This gives the property observed by Long (1964) a certain fundamental importance in computational hydraulics, probably necessitating, in particular, a certain reformulation of the notion of consistency as introduced in Section 4.5.

5 The foundations of computational hydraulics

5.1 Introduction

The computational hydraulics covered in this book introduce many new concepts, corresponding to the new opportunities of digital computation and the new demands of 'high information' methods of engineering construction. However, the fundamental notions that underlie this new hydraulics are really of very early origin, and were, in point of fact, central issues during the foundation of mechanics and analysis during the seventeenth and eighteenth centuries. As will now appear, computational hydraulics reclaims a heritage – that of the discrete view of the world – that had been largely lost for two centuries. The problems that so preoccupied Galileo, Newton, d'Alembert, Maupertuis, Carnot and other early physicists were far from trivial, but they were put aside during the reign of analysis, a reign that has held hydraulics in its dominion until the present time. The aim of this chapter is to present the discrete view of hydraulic processes, with its non-equivalence of momentum and energy formulations, its irreversibilities, its compatibilities and its sequential–recursive procedures, and to relate this view to that of the founders of modern mechanics and analysis, primarily Galileo and Newton. This chapter will further be concerned to show how the mechanical non-equivalences are related to the set-theoretic non-equivalence of countable sets and sets of the power of the continuum, as presaged by Galileo.

5.2 Historical background

During the seventeenth and eighteenth centuries, the prototype of continuous motion was provided by the free fall of a body, the law of which was established by Galileo in the form that 'the descent of bodies varies as the square of the time and that the motion of projectiles is in the curve of a parabola' (Newton, 1687). Most of the early laws of hydraulics were expressed in terms of Galileo's law of falling bodies, whether to relate the velocity at which a jet issued from an orifice under a given head or to relate the celerity at which a wave propagated in water of a given depth (Rouse and Ince, 1957). The law provides what is now called

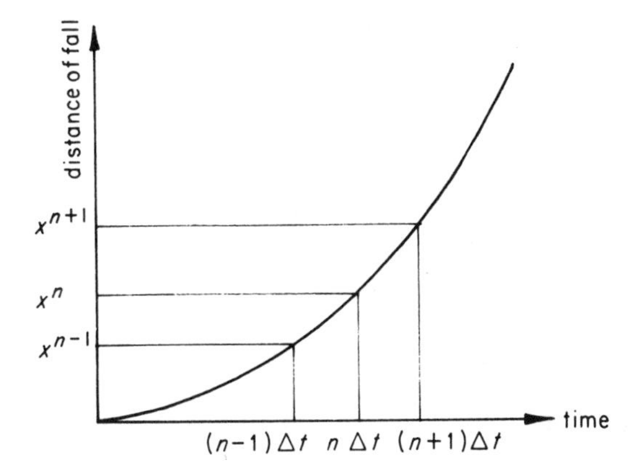

Fig. 5.1. Discretization of Galileo's law of falling bodies

'Newton's second law of motion', whether by simply differentiating, or by taking the limit of the following discrete construction (Fig. 5.1):

$$x^{n+1} = k(n+1)^2 \Delta t^2$$

$$x^n = kn^2 \Delta t^2$$

$$x^{n-1} = k(n-1)^2 \Delta t^2$$

with k = constant, giving

$$x^{n+1} - 2x^n + x^{n-1} = 2k\,\Delta t^2 \tag{5.2.1}$$

with a limit, in Leibnitz's notation, of

$$\mathrm{d}^2x = 2k\,\mathrm{d}t^2 \tag{5.2.2}$$

When the constant $2k$ is taken as 'the acceleration due to gravity', this is the same form of the second law as that given in Section 1.1, the form that will now be called 'form C' (= continuous). However, even at the time of Galileo (1590, 1600), such a limit process presented certain conceptual difficulties. Thus, Equation (5.2.1) could as well be written as

$$\frac{x^{n+1} - x^n}{\Delta t} - \frac{x^n - x^{n-1}}{\Delta t} = 2k\,\Delta t \tag{5.2.3}$$

where it corresponded to a statement that the change in velocity from 'state $n - \frac{1}{2}$' to 'state $n + \frac{1}{2}$' of a unit of mass was given by an impulse proportional to the time separating these states. The motion could then as well be conceived, to some approximation, as being driven by finite 'blows' of impulse-magnitude $2k\,\Delta t$, driving the body along polygonal element paths jointing the 'grid-points' of Fig. 5.1. Galileo observed that the set of times, scaled to integers $\{1, 2, 3, \ldots\}$, was then set into correspondence with the set of displacements scaled to

{1, 4, 9, . . .}, the latter set being a proper subset of the former set. In the limit, both sets would have to tend to infinite sets, presumed at that time to be 'of equal number of elements', which then appeared to be patently impossible when one set was a proper subset of the other and where, indeed, the one appeared to have 'infinitely more elements' than the other.

For any choice of a *fixed* Δt in (5.2.3), so for any discrete process, (5.2.3) corresponds to the form of the second law of motion as given by Newton in his *Principia*: 'The change in motion is proportional to the motive force impressed and is made in the right line in which that force is impressed'.

The first, historically, of the questions that were posed, was that of whether a sequence of impacts, the magnitude of which decreases as their frequency increases towards some limit, would in fact provide the motion predicted by the continuous form of the second law of motion.

In the case considered above, where the limit is a smooth continuous motion, it could be anticipated that the continuous form would indeed be provided, at least so long as the problem of conceiving of the limit was put aside. However, consider the law of momentum conservation of the standing hydraulic jump, from Equation (1.4.1):

$$\rho[u^2 h + gh^2/2]_1^2 = 0 \quad \text{or} \quad p(u_1 - u_2) = \left(\frac{\rho g h_2^2}{2} - \frac{\rho g h_1^2}{2}\right) \tag{5.2.4}$$

This is also a statement of the second law of motion, but now applied to the case of impact or percussion, associated with rapidly varying, irreversible processes. It clearly corresponds more to the form of the second law of motion given by Newton in his *Principia*. This form, called 'D' (= discrete), does not contain time as an explicit variable and it is difficult, as will shortly appear, to derive Equation (5.2.4) or any other of its consequences from the partial-differential form of Equation (5.2.2), whether from a conservation or from an Eulerian form. Referring back to Chapter 1, it will be seen that the formulation of Equation (5.2.4) proceeded using a discrete view of processes. On the other hand, the process of passing from the discrete to the continuous was demonstrated in Chapter 3, where Equation (5.2.4) was used to derive the method of characteristics on the assumption that the flows were continuous in a classical sense, the magnitude of impacts tending to zero as their number tended to infinity over any finite interval of time and space. This led again to the characteristic relations, that were no explicit functions of time. The resulting relations derived in this way from the discrete form of Newton's second law were seen to be the same as those provided by the use of the corresponding continuous partial-differential form of Equation (5.2.2). Thus it has already been shown that the discrete form ('D') leads to the results of the continuum form ('C') when the flow is smoothly varying and the processes reversible, but it still remains to show how, or in what sense, the partial-differential form ('C') leads to the results of the discrete form ('D') when the flow is rapidly varying and the process is irreversible. Moreover, it may already be anticipated that the irreversibility of numerical schemes, as manifest by energy variations when they are built upon mass and momentum conservation laws, is intrinsically related to their discrete nature and that these schemes model more a law of motion of type 'D' than they do one of type 'C'.

This greater generality of the form 'D' could well have influenced Newton when he formulated his second law explicitly in the form 'D' so that it applied, in the first instance, to problems of impact. Newton later *extended* the law to continuous motions, by introducing a time into his definitions, but a law of impact, or discontinuous motion, initiated his chain of definitions.

Now in a paper on the history of Newton's second law, in which the above problems are introduced, Hankins (1967) has drawn special attention to the claim by Euler, in 1750, to have discovered a 'new principle of dynamics' which could be 'regarded as the unique foundation of all mechanics and of all the other sciences which treat the motion of any kind of bodies whatsoever', so that 'all other principles ought to be established on this simple principle whether they are already accepted in mechanics and hydraulics and used to determine the movement of solid and fluid bodies; or whether they still have to be developed'.

This new principle was written in the form

$$2M\,\mathrm{d}^2x = \pm F\,\mathrm{d}t^2$$

or

$$F = \pm 2M\frac{\mathrm{d}^2x}{\mathrm{d}t^2}$$

with one equation for each of the three spatial axes. (The ± implied a vector equation and the 2 was introduced through Euler's choice of units for force and time.) As Hankins remarks, 'Euler's new principle may be a fair statement of the second law of motion as it is commonly used today, but it is not the law that Newton wrote in the *Principia*.' Euler, who was thoroughly familiar with the *Principia* and was always fastidious in acknowledging his sources, clearly recognized his formulation of the second law as something other than the formulation of Newton. For Maupertuis (Hankins, 1967), Euler 'deduced these laws from the principle discovered by Galileo', as comparison of the present equations (5.2.1) and (5.2.2) will indicate.

Euler chose to put aside the difficulties inherent in the dualism of discrete–continuum or irreversible–reversible forms of the second law of motion and to express the behaviour of a fluid system through systems of *differential* equations. In the subject of *hydrodynamics*, that he thus initiated, it was thenceforth impossible to describe irreversible processes at the order of the dynamic equations themselves. In hydraulics, for example, these irreversible processes appeared primarily as zero-order resistance terms in first order, otherwise reversible differential equations. Zero-order terms could never, of course, describe such solution-influencing processes as hydraulic jump formation. From the point of view of hydrodynamics, any one differential equation is an 'ultimate' form, while the infinity of difference equations that are consistent with it are mere approximations. From the point of view of computational hydraulics, the difference equations are more general, and potentially closer to the physical reality, while the virtue of the differential equation resides more in its uniqueness, whereby it can serve as a fixed point of reference in the formulation of the difference equations, at least for lower-order equations homogeneous in the order of magnitude of their terms (Section 2.7).

Euler cut the Gordian knot of the continuum–discrete controversy, and thereby released the whole science of hydrodynamics (*see* Bourbaki, 1960, pp. 216–221), but this science developed only at the cost of a deepening rift between the mathematics and the physics of irreversible processes. Today discontinuities are treated mathematically in the 'theory of distributions' (of Sobolev, 1936; Schwartz, 1957, 1959; and Mikusinsky, 1959) while the physics of discontinuity is treated far away in 'irreversible thermodynamics' (of de Groot, 1951; and Prigogine, 1955). It was the achievement of Lax (1954) that he brought these two great scientific tendencies back together and demonstrated their natural unity *in numerical analysis*, through the study of weak solutions of systems of conservation laws.

5.3 Weak solutions of systems of conservation laws

Once again, all energy-diffusing terms such as bed-slopes, resistances, wind stresses, etc. are put aside, as also are the Coriolis accelerations. The conservation forms then become, from Equations (2.1.8) and (2.4.7):

$$\frac{\partial f}{\partial t} + \frac{\partial g(f)}{\partial x} = 0 \tag{5.3.1}$$

for one-dimensional flows and

$$\frac{\partial f}{\partial t} + \frac{\partial g_1(f)}{\partial x} + \frac{\partial g_2(f)}{\partial y} = 0 \tag{5.3.2}$$

for two-dimensional (plane) flows.

Now it has been shown in Section 3.4 that no single-valued non-trivial solution of these equations will exist for all time. At some time, characteristics of the same family will intersect, to form, physically, a hydraulic jump, or, as introduced in Section 3.7, characteristics of different families may intersect to form fronts, Jeffreys–Vedernikov instabilities and so forth. If, following Lax (1954), it is supposed that the resulting discontinuity in the flow is restricted, for one-dimensional flow, to a single point, the physical situation can be described by a certain generalized solution of Equation (5.3.1) as follows:

A solution f is called a 'weak solution' of Equation (5.3.1) with initial values $\phi(x)$ if the integral relation

$$\int_0^\infty dt \int_{-\infty}^{\infty} dx \left(\frac{\partial w}{\partial t} \cdot f + \frac{\partial w}{\partial x} \cdot g(f)\right) + \int_{-\infty}^{\infty} dx \left(w(x,0) \cdot \phi(x)\right) = 0 \tag{5.3.3}$$

holds for every test function w that tends rapidly to zero as $|x| + t$ increases and which has continuous first derivatives. The definition extends easily to more dimensions, the point of discontinuity becoming a line or surface of discontinuity and the space of definition of the integrals being suitably extended.

It will now be shown that (a) Equation (5.3.3) is equivalent to Equation (5.3.1) for continuous flows, but that (b) when the flows are discontinuous, the following theorem obtains:

Theorem I:

If

$$f = f(x, t) = \alpha_0 - \alpha_1 H(x - ct)$$

where H is the Heaviside step function defined by

$$H(x) = \begin{cases} 0, & x < 0 \\ 1, & x \geqslant 0 \end{cases}$$

then

$$g = g(f) = c[\alpha_0 - \alpha_1 H(x - ct)]$$

satisfies Equation (5.3.3), where c is the celerity of the step discontinuity.

In fact:

(a) One multiplies Equation (5.3.1) by w on the left and integrates the resulting equation, separating by parts to give

$$\int_0^\infty \mathrm{d}t \int_{-\infty}^\infty \mathrm{d}x \left[\left(\frac{\partial}{\partial t}(wf) - f\frac{\partial w}{\partial t}\right) + \left(\frac{\partial}{\partial x}(wg) - g\frac{\partial w}{\partial x}\right)\right] = 0$$

Then by virtue of the rapidly decreasing or 'tempered' properties of the test functions

$$\int_0^\infty \mathrm{d}t \int_{-\infty}^\infty \mathrm{d}x \frac{\partial}{\partial x}(wg) = 0, \quad \int_{-\infty}^\infty \mathrm{d}x \int_0^\infty \mathrm{d}t \frac{\partial}{\partial t}(wf) = -\int_{-\infty}^\infty \mathrm{d}x(wf)\bigg|_{t=0}$$

so that Equation (5.3.3) is simply Equation (5.3.1) rewritten when first derivatives of Equation (5.3.1) exist.

(b) The terms of Equation (5.3.3) are evaluated:

$$\int_0^\infty \frac{\partial w}{\partial t} \cdot f \,\mathrm{d}t = \int_0^\infty \frac{\partial w}{\partial t}[\alpha_0 - \alpha_1 H(x - ct)]\,\mathrm{d}t$$

$$= \alpha_0 \int_0^\infty \frac{\partial w}{\partial t}\,\mathrm{d}t - \alpha_1 \int_0^\infty \frac{\partial w}{\partial t} \cdot H(x - ct)\,\mathrm{d}t$$

$$= \alpha_0 \int_0^{\infty} \frac{\partial w}{\partial t}\, dt - \alpha_1 \int_0^{(x-ct=0)} \frac{\partial w}{\partial t} \cdot H(x)\, dt$$

$$= -\alpha_0 w(x,0) - \alpha_1 [w\big|_{x-ct=0} - w(x,0)] \cdot H(x)$$

$$= -[\alpha_0 - \alpha_1 \cdot H(x)] \cdot w(x,0) - \alpha_1 \cdot w\big|_{x-ct=0} \cdot H(x)$$

Thence

$$\int_{-\infty}^{\infty} dx \int_0^{\infty} \frac{\partial w}{\partial t} \cdot f\, dt = \int_{-\infty}^{\infty} -[\alpha_0 - \alpha_1 \cdot H(x)]\, w(x,0)\, dx - $$

$$- \alpha_1 \int_{-\infty}^{\infty} w\big|_{x-ct=0} \cdot H(x)\, dx \tag{5.3.4}$$

Similarly,

$$\int_{-\infty}^{\infty} \frac{\partial w}{\partial x} \cdot g(f)\, dx = \int_{-\infty}^{\infty} \frac{\partial w}{\partial x} [\alpha_0 - \alpha_1 \cdot H(x-ct)]\, c\, dx,$$

$$= \alpha_0 c \int_{-\infty}^{\infty} \frac{\partial w}{\partial x}\, dx - \alpha_1 \int_{(x-ct=0)}^{\infty} \frac{\partial w}{\partial x} \cdot c\, dx$$

$$= 0 + \alpha_1 \cdot w\big|_{x-ct=0} \cdot c$$

so that

$$\int_0^{\infty} dt \int_{-\infty}^{\infty} \frac{\partial w}{\partial x} \cdot g(f)\, dx = \alpha_1 \int_0^{\infty} w\big|_{x-ct=0} \cdot c\, dt \tag{5.3.5}$$

Integrating w along $x - ct = 0$ from $x = ct = 0$ to ∞ causes the second integral of Equation (5.3.4) to cancel with the integral in Equation (5.3.5) while the first integral of Equation (5.3.4) cancels with the second term of Equation (5.3.3), so that (5.3.3) is satisfied identically. (*See also* Aramanovich *et al.*, 1965, p. 215.)

It follows from Theorem I that, if f_1 and f_2 are two solutions of Equation (5.3.1) whose domains of definition in the x–t plane are separated by a smooth curve, so that Theorem 1 applies locally, then the two taken together will constitute a weak solution if, and only if, the local slope $dx/dt = c$ of the separating curve and the values of f_1 and f_2 on either side of it satisfy the condition

$$c[f]_1^2 = [g]_1^2 \tag{5.3.6}$$

where the notation $[\ \ ]_1^2$ again denotes the jump in the bracketed quantities from one side of the dividing curve to the other. This result is, of course, identical to that obtained directly through the discrete calculation of Chapter 1.

From Equation (5.3.6) it again follows, however, that *weak solutions of Equation (5.3.1) depend upon the form in which (5.3.1) is written*. Thus if

$$f = \begin{bmatrix} h \\ uh \end{bmatrix}, \qquad g(f) = \begin{bmatrix} uh \\ u^2h + gh^2/2 \end{bmatrix} \tag{5.3.7}$$

are used, the jump must satisfy

$$c\begin{bmatrix} h \\ uh \end{bmatrix}_1^2 = \begin{bmatrix} uh \\ u^2h + gh^2/2 \end{bmatrix}_1^2$$

If, however, any non-linear transformation of Equation (5.3.1) with (5.3.7) is used, such as one to

$$f = \begin{bmatrix} h \\ u \end{bmatrix}, g(f) = \begin{bmatrix} uh \\ u^2/2 + gh \end{bmatrix} \tag{5.3.8}$$

the jump conditions are

$$c\begin{bmatrix} h \\ u \end{bmatrix}_1^2 = \begin{bmatrix} uh \\ u^2/2 + gh \end{bmatrix}_1^2$$

providing a totally different value for c, and accordingly defining a dividing curve quite different from that defined by Equation (5.3.7). It follows from the very definition of these differential forms in Chapter 2 that the mass conservation equation and Newton's laws of motion (the second law being taken in the form of Euler's (1750) formulation) are not sufficient to define the solutions of mixed continuous–discontinuous flows.

A further principle has to be invoked – the second law of thermodynamics – and it is this law that insists that the mass–momentum formulation (5.3.7) be chosen and all of its non-linear transformations, such as formulation (5.3.8), be rejected. Such solutions that simultaneously satisfy conservation laws for continuous and discontinuous flows are called *genuine weak solutions.* It is again remarked that the use of Newton's original formulation of the second law provides genuine weak solutions without explicit resort to the second law of thermodynamics – and it is accordingly just this formulation that is used in nearly all books on hydraulics when describing irreversible processes. It is this form that has been used when passing from the results of Chapter 1 to those of Chapter 3.

5.4 Dissipative difference schemes

As demonstrated in Chapter 1, the above results follow from the non-equivalence of such measures of state of motion as momentum and energy in discontinuous flows. However, difference schemes also represent discontinuous (discretized) processes and may therefore provide a similar non-equivalence. Consider, as an idealized example, the following difference form of Equation (5.3.1):

$$\frac{f_j^{n+1} - \frac{1}{4}(f_{j+1}^n + 2f_j^n + f_{j-1}^n)}{\Delta t} + \frac{g_{j+1}^n - g_{j-1}^n}{2\Delta x} = 0, \quad j = 1, \ldots jj - 1 \qquad (5.4.1)$$

operating on the flow

$$f_j^n = f_0 + (-1)^j \Delta f, \quad j = 0, \ldots jj \qquad (5.4.2)$$

implying that

$$g_j^n = g_0 + (-1)^j \Delta g, \quad j = 0, \ldots jj$$

Then

$$f_j^{n+1} = f_0$$

so that, if f has the form of Equation (5.3.7), there occurs an energy loss per unit support

$$0(h\,\Delta u^2, g\,\Delta h^2) \qquad (5.4.3)$$

where Δu and Δh are the increments providing Δf and Δg. It is seen that the magnitude of the term (5.4.3) is independent of Δx and Δt.

As has been seen in Chapter 4, the example (5.4.1)–(5.4.3) is a case of non-equivalence brought about by amplification errors, but even in schemes without amplification error, the phase error will still ensure non-equivalence. For translating a component such as the perturbation in Equation (5.4.2) relative to a component of longer wave length will have the effect of changing its mean elevation and its mean velocity so as to give a further term of the order of (5.4.3). In general, then, the non-equivalence of momentum and energy is manifest by a steady dissipation of energy when momentum is conserved in dissipative difference schemes and by a variation of the energy about some mean in non-dissipative but stable schemes. Indeed, it is this non-equivalence that serves as the basis for the so-called 'energy method' of stability analysis (Richtmyer and Morton, 1967; Rozhdestvensky and Yanenko, 1968; Gustafsson, 1971) and is implicit in variational finite element methods (e.g. Norrie and de Vries, 1973). It is also just this non-equivalence of momentum and energy in difference schemes that allows the generation of genuine weak solutions as filtered solutions of the difference forms of mass and momentum conservation laws, as will be exemplified in the sequel.

5.5 Generalized difference forms of conservation laws

Difference schemes are considered centred at $j\,\Delta x$, and having range $[(j-i)\,\Delta x, (j+i)\,\Delta x]$ in x. By extending the notion of generalized consistency suggested by

Lax and Wendroff (1960), it is possible to describe a large class of schemes by introducing a vector-valued function f^* of $2i+1$ arguments $\{f_k, k=j-i, \ldots j+i\}$ related to f only by the requirement that when all $2i$ arguments are equal f^* reduces to f:

$$\underbrace{f^*(f,f,\ldots f)}_{2i+1 \text{ times}} = f$$

In a similar way a $g^* = g^*\{f_k, k=j-i, \ldots j+i\}$ is introduced under the requirements that when all $2i+1$ arguments are equal, g^* reduces to g. Moreover

$$g^*_{j+\frac{1}{2}} = g^*(f_{j-i+1}, \ldots, f_{j+i})$$

$$g^*_{j-\frac{1}{2}} = g^*(f_{j-i}, \ldots, f_{j+i-1})$$

and the problem may be conveniently posed as a pure initial-value problem by allowing $-\infty < j < \infty$. Then a wide range of two-level schemes, generalized consistent with Equation (5.3.1) and also with its corresponding Eulerian and algorithmic forms can be written as

$$\frac{f_j^{n+1} - f_j^{*n}}{\Delta t} + \theta\left(\frac{g^{*n}_{j+\frac{1}{2}} - g^{*n}_{j-\frac{1}{2}}}{\Delta x}\right) + (1-\theta)\left(\frac{g^{*n+1}_{j+\frac{1}{2}} - g^{*n+1}_{j-\frac{1}{2}}}{\Delta x}\right) = 0 \quad 0 \leqslant \theta \leqslant 1 \tag{5.5.1}$$

The following 'theorem of generalized consistency' then follows:

Theorem II: *Denote by $\bar{f} = \bar{f}(x, t, \Delta x, \Delta t)$ the solution of the difference equation (5.5.1) with initial data ϕ and assume that as Δx and Δt tend to zero f converges boundedly almost everywhere to some function $f = f(x, t)$. Then this f is a weak solution of (5.3.1) with initial values ϕ*

In order to prove generalized consistancy, multiply (5.5.1) by any test function w and integrate with respect to x, with f_j^n representing the value of $f^n(x)$ over $j\,\Delta x \leqslant x < (j+1)\,\Delta x$, but now summing over all values of t which are integral multiples of Δt and multiplying by Δt. The first term is summed by parts to obtain,

$$\sum_{n=0}^{\infty}\int_{-\infty}^{\infty} \mathrm{d}x \left(\frac{w_j^{n+1} + w_j^n}{2}\right)\left(\frac{\bar{f}_j^{n+1} - \bar{f}_j^{*n}}{\Delta t}\right)\Delta t$$

$$= \int_{-\infty}^{\infty} \mathrm{d}x \sum_{n=0}^{\infty}(w_j^{n+1}\bar{f}_j^{n+1} - w_j^n\bar{f}_j^{*n}) - \int_{-\infty}^{\infty}\mathrm{d}x \sum_{n=0}^{\infty}\left(\frac{\bar{f}_j^{n+1} + \bar{f}_j^{*n}}{2}\right)\left(\frac{w_j^{n+1} - w_j^n}{\Delta t}\right)\Delta t$$

$$= -\int_{-\infty}^{\infty}\mathrm{d}x(w_j^0\bar{f}_j^0) - \int_{-\infty}^{\infty}\mathrm{d}x\sum_{n=0}^{\infty}\left(\frac{w_j^{n+1} - w_j^n}{\Delta t}\right)\left(\frac{\bar{f}_j^{n+1} + \bar{f}_j^{*n}}{2}\right)\Delta t +$$

$$+ \int_{-\infty}^{\infty}\mathrm{d}x \sum_{n=0}^{\infty}(w_j^n\bar{f}_j^n - w_j^n\bar{f}_j^{*n})$$

while for the second term the variable of integration is adjusted to obtain

$$\int_{-\infty}^{\infty} dx \left(\frac{w_j^{n+1} + w_j^n}{2}\right) \left[\theta \, \frac{(g^*_{j+\frac{1}{2}} - g^*_{j-\frac{1}{2}})^n}{\Delta x} + (1-\theta)\, \frac{(g^*_{j+\frac{1}{2}} - g^*_{j-\frac{1}{2}})^{n+1}}{\Delta x}\right] =$$

$$= \int_{-\infty}^{\infty} dx \left[\frac{\theta w_j^n}{\Delta x} (g^*_{j+\frac{1}{2}} - g^*_{j-\frac{1}{2}})^n + \frac{(1-\theta) w_j^{n+1}}{\Delta x} (g^*_{j+\frac{1}{2}} - g^*_{j-\frac{1}{2}})^{n+1}\right] + 0(\Delta w)$$

$$= \int_{-\infty}^{\infty} dx \left[\theta \, \frac{(w_{j-\frac{1}{2}} - w_{j+\frac{1}{2}})^n}{\Delta x} g_j^{*n} + (1-\theta)\, \frac{(w_{j-\frac{1}{2}} - w_{j+\frac{1}{2}})^{n+1}}{\Delta x} g_j^{*n+1}\right] + 0(\Delta w)$$

where $0(\Delta w), = 0(\int_{-\infty}^{\infty} dx \; [(w_j^{n+1} - w_j^n)(g^{*n}_{j+\frac{1}{2}} - g^{*n}_{j-\frac{1}{2}})]), \to 0$ as $\Delta t \to 0$

because $w(x, t) \to 0$ rapidly with increasing $|x| + t$.

Thus Equation (5.5.1) implies that

$$\sum_{n=0}^{\infty} \Delta t \int_{-\infty}^{\infty} dx \left[\frac{(w_j^{n+1} - w_j^n)}{\Delta t} \frac{(\bar{f}_j^n + \bar{f}_j^{*n})}{2} + \theta \, \frac{(w_{j+\frac{1}{2}} - w_{j-\frac{1}{2}})^n}{\Delta x} g_j^{*n}\right.$$

$$\left. + (1-\theta)\, \frac{(w_{j+\frac{1}{2}} - w_{j-\frac{1}{2}})^{n+1}}{\Delta x} g_j^{*n+1}\right] - \int_{-\infty}^{\infty} dx \sum_{n=0}^{\infty} w_j^n (\bar{f}_j^n - \bar{f}_j^{*n}) -$$

$$- \int_{-\infty}^{\infty} dx \; [w(x, 0)\, \varphi(x)] = 0(\Delta w) \qquad (5.5.2)$$

and as $\Delta x, \Delta t \to 0$ in all points of continuity of f, $f^*(f^n_{j-i}, \ldots, f^n_{j+i})$ at $(x, n\,\Delta t)$ tends boundedly to $f^*(f, \ldots, f) = f$ at (x, t) while $g^*(f_{j-i}, \ldots, f_{j+i})$ at $n\,\Delta t$ and $(n+1)\,\Delta t$ tend boundedly to $g^*(f, \ldots, f) = g(f)$ at (x, t). Hence the limit of Equation (5.5.2) is the integral relation (5.3.3).

That Equation (5.5.1) provides the discontinuous solution (5.3.6) in the same well-behaved limit, $\Delta x, \Delta t \to 0$, follows easily when g^* is a linear function of the g (i.e. $g_j^* = \sum_{k=j-i}^{j+i} \alpha_k g_k$, so that $\sum_{k=j-i}^{j+i} \alpha_k = 1$) and f^* is a linear function of the f (i.e. $f_j^* = \sum_{k=j-i}^{j+i} \beta_k f_k$ so that $\sum_{k=j-i}^{j+i} \beta_k = 1$), as by using the summation technique described in section 4.6 and the transformation of section 1.4. (See also Hamming, 1977). (When the $g^* = g^*(g)$ and $f^* = f^*(f)$ are non-linear relations, the situation becomes considerably more complicated: *see*, for example, Rosinger, 1982.

This last theorem provides the assurance that if the solutions of the difference scheme (5.5.1) converge at all to some solution $f = f(x, t)$ as $\Delta x, \Delta t \to 0$, then that solution f is a weak solution of Equation (5.3.1). Before continuing to the study of means of obtaining convergence and the more general 'information . . . transmission' properties of schemes, however, some notions of functional analysis now have to be introduced.

5.6 Some elements of functional analysis

Once again, there is space in this introductory hydraulics text for only a cursory glance at a very extensive, general and powerful theory. A more extensive knowledge of this theory is, in fact, essential if one is to understand the limitations of the present methods and the possible ways to improve upon them, but this knowledge appertains more to mathematical modelling in general and can better find its place in a more general text. Only a few key definitions and results can be presented here, directed to a few immediate applications.

The notion of a *set*, composed of a number of *elements*, has already been used quite extensively in the earlier chapter. It is so general and intuitive that it is left undefined. Any collection of elements selected from out of a set is called a *subset*. When the subset is not the whole set it is called a *proper subset*. One of the first problems of set theory is to define the relation of 'equal numbers of elements' between two sets. This is done by introducing the notion of *equivalence*: two sets are said to be *equivalent* if it is possible to establish a correspondence between the elements of the sets by assigning to each element of one of the sets one and only one element of the other set, and conversely. This definition has the property that, while it corresponds to the notion of 'same number' for finite sets, it can be extended to infinite sets.

Among all possible infinite sets, the simplest is the set of natural numbers

$$1 \quad 2 \quad 3 \quad 4 \quad 5 \ldots$$

Any set that is equivalent to this prototype set is called a *denumerable* (or enumerable or *countable*) set. For example, the set of integers and the set of numbers 2, 4, 8, . . . 2^n, . . . are denumerable sets, since the following correspondences can be made:

$$0 \quad -1 \quad 1 \quad -2 \quad 2 \ldots$$

$$1 \quad 2 \quad 3 \quad 4 \quad 5 \ldots$$

$$2 \quad 4 \quad 8, \ldots 2^n, \ldots$$

$$1 \quad 2 \quad 3, \ldots n, \ldots$$

The essential property of an infinite set then appears, that it is equivalent to a proper subset of itself (Dedekind, 1872).

Any function that associates a real number s_n with each integer $n \geqslant 1$ is called a *sequence*.

The central theorem of computational hydraulics is just the following central theorem of set theory (e.g. Cantor, 1895–1897; Kamke, 1950; Kolmogorov and Fomin, 1957; Weir, 1973):

Theorem III *The set of real numbers in the closed interval (0, 1) is non-denumerable* This may be shown (Weir, 1973) by letting $\{s_n\}$ be an arbitrary sequence of real numbers in the open interval (0, 1). The proof consists in showing that there is at least one real number in (0, 1) which does *not* appear as one of these s_n values. Now the s_n values can be expressed without ambiguity as non-terminating decimals; for example, choosing 0.1999 . . . to represent $\frac{1}{5}$ rather than 0.2. Let

$$s_1 = 0 \,.\, a_{11}a_{12}a_{13} \ldots$$

$$s_2 = 0 \,.\, a_{21}a_{22}a_{23} \ldots$$

$$s_3 = 0 \,.\, a_{31}a_{32}a_{33} \ldots$$

If $a_{nn} \neq 1$, let $b_n = 1$ and if $a_{nn} = 1$, let $b_n = 2$. This defines b_n for $n \geqslant 1$. Now the non-terminating decimal expansion $0 \,.\, b_1b_2b_3 \quad . \quad . \quad .$ converges to a real number b in (0, 1) which differs from that of s_n in the nth place, by virtue of the construction. By setting up a centred projection it is easily seen that the set of all points on an arbitrary segment (a, b) are equivalent to the set of all points on the segment (0, 1). It may as well be shown that the set of all points in the plane, in space, on the surface of a sphere, in the interior of a sphere and so on, are all equivalent to this same prototype set. In the same way as the set of natural numbers forms the prototype for the countable sets, so the set of all real numbers in the closed interval (0, 1) forms the prototype of sets that are said to be *of the power of the continuum.* Thus the mathematical analysis that is used in classical hydraulics and hydrodynamics is defined upon sets of the power of the continuum while the numerical analysis that is used in computational hydraulics is defined upon countable sets, and these two distinct sets of definition of the continuum and discrete approach are not equivalent. (See also Manin, 1977.)

Sets may be subject to operations, such as addition of elements of two sets in some order, multiplication of all their elements by a common scalar, convolution of one against the other and so on. A class of sets may then be defined that are additive in some sense and which can be multiplied by scalars. More precisely, a *linear space* is defined as a set R of elements $f, g, \ldots$ with the following properties:

(1) *Addition:* For any two elements $f, g \in R$ (read as 'f and g elements of R'), there exists a unique element $h = f + g$ called their sum such that

(a) $f + g = g + f$ (commutativity);
(b) $f + (g_1 + g_2) = (f + g_1) + g_2$ (distributivity);
(c) there exists an element 0 such that $f + 0 = f$ for all $f \in R$;
(d) for every $f \in R$ there exists an element $-f$ such that $f + (-f) = 0$

(2) *Multiplication by scalars:* For any arbitrary number α and element $f \in R$ there is defined an element αf such that

(a) $\alpha(\beta f) = (\alpha\beta)f$
(b) there exists an element 1 such that $1 \cdot f = f$

(3) *Combined addition and multiplication by scalars:*

(a) $(\alpha + \beta)f = \alpha f + \beta f$
(b) $\alpha(f + g) = \alpha f + \alpha g$

It is easily seen that physical Euclidean space is a linear space of the power of the continuum. Consider the countable (even finite) set of dependent variables that describe the state of a fluid system at any time, such as the vector $\{h_j^n; u_j^n; j = 0, 1, \ldots jj\}$. Then it is easily verified that the set of all possible vectors of this configuration is also an example of a linear space. The last space is sometimes called a *configuration space* or even a *phase space*. It should be remarked that among the possible vectors are those containing negative h values, in order that expressions such as $h_j^{n+1} - h_j^n = h_j^{n+1} + (-h_j^n)$ can be accommodated conceptually.

As a computation progresses from time level to time level, so it generates a sequence of points in the configuration space, and in the limit, $\Delta t \to 0$, it generates a line in the space. The sequence of points or its limiting line are called *trajectories* of the computation. It is possible to describe many interesting properties of solutions in terms of their trajectories in phase space or configuration space. For example, the principle of contraction mapping (e.g. Kolmogorov and Fomin, 1957; Liusternik and Sobolev, 1961) provides an elegant theory of numerical stability in these spaces. On the other side, it is possible to develop theorems on the divergence of systems of trajectories in phase space, and especially to show that linearized equations of the type of (2.6.15), (2.6.16) generate a 'flow' of trajectories that is non-diverging (e.g. Welander, 1955).

In order that one can introduce limit processes into a linear space, it is usual to introduce a *norm* into the space, this being a real non-negative number written $\|f\|$, defined for all vectors $f \in R$ and satisfying the following conditions:

$$\|f\| = 0 \quad \text{if and only if} \quad f \equiv 0$$

$$\|\alpha f\| = |\alpha|\,\|f\|$$

$$\|f + g\| < \|f\| + \|g\| \quad \text{(the 'triangle inequality')}$$

The most common norms of computational hydraulics are defined in terms of the total system mass $\left(\text{e.g. } \sum_{j=C}^{jj} \rho_j h_j \Delta s\right)$, the total system momentum $\left(\text{e.g. } \sum_{j=0}^{jj} \rho_j u_j h_j \Delta s\right)$ and the total system energy $\left(\text{e.g. } \sum_{j=0}^{jj} \rho_j h_j (u_j^2/2 + g h_j/2)\ \Delta s\right)$.
Obviously more 'accurate' measures of total mass, momentum and energy can be used, corresponding to higher-order interpolations, and it may be more consistent to use these when using high-order-accurate difference schemes.

Now introducing limit processes in such a space, using $\|f - f_i\| \to 0$ as a sequence of element vectors $\{f_i\}$ converges towards some limit vector f (in which case the sequence $\{f_i\}$ is said to be a *Cauchy sequence* or *fundamental sequence*), it may happen that all possible limits are also elements of the space or it may happen that some limits are not elements of the space. For example, all limits of a finite configuration space of vectors $\{h_j; u_j; j = 0, 1, \ldots jj\}$ are elements of the configuration space while, on the other hand, limits of the space of all continuous functions may very well have limits, in some norms, that are discontinuous functions. In the event that all converging sequences in a space have limits within that space, the space is said to be *complete*. In the contrary case, that some limits are not elements of the space, the space is said to be incomplete. A complete normed linear space is called a *Banach space*. Many of the processes of theoretical numerical analysis are most conveniently described in terms of Banach spaces (e.g. Richtmyer and Morton, 1967).

The elements of a Banach space may as well be continuous functions as finite vectors in the familiar sense. In either event the elements can still be conveniently referred to as 'vectors'. Within such a space it is then meaningful to define a *scalar product* or *inner product* of two vectors f and g over a given support, written as (f, g) and satisfying the following conditions:

(a) $(f, g) = (g, f)$
(b) $(f_1 + f_2, g) = (f_1, g) + (f_2, g)$
(c) $(\alpha f, g) = \alpha(f, g)$
(d) $(f, f) \geqslant 0, \quad \text{if } f \neq 0.$

The scalar product of two functions $f = f(x)$ and $g = g(x)$ is usually defined through $(f, g) = \int f(x) g(x)\, dx$ and the scalar product of two finite vectors f and g through $(f, g) = \sum_{ji} f_j g_j$. It is then seen, in passing, that the norm $\|f\|$ can be defined through $\|f\|^2 = (f, f)$, which provides the explicit definitions of the more common norms used in nearly horizontal flow hydraulics:

$$\|f\|^2 = \sum_{j=0}^{jj} \rho_j h_j \, \Delta s \qquad \textbf{total system mass}$$

or

$$\|f\|^2 = \sum_{j=0}^{jj} \rho_j (u_j h_j) \, \Delta s \qquad \textbf{total system momentum}$$

or

$$\|f\|^2 = \sum_{j=0}^{jj} \rho_j h_j (u_j^2/2 + g h_j/2) \, \Delta s \quad \textbf{total system energy}$$

Two vectors f and g are further said to be *orthogonal* if $(f, g) = 0$. For example, the functions $f_n = \exp inx$ and $f_m = \exp imx$ are orthogonal so long as $n \neq m$ when the norm is defined over the range $R = [0, 2\pi]$ for complex numbers. In fact

$$(f_n, f_m) = \int_0^{2\pi} (\exp inx)(\exp - imx)\, dx = 0, \quad n \neq m$$

$$= 2\pi, \quad n = m$$

In this example, dividing by $\sqrt{2\pi}$ throughout the elements of the set, so as to use the *basis vectors* $\{(\exp inx)/\sqrt{2\pi}; n = 0, 1, \ldots jj\}$ one could get

$$(f, g) = \delta_{nm}$$

where δ_{nm} is the Kronecker delta which takes the value one for $n = m$ and the value zero for $n \neq m$. Consider now any system of non-zero vectors $\{e_1, e_2, \ldots e_n\}$ of a normed linear space. They are said to form an *orthogonal basis* if they are pair-wise orthogonal, and an orthonormal basis if, in addition, each has a unit length. That is to say, the $\{e_j\}$ form an orthonormal basis if $(e_i, e_j) = \delta_{ij}$, for every i, j.

Now in normal kk-dimensional Euclidean space R, any vector $f \in R$ can be expressed in terms of the orthogonal normalized basis $(e_1, e_2, \ldots e_{kk})$ of 'unit vectors' in the form

$$f = \sum_{k=1}^{kk} C_k e_k$$

where, then,

$$C_k = (f, e_k)$$

through the usual notion of the scalar product between any two vectors as the product of the magnitude of the one and the magnitude of the projection of the other orthogonally onto it. Extending this, at first by analogy, to any vectors, including functions, consider the series $\Sigma C_k e_k$ defined on an orthonormal function basis $\{e_k\}$ and pose the following *approximation problem*:

For a given value of kk, select the coefficients $\{C_k, k = 1, 2, \ldots kk\}$ in such a way that the 'distance' between f and the sum $S_{kk} = \sum_{k=1}^{kk} C_k e_k$, defined by $\|f - S_{kk}\|$, be as small as possible.

By analogy with the finite case, set numbers $\xi_k = (f, e_k)$. Then

$$\|f - S_{kk}\|^2 = \left(f - \sum_{k=1}^{kk} C_k e_k, f - \sum_{k=1}^{kk} C_k e_k\right)$$

$$= (f, f) - 2\left(f, \sum_{k=1}^{kk} C_k e_k\right) + \left(\sum_{k=1}^{kk} C_k e_k, \sum_{k=1}^{kk} C_k e_k\right)$$

$$= \|f\|^2 - 2\sum_{k=1}^{kk} C_k \xi_k + \sum_{k=1}^{kk} C_k^2$$

$$= \|f\|^2 - \sum_{k=1}^{kk} \xi_k^2 + \sum_{k=1}^{kk} (C_k - \xi_k)^2$$

Clearly this expression takes its minimum value when the last term is equal to zero, i.e. for

$$C_k = \xi_k, \quad k = 1, 2, \ldots kk$$

In this case

$$\|f - S_{kk}\|^2 = \|f\|^2 - \sum_{k=1}^{n} \xi^2 \tag{5.6.1}$$

The numbers $\xi_k = (f, e_k)$ are called the *Fourier coefficients* of the function f on the orthonormal system $\{e_k\}$ and the series

$$\sum_{k=1}^{\infty} \xi_k e_k$$

is called the *Fourier series* of the function f on $\{e_k\}$.

Since $\|f - S_{kk}\|^2$ is necessarily non-negative, it follows from Equation (5.6.1) that

$$\sum_{k=1}^{kk} \xi_k^2 \leqslant \|f\|^2$$

Since kk is arbitrary and the right-hand side is no function of kk, this can be extended to

$$\sum_{k=1}^{\infty} \xi_k^2 \leqslant \|f\|^2$$

This is called *Bessel's inequality*.

It may be shown (e.g. Kolmogorov and Fomin, 1961, pp. 112ff.) that, for Banach spaces of the type used in computational hydraulics, the inequality is in fact reduced to an equality as $kk \to \infty$ so that

$$\sum_{k=1}^{\infty} \xi_k^2 = \|f\|^2 \tag{5.6.2}$$

This is called *Parseval's equation*. It states, in words, that the sum of the squares of the Fourier coefficients is equal to the square of the norm of the function f. It clearly underlies all the stability considerations of Chapter 4. In particular, referring to the proof of stability used in Section 4.7, where h was taken as a variation of the average depth, it is seen that the Fourier decomposition can be regarded as a superposition of elementary flows corresponding to the basis function forms as schematized in Fig. 3.44, and Parseval's equation can then be regarded simply as a statement of the principle of superposition of component energies to form a total flow energy. The analysis of Section 4.7 then showed that, in the *linear analysis at least*, the scheme considered would not only conserve the totals of mass and momentum, upon which quantities it was built, but it would also conserve energy. As shown in Section 5.4, of course, this would not say that energy could not vary from time step to time step in a real, finite computation, but it does suggest that, even then, total system energy would be approximately conserved, albeit incorrectly distributed relative to the exact solutions through the agency of the phase errors.

5.7 Construction of weak solutions by filtering

It is now possible to see how the predictions of Section 5.5 work out in practical computations, by actually constructing a sequence of solutions corresponding to continually reduced values of Δt and Δx. If such a sequence converges to a (necessarily unique) limit, it will constitute a fundamental sequence. A process for calculating the behaviour of the limit of calculations through the construction of elements of such a fundamental sequence provide a weak solution. In the case o: discontinuous flows, the problem has two parts:

(a) to determine a norm that will resolve the approach of the elements of a fundamental sequence of continuous solutions to the discontinuous solution;
(b) to construct a sequence of solutions that actually is fundamental in this norm.

The problem was solved by Abbott, Marshall and Ohno (1969) for the fixed points of the Lax and Lax–Wendroff schemes of Section 4.6. (By a 'fixed point' under a mapping A is meant that point (vector) $\bar{x}$ such that $\bar{x} = A\bar{x}$. Physically, these are the set of all stationary hydraulic jumps.) They introduced the norm

$$\|f\| = \int g(f)\, \mathrm{d}x$$

and gave the following argument, here restricted to the mass equation of the Lax scheme.

1. The Taylor-series expansion of the Lax scheme for a steady state is

$$\sum_{n=1}^{\infty} \left\{ \frac{\mathrm{d}^{2n} f}{\mathrm{d}x^{2n}} \cdot \frac{\Delta x^{2n}}{2n!\,\Delta t} + \frac{\mathrm{d}^{2n-1} g(f)}{\mathrm{d}x^{2n-1}} \cdot \frac{\Delta x^{2n-2}}{(2n-1)!} \right\} = 0$$

This can be transformed to the variable $\eta = (x - x_0)/\Delta t$ and multiplied through by Δt to obtain the (presumably convergent) Taylor series

$$\sum_{n=1}^{\infty} \left\{ \frac{\mathrm{d}^{2n} f}{\mathrm{d}\eta^{2n}} \cdot \frac{\rho^{2n}}{2n!} + \frac{\mathrm{d}^{2n-1} g(f)}{\mathrm{d}\eta^{2n-1}} \cdot \frac{\rho^{2n-2}}{(2n-1)!} \right\} = 0$$

where $\rho = \Delta t/\Delta x$ is then the only parameter. Thus, in the variable η there is a unique function $\{f, g(f)\}$ for given ρ and given boundary conditions.

2. The sum is taken over the difference scheme from the boundary at 0 to the inflexion point in h at $J\,\Delta x$, to obtain

$$\left[\frac{(uh)_J + (uh)_{J-1}}{2} - \frac{(uh)_1 + (uh)_0}{2}\right] \doteq \frac{\Delta x}{\Delta t}\,[h_J - h_{J-1}]$$

It follows from (1) however that $(h_J - h_{J-1})$ and $[(uh)_J - (uh)_{J-1}]$ will depend on the ratio $\Delta t/\Delta x = \rho$ so that when this is fixed, $(h_J - h_{J-1})$ and $[(uh)_J - (uh)_{J-1}]$ are constant for all $\Delta t(\Delta x)$. Thus $(uh)_J$ retains a constant finite value. A numerical demonstration of this argument is shown in Fig. 5.2. Now in the step function limit, (uh) takes one constant value everywhere except at the step co-ordinate x_0. Since at this step it retains a finite magnitude, the integral of the variation from the boundary values of (uh) will be zero. Thus if a sequence of such integrals converges to zero, it is justified to suppose that the sequence of solutions converges to the true (discontinuous) weak solution. Figure 5.3 shows that this is in fact realized with the Lax scheme. (See also Shokin, 1979).

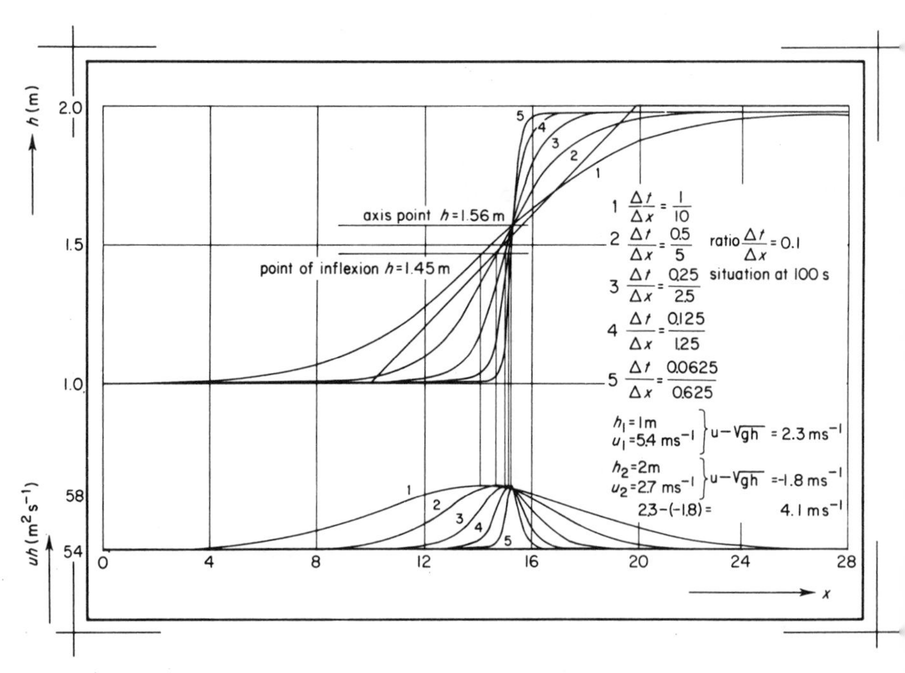

Fig. 5.2. Representation of a filter construction in which the approach to a step function limit of a level function as $\Delta t \to 0$ with $\Delta t/\Delta x$ fixed is characterized by the condition that the area under the corresponding flux density function should tend to zero. The value of $(uh)_J$, J being the address of the inflexion point in h, is seen to remain constant as $\Delta t \to 0$ (curve 1 has the largest value of Δt, curve 5 the smallest)

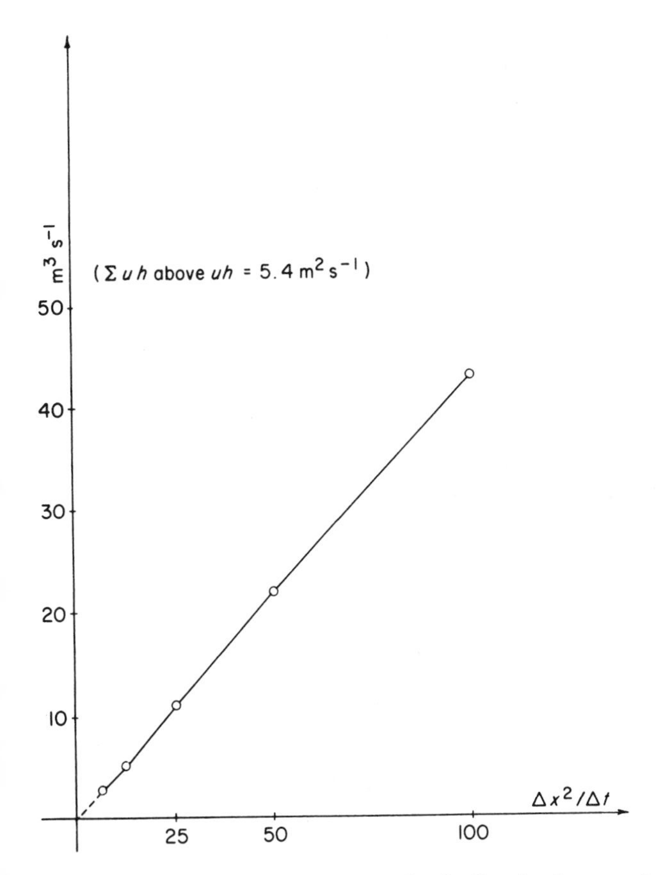

Fig. 5.3. The approach of the area under the flux density curve to zero as $\Delta t \to 0$ with $\Delta t/\Delta x$ fixed

5.8 Applications of the energy norm

From Section 5.5 it appeared that it would be desirable to use the mass level and the momentum level as dependent variables in a wide range of applications. By using a conservation form of difference method, the total system mass and total system momentum could be conserved throughout a calculation. In this case the only function of the mass and momentum norms is to ensure that the conservation is properly maintained. Much more interesting in this case is the behaviour of the energy norm, and it is this that will be exclusively considered in the sequel. Thus, in one dimension,

$$\|f\|^2 = \sum_j \left[\left(\left(\frac{h}{2}\right)^{\frac{1}{2}} u_j\right)^2 + \left(\left(\frac{g}{2}\right)^{\frac{1}{2}} h_j\right)^2\right]\Delta x \tag{5.8.1}$$

A dissipative interface on f, $f^* = f^*(f)$, is then defined as any transformation of f over some range $2i + 1$, $f^* = f^*(f_{j-i}, \ldots f_j, \ldots f_{j+i})$, satisfying the generalized consistency condition that f^* reduced to f when all $2i + 1$ arguments of f^* are identical (Hamming, 1977) and further such that

$$\|f^*\| < \|f\| \tag{5.8.2}$$

The centred, forward and backward linear dissipative interfaces given in Section 4.6 are seen to satisfy this condition, but obviously the range of dissipative interfaces is much wider than these linear three-element prototypes. For example, the 'pseudo-viscosity' introduced into the Euler equation of gas dynamics by von Neumann and Richtmyer (1950) and adapted to hydraulics by Preissmann and Cunge (1961) used an interface of the form

$$u_j^* = u_j - \frac{2\eta^2 \Delta t}{\Delta x}(u_{j+i} - u_{j-1})(u_{j+1} - 2u_j + u_{j-1})$$

where η is a pseudo-viscosity parameter. This is seen to be more selective than the interfaces of Section 4.6 in that the extent of the dissipation is accentuated as the slope itself increases (*see*, for example, the note of Richtmyer and Morton, 1967, p. 312). Like most such non-linear schemes, however, it may influence the conservation of momentum. The selectivity can be increased still further by weighting the linear dissipative interfaces with higher powers of local first and second differences. As the exponent of the weighting increases, so the weight of the greatest differences is emphasized against all other such differences and in the limit, as the exponent tends to infinity with f_j suitably bounded, these interfaces reduce to:

$$f_j^* = \begin{cases} f_j + \beta(f_{j+1} - 2f_j + f_{j-1}); \text{ for } j \text{ such that } |e_{j+1} - 2e_j - e_{k-1}| \\ \qquad \geqslant |e_{k+1} - 2e_k + e_{k-1}| \text{ for every } k \in K \\ f_j \qquad ; \text{for all other } j \end{cases}$$

where β = constant, K = the set of integers 0, 1, 2, . . . , jj and here e is any element of f (see the description of the 'Polya algorithm', by Rice, 1964).

Now the generation of weak solutions through the use of dissipative schemes, so easily realized through the use of dissipative interfaces, leads to the following paradox: that the solution of a difference scheme may depart from the solution of the differential equation with which it is consistent in such a way that it becomes more realistic physically even though the physical processes thereby simulated occur at scales far below the resolving power even of the differential equation with which the difference scheme is consistent.

Thus, for example, a difference scheme that is generalized in a way which is consistent with the equations of nearly horizontal flow may provide the physically correct energy loss to the nearly horizontal flow at a jump, even though the nearly horizontal flow differential equations with which it is consistent cannot resolve the undulations and turbulence of the jump.

The paradox is in part resolved by the observation that the difference scheme may express the discrete form of the second law of dynamics, so that it provides that overall balance of velocity change of convected mass against force which holds across any discontinuity as a vector relation, regardless of the inner processes of the fluid. In mathematical terms, to any one partial differential equation, defined at almost every point of the continuum, there corresponds an infinity of consistent difference equations, that may be defined through Taylor-series expansions over this same continuum. These Taylor-series expansions consist of the differential form that is 'approximated', plus 'error terms' consisting of products of higher derivatives with powers of the grid intervals. Then, when the true solution is continuous, the higher derivatives remain bounded as the grid intervals tend to zero, so that the error products themselves tend to zero. The situation is quite different, however, in the case of discontinuous solutions, for then, as the grid becomes progressively finer, the solution defined on this grid is capable of locally resolving larger and larger values of the derivatives. Thus, as a discontinuity forms up at any point, the derivatives at that point are able to increase at

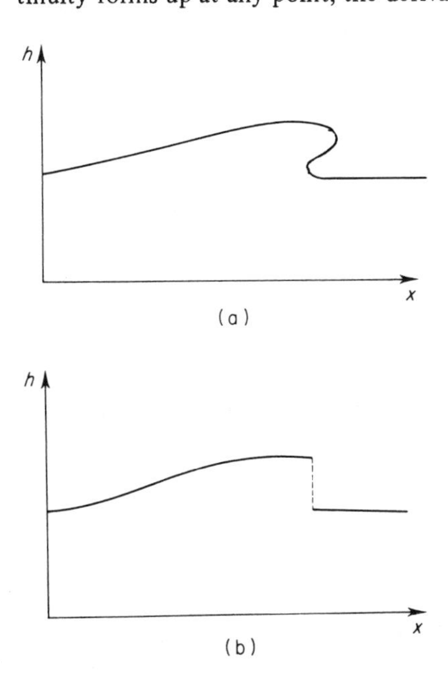

Fig. 5.4. Continuum and discrete solutions for nearly horizontal flows following the formation of a discontinuity. The solution of the differential equation, defined on the continuum, is a multi-valued function with continuous elements (a), while the solutions of the difference equation tend in the limit, as $\Delta t \to 0$ with $\Delta x/\Delta t$ fixed, to a function that is single-valued except at a single discontinuity (b). As the discontinuity of the limit of genuine discrete solutions travels with the celerity of the jump, these discrete solutions are altogether more realistic physically than is the multi-valued solution of the differential equation

the same rate as the grid intervals tend to zero and the error products remain finite.

This process of resolving discontinuities is clearly restricted to difference equations since it is realized through the truncation error of the schemes and need leave no trace upon the differential equations with which the difference equations concerned are consistent.

Thus, as discontinuities form up, the solution of the difference equation is quite different from the solution of the differential equation, and this difference can be maintained even in the limit, as $\Delta x \to 0$ and $\Delta t \to 0$. The solution of the conservation laws on the countably infinite set is entirely different from the solution of these same laws on the continuum. This vital difference is schematized in Fig. 5.4.

The key to the above paradox of resolution, as to the related 'stability paradox' of Richtmyer and Morton (1967) given in Section 4.5 is thus to be found ultimately in Cantor's proof, of 1873, that the set of the power of the continuum is not equivalent to any countable set. At the cost of repetition, it should be emphasized that analysis *per se* deals with operations on the continuum, while numerical analysis deals with operations on countable sets (and usually, for theoretical purposes, on infinite countable sets). Computational hydraulics is concerned with the description of hydraulic phenomena with countable sets (of numbers and operations on numbers).*

Now because of the historical prejudice that the differential equation was

* The tenuous thread that strings together the elements of computational hydraulics, passing from the second law of thermodynamics to Cantor's celebrated theorem, has been traced in this book through an opposing of the discrete view of processes to the more familiar continuum view. It is necessary, however, to distinguish between the method used here to promote understanding and the promotion of any kind of so-called 'world system'. It makes no sense, at least since Kant, to say that 'the world is essentially discontinuous' or 'essentially continuous'. The real world, seen as an interacting assemblage of 'things in themselves', has nothing whatsoever to do with these abstractions, abstractions which only arise in the mind, through the agency of reason. 'Reason only perceives that which it produces after its own design': the discrete view of the world is a way of perceiving the world that presented human-conceptual advantages in the earlier period of the scientific enlightenment and which now presents machine-conceptual advantages in the current period of digital computer application. Even in this current period, moreover, it would be naive to reject the notion of continuity, as by refusing to work with derivatives of functions on the basis that, in the discontinuous picture, these are δ-like and dipole-like and so on, and so are not resolvable in any finite description. Even in applications of hyperbolic equations, the notion of analytic continuation is often useful and productive, as exemplified at various places in this book, while in applications of first order equations and second order equations of parabolic and elliptic type, where certain directions of analytic continuation are explicitly assured, it has an obvious utility. To summarize: the physical world is in itself entirely 'indifferent' to our descriptions, the relative values of these descriptions being only judged in terms of the help they provide to human and 'artificial' intelligence in its various attempts to perceive the world. These more philosophical aspects of the problems of mathematical perception and description were discussed at length by Wittgenstein (1956, 1969).

universally 'more correct' than any consistent difference equation, the departure of the solution of the difference equation from that of the differential equation was ascribed to 'truncation errors'. Since, however, a considerable part of this 'error' may go to making the difference equation more physically realistic than the differential equation, this is a terminology that should be modified, at least for the present purposes. Accordingly the notion of the 'information loss' through a computation will now be introduced, in such a way that when the flow is smooth and continuous this information loss may be related to error in a conventional sense, but that when the flow tends to discontinuity, so that irreversible processes intervene, the information loss becomes a measure of the 'physical loss of information or negentropy'; that is to say, of the actual production of entropy (Shannon and Weaver, 1949; Brillouin, 1956).

It should be remarked that in Shannon's information theory, a 'loss' of information is not allowed and the noise that disturbs Shannon 'signals' in fact increases the information arriving at the receiving end of a channel. However, as there seems little point in pressing analogies with Shannon's theory too far, at least at this early stage (where it is still by no means certain that such an analogy will even be productive), the more popular usage seems preferable (*see* Brillouin, 1956, p. 161).

Several such measures of information loss have been investigated, mainly those based upon Fourier decompositions of the solutions, but the following appears best to satisfy the requirements of such a measure, here defined again for a closed system:

$$L_p^q = \frac{1}{T_p^q} \sum_{n=p}^{q-1} \frac{\left| \|f^{n+1}\|^2 - \|f^n\|^2 \right|}{\|f^{n+1}\|^2 + \|f^n\|^2} \tag{5.8.3}$$

with $p = 1, 2, \ldots ; q = 1, 2, \ldots$, and $p < q$. As before, it is the energy levels that are integrated or summed to give the square of the norm, while the mass and momentum provide the elements of f appearing in the difference equations. The T_p^q is the time elapsing between state $\{f^p\}$ and state $\{f^q\}$ so that L_p^q is a measure of the mean rate of information loss between state $\{f^p\}$ and state $\{f^q\}$.

Clearly

$$L_p^r T_p^r = L_p^q T_p^q + L_q^r T_q^r$$

that is, the actual (meaned) information losses are additive.

For a continuous flow, over fixed time T_p^q, the information loss $L_p^q T_p^q$ tends to zero as the net size is reduced; that is, as $(q-p) \to \infty$.

In fact, over each time interval in which $\|f\|$ varies monotonically,

$$L_p^q T_p^q \to \left| \int_p^q \frac{\|f\| \, \mathrm{d}\|f\|}{\|f\|^2} \right| = \left| \int_p^q \frac{\mathrm{d}\|f\|}{\|f\|} \right| = \left| \ln\left(\frac{\|f^q\|}{\|f^p\|} \right) \right|$$

In a closed system, therefore, the equivalence of differential forms implies that

$$L_p^q T_p^q \rightarrow |\ln 1| = 0$$

This 'filtered' situation corresponds, in a certain sense, to the best possible resolution of the information loss.

In the other extreme, when $q = p + 1$ Equation (5.8.3) gives

$$L_p^{p+1} = \frac{1}{T_p^{p+1}} \frac{|\|f^{p+1}\|^2 - \|f^p\|^2|}{(\|f^{p+1}\|^2 - \|f^p\|^2)}$$

providing a 'worst possible' resolution of the information loss. In this single-step situation

$$0 \leqslant L_p^{p+1} T_p^{p+1} < 1$$

The lower limit, corresponding to $\|f^{p+1}\|^2 = \|f^p\|^2$, may arise from a fortuitous cancellation of variations of the norms of subsystems within the total system or by applications of schemes with zero rounding error to a steady state. The upper limit point corresponds to the transformation of any function-described state to a state where u is zero everywhere and the fluid is supposed spread out in an infinitesimal sheet from $-\infty$ to ∞. (More strictly, of course, this distribution of fluid mass corresponds to the ϵ-distribution, introduced in Section 1.5 and defined there distributionally through

$$\int_{-\infty}^{\infty} \epsilon(x)\, w(x)\, \mathrm{d}x = \overline{w(x)}$$

where $\overline{w(x)}$ is the mean of $w(x)$ between $-\infty$ and ∞ and $w(x)$ is any test function possessing such a mean.)

Consider now the situation where a discontinuity connects two steady states. Then introducing

$$L_n^{n+1} = \frac{1}{T_n^{n+1}} \frac{|\|f^{n+1}\|^2 - \|f^n\|^2|}{(\|f^{n+1}\|^2 + \|f^n\|^2)}$$

provides

$$(\|f^{n+1}\|^2 + \|f^n\|^2) L_n^{n+1} = \frac{|\|f^{n+1}\|^2 - \|f^n\|^2|}{T_n^{n+1}} \tag{5.8.4}$$

Then, if the square of the norm is taken as the total energy level with the elements of f in the difference equations taken as the mass and momentum levels, the term on the right of Equation (5.8.4) is the rate of energy loss. Since in this special case no information is lost outside of the discontinuity, and always supposing that the process can be taken as isothermal, the term on the right-hand side of Equation (5.8.4) is proportional to the rate of entropy production of the system considered. Towards the limit, $T_n^{n+1} = \Delta t \rightarrow \mathrm{d}t$, $\|f^{n+1}\|^2 - \|f^n\|^2 \rightarrow \mathrm{d}\|f\|^2$ and L is proportional to the instantaneous rate of entropy production.

The definition (5.8.4) can be linked to definitions in terms of Fourier coefficients $\{\xi_m\}$ on any orthonormal basis $\{\psi_m\}$, their amplification factors $\{A_m\}$, celerity rations $\{Q_m\}$ and related quantities through Parseval's equation

$$\sum_m \xi_m^2 = \|f\|^2$$

Here m is taken over the infinity of components, or wave numbers, as Δt, Δx become infinitesimal. Various analytical representations of information loss rates are thereby provided, even for the countably infinite case.

The operation of the dissipative interface can also be related to the entropy production concept of classical information theory, through the following analogy. The centred dissipative interface, to take the simplest example,

$$f_j^* = \alpha f_{j+1} + (1-2\alpha)f_j + \alpha f_{j-1} \tag{5.8.5}$$

can be regarded as a modelling, at some other scale, of the random walk

$$p(0, \delta t) = p_{j+1}^j p(\delta x, 0) + p_j^j p(0,0) + p_{j-1}^j p(-\delta x, 0), \quad p_{j+1} = p_{j-1} \tag{5.8.6}$$

where $p(j\,\delta x, n\,\delta t)$ is the probability of an element being at distance $j\,\delta x$ at time $n\,\delta t$ and p_j^k is the probability that the element moves from $j\,\delta x$ to $k\,\delta x$ at each time increment of the walk. The process is assumed to be Markovian and the events mutually exclusive. Then each application of the interface can be viewed as the outcome of multiple applications of a finite scheme:

Scale	Event		
	Element moves from $j+1$ to j	Element remains at j	Element moves from $j-1$ to j
microscopic	p_{j+1}^j	p_j^j	p_{j-1}^j
macroscopic (model)	α	$1-2\alpha$	α

The entropy of this finite scheme (Khinchin, 1957),

$$-(p_{j+1}^j \ln p_{j+1}^j + p_j^j \ln p_j^j + p_{j-1}^j \ln p_{j-1}^j)$$

is modelled by

$$-(\alpha \ln \alpha + (1-2\alpha)\ln(1-2\alpha) + \alpha \ln \alpha)$$

It follows that the application of the centred dissipative interface can be viewed as the addition to the equations of mass and momentum conservation of an equation of diffusion of mass and momentum, in effect as an extra stage in a multi-stage difference scheme.

However, being of the order of the truncation error, this extra stage is introduced instantaneously after every time step, or, mathematically, it is 'δ-like' in time with minimum period Δt. Its effect is a function only of Δx and no function of Δt. Since this effect occurs after every Δt or some multiple of Δt, it can be conceived as being spread out over Δt or the multiple of Δt used in the application. When viewed in this last way, and over Δt, it introduces an effective dispersion coefficient

$$D = \frac{\alpha \Delta x^2}{\Delta t}$$

by Taylor-series expansion of Equation (5.8.5). The unbiased random-walk equation (5.8.6) provides a diffusion equation in much the same way.

Evidently when $f = u$ the form of dispersion equation

$$\text{`}\frac{\partial f}{\partial t} - D\frac{\partial^2 f}{\partial x^2} = 0\text{'}$$

that corresponds to this view of the dissipative interface in a finite scheme itself corresponds to the behaviour of a fluid in which stress is proportional to strain rate; that is, to the 'Newtonian fluid' of hydrodynamics. And, of course, the Newtonian behaviour of a perfect gas is derived in the kinetic theory of gases from just such a random-walk model as Equation (5.8.6) (e.g., Jeans, 1940).

Several other approaches to the modelling of turbulent mass and momentum diffusion are currently being considered, and indeed the subject is still largely at the research stage. Most of these approaches devolve around the introduction of differential equations that describe the turbulent flow process in such a way as to 'close' the total equation system (e.g. Launder and Spalding, 1972; Falconer, 1976; Rodi, 1978). Most such equations describe turbulent diffusion processes. Some attention should, however, be given to a form of diffusion that can be introduced by computational grid-scale processes. As shown in Section 4.5, solutions of the complete equations of nearly horizontal flow, thus including convective terms, demonstrate the breeding of higher-wave-number components. In a discrete, two-dimensional system, with resistance terms proportional to the nth power of the velocity with $n > 1$, this process provides a rotational velocity field in which circulations at one scale break down into circulations at a smaller scale and energy cascades through these successively smaller scales very much in the manner observed in the physical reality. Of course, the breeding process proceeds somewhat differently in the discrete modelling case as compared with the more continuous physical case. For example, in the former case it is arrested at the scale of the discretization, as also described in Section 4.5. However, the grid-scale turbulence arising in the discrete case transfers momentum horizontally through the fluid, and refining the grid, although it resolves further scales in the turbulent cascading process, may otherwise have little influence on the total momentum diffusion process. In terms of a Taylor-series representation of the difference scheme, this means that some terms of the truncation error remain finite as the discretization increments tend to zero, and, in this characterization, such turbulent flows are closely associated with the strongly discontinuous flows discussed earlier in this chapter. In order to illustrate some properties of such grid scale flows, consider a sequence of flows at smaller and smaller scales, of order $\Delta x = \Delta y = \Delta s$, in each of which the flow is confined to

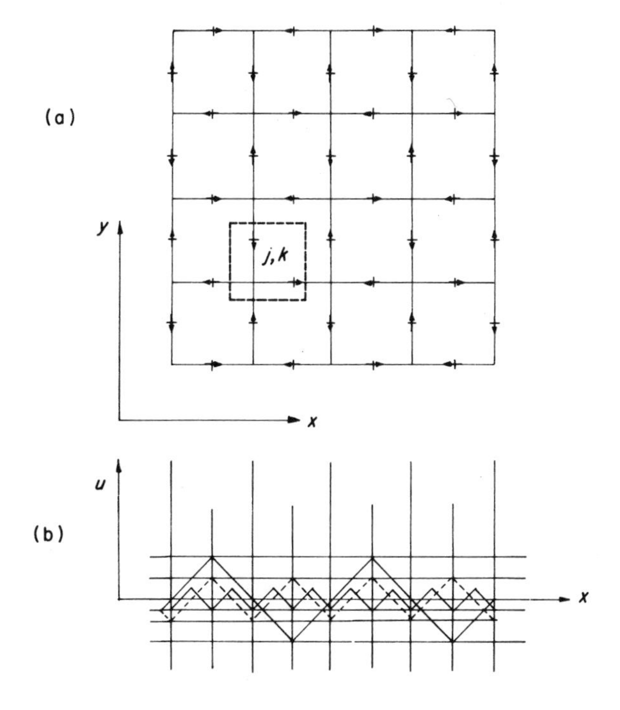

Fig. 5.5. Schematization of an idealized numerical-grid-scale turbulent flow, (a) in plan on x–y, and (b) in elevation, with velocity against x, for a sequence of such flows, with decreasing Δx

the representation of highest resolvable wave number. A typical flow is illustrated in Fig. 5.5a and notated, using the addressing of Fig. 5.5, by

$$u_{j,k} = (-1)^{(j+k)} \Delta u$$
$$v_{j,k} = (-1)^{(j+k+1)} \Delta u$$

This is clearly an entirely legitimate solution for a whole class of difference schemes on (2.6.19) – (2.6.21).

In general the Taylor series for a first difference would appear, formally, as

$$\frac{u_{j+1} - u_j}{\Delta x} = \frac{\partial u}{\partial x} + \left(\frac{\partial^2 u}{\partial x^2}\right)\frac{\Delta x}{2!} + \text{(higher-order terms)}$$
$$= \frac{\partial u}{\partial x} + \left(\frac{u_{j+1} - 2u_j + u_{j-1}}{\Delta x^2}\right)\frac{\Delta x}{2} + \text{(higher-order terms)}$$
$$= \frac{\partial u}{\partial x} + \frac{1}{2}\left(\left(\frac{u_{j+1} - u_j}{\Delta x}\right) - \left(\frac{u_j - u_{j-1}}{\Delta x}\right)\right) + \text{(higher-order terms)}$$

In the case of a function $u = u(x)$ continuous in first derivatives, the leading term of the truncation error tends to zero as Δx tends to zero, but in the type of flow illustrated in Fig. 5.5 this 'truncation error' may well remain finite as Δx tends to zero, even as the variation of u from a uniformly continuous function also tends to zero, as illustrated in Fig. 5.5b for the case where the leading term of the truncation error remains constant.

This type of situation can arise whenever the resolution of derivatives can improve at about the same rate as the increments, that these derivatives multiply in the truncation error, diminish. Thus although a stable solution of a system of difference equations may converge to a limit in a turbulent-resisted, free-surface, two-dimensional flow, this limit may demonstrate a momentum diffusion that does not appear, or cannot be resolved, in the solution of the corresponding system of differential equations. The related situation that arises at the hydraulic jump is used here to demonstrate the differences between the approaches of computational hydraulics and the analytical approaches of classical hydrodynamics, this case being much the most familiar and by far the most thoroughly studied, but some of these differences appear to carry over to turbulent diffusion as well. Thus, just as the difference equations 'provide their own closure' in the case of the hydraulic jump, so they may well do so also in the case of turbulent-diffused flows. Notions similar to these were set out long ago by von Neumann, referring especially to earlier work of Burgers (*see* von Neumann, 1949/1963).

It is, however, difficult even to talk about a truncation error – which presumes differentiability to any order – in the case of flows such as those illustrated in Fig. 5.5. This figure illustrates legitimate solutions of difference equations that are strongly consistent with differential equations such as (2.6.19)–(2.6.21), but it is difficult to regard them, for all Δs, as solutions of the differential equations themselves. Consider, for example, the continuum description of the superposition of all such flows below a certain unit scale. Let $\{x\}$ denote the distance between x and its nearest integer. Then such a superposition can be described in one coordinate direction, say in the x-direction, by a function of the form

$$u = u(x) = u_0 \sum_{n=0}^{\infty} \left(\frac{\{2^n x\}}{2^n} - \frac{1}{2^{n+2}} \right)$$

However, this is an example, classical in the theory of functions and functional analysis, of a continuous function without a derivative (*see* e.g. Riesz and Nagy, 1955).

At a more empirical level, it is possible to introduce just that degree of numerical dissipation which would withdraw energy at a rate corresponding to the dispersion observed in nature. This procedure corresponds to the assumption of a Newtonian fluid; that is, one in which shear stresses across verticals are proportional to strain rates in the horizontal plane. For example, the Kolmogorov similarity hypothesis can be used with Ozmodov's calibration (e.g. Nihoul, 1975) to provide a degree of dissipation in the α-algorithm given over a wide range of scales by Abbott (1976):

$$\alpha \approx \frac{10^{-8/3}\,\Delta t}{\Delta x^{2/3}} \text{ (meter-second units)}$$

It is easily verified that this gives α values one or two orders of magnitude lower than those commonly used (intentionally or otherwise) just to stabilize difference schemes as described in Chapter 4. (*See also* Leonard, 1974; Flokstra, 1977; Leslie and Quarini, 1979).

5.9 The description problem: third remark

It is often said that all hydraulics should be based upon hydrodynamics, so upon differential and other continuum descriptions. If computational hydraulics were to be so based, then it would only have to concern itself with approximating strong solutions of differential equations, as through the notions of consistency, stability and convergence. This computational hydraulic programme now appears to be too restrictive, however. In hydrodynamics one writes differential equations whose range of description is limited, by and large, to well-ordered processes and totally disordered (statistically non-transient, ergodic) states or processes. One then tries to find functions from the set of closed linear closures of polynomials which, when substituted for the dependent variables of the differential equation, make the left hand side of that equation equal to its right hand side, ultimately in the sense of equality of polynomial coefficients. These functions one calls 'solutions' of the differential equation and one claims that they describe a certain physical reality, while they can only possibly be computed, in the last analysis, through the computation of polynomials.

However, it is already observed in computational hydraulics that one can simulate certain phenomena, such as the hydraulic jump, the salt front and turbulent mass and momentum transfers for which there is apparently no differential formulation. Correspondingly, as soon as one actually tries to construct the solutions of differential equations numerically, using the notions of consistency, stability and convergence, one meets certain problems of resolution, of consistency and convergence, that appear to be exceedingly difficult to resolve, if not insurmountable. In certain cases, exemplified above, the solution of the difference scheme is simply not converging to the solution of the differential equation with which it is strongly consistent.

This difference in behaviour is, of course, to some extent predictable in terms of the theory of analytic continuation. To press the matter figuratively, a knowledge of an elevation and all derivatives at a point on a table will give no information at all about the positions of the edges of the table! Analytic continuation, as Hadamard (1923) explained, is a very special and peculiar property of certain differential equations, Taylor series, polynomials and certain of their closed linear closures, but it has no 'physical reality' as such and its relation to countable schemes is at best tenuous and in some cases non-existent.

Moreover, an essential, integrated concern of computational hydraulics is the economic-utilitarian one of 'best description', such as the finding of minimal sets of descriptive numbers providing a given resolution and minimal sets of operations providing a given accuracy. These game-theoretic aspects, that are central to computational hydraulics and that have been exemplified frequently in the preceeding chapters, are again quite foreign to hydrodynamics, or at least to hydrodynamics in its algebraic, non-arithmetic form.

It follows that the programme of computational hydraulics should not be circumscribed by the programme of hydrodynamics, but, even while drawing upon hydrodynamics, it should share the programme of other branches of science and engineering that are currently being rethought from the digital-computational point of view. It should now be clear that this programme must propose certain problems of a fundamental nature quite unrestricted by analytic constraints. For example:

1) To what extent can one describe the world at all with numbers and operations on numbers and, in so far as one can, how can such descriptions be constructed most economically for a given resolution or accuracy?

2) The physical reality is described on the one side by the behaviour of numerical quantities under elementary operations and on the other side by observation-numbers. How are the economy-resolution characteristics of these two sides related?

The *reasons* for this change in emphasis, from the hydrodynamic to the computational hydraulic, have of course to be sought in social-economic needs and possibilities, as introduced in the preface and as exemplified further in Chapter 6.

6 Applications of computational hydraulics

6.1 Introduction

Computational hydraulics, like most branches of science, is a subject of aesthetic productivity and contemplation, which makes it interesting, and an economic benefit as well, which can make it doubly interesting. In this introductory work, the aesthetic aspects have predominated almost entirely so that a final chapter on applications is essential in order to redress some of the balance between the aesthetics and the ethics of the subject.

There is obviously a considerable gap between the fundamental considerations of this introductory work and the design, development and application of a modelling system, but it is precisely this gap that should be closed by the rest of the present book series. The object in this final chapter is only to sketch out the path to the applications and to show how this path is laid over the foundations described in the earlier chapters of this work.

Every mathematical model has to reconcile the interests and demands of its *external* and *internal specifications*, usually within a certain budget frame. The external specification lays down what the model should achieve, in terms of range of applications, accuracy, ease of handling, running speed (turnaround time and machine costs) and other such factors. The external specification is decided on the basis of *market research*, so in terms of what the market will need and can support by the time that the model is available. The internal specification lays down the means available for meeting the requirements of the external specification in terms of basic physical and numerical understanding, available staff, known schemes and algorithms, specific code procedures, service programs and so on, all related to available machine and other facilities. The internal specification is a result of *product research*, which must often precede the confrontation with the reality of the external specification by many years if it is to have full scope for its development.

It follows that every non-trivial mathematical model has a more or less long period of gestation, during which the research of the modelling organization has to proceed in directions that are relevant to its future modelling needs without being too constrained by immediate technical requirements. It is this research that gives mathematical modelling its special commercial character, as one with quite inordinate ratios of research expenditure to total added value, with the corollary, in classical management theory, of incipient commercial instability (e.g. Mahieux, 1972; Abbott, 1979). Typical examples of long-range research are provided by the works of Daubert and Graffe (1967), Cunge (1969) and Abbott and Vium (1977), referred to in the earlier chapters. This research rarely translates into a specific difference scheme or algorithm, but provides more a basic approach that may generate whole families of schemes – or nothing of immediate practical value at all. For example, during the redesign of one system, ten different schemes were investigated and six of these were programmed and test-run (Abbott, 1976). All of these were, of course, linearly stable, so that attention was entirely directed to problems of non-linear stability and accuracy. The last four of the proposed schemes converged to a 'production' scheme, that was then subjected to further continuous testing and development, extending over more than two years, later continued alongside and in conjunction with its application in engineering practice.

6.2 Rig testing and field testing

The testing of a modelling system is first carried out on *test rigs* in which idealized bathymetries are used that isolate or accentuate one or another problem of specific interest. Very many different test rigs are employed and each of these is run over a wide range of flow conditions and discretization conditions during the development process. The test topographies range from simple one-dimensional horizontal bed tests, for amplitude error and phase error testing, through 'L-tests', where the flow is taken around L-bends of different orientations to test the model symmetries, to very special bathymetric situations where specific forms of ill-conditioning can be isolated and corrected. From out of this test program, an operational range of system modelling can be demarcated, relating accuracy to resolution and thence to program running speeds and machine costs. The values of the parameters that delineate the accuracy–operational range of the system models together constitute the *performance envelope* of the system.

Even while the test-rig program is continuing, a first series of tests will usually be made on some typical, but individually disparate, real-world bathymetries, as a guide to further areas of difficulty and so as a basis for further development. This test series constitutes the so-called *field testing* of the system. A standard exhortation of production engineering is 'never test on the client', but it is often difficult to avoid doing this in practice. If the exhortation cannot be heeded, then at least the client should be aware of the trials and tribulations that he may face, so that he can make contingency plans and obtain compensation for his difficult situation. It is far better for the modelling organization, however, if the field testing can continue outside of the strict time schedules and other exigencies of the commercial application – while the last thing the commercial side wants is to be troubled with a string of development problems, each with its trail of expensive re-runs, lost engineer time and schedule slippage.

By the time a modelling system is introduced for commercial exploitation, it should be reliable, flexible and economically viable. It cannot be expected that all problems and difficulties will have been eliminated, of course, but it should always be possible to circumvent these in practice by a trade-off against other facilities or, at worst, by minor changes at the level of the code. Thus, upon its commercial introduction, it should always be possible to keep the system running. Development then proceeds alongside the applications, interacting with the applications so as to improve the performance characteristics, extend the range of the system and make it generally more useful and accessible.

6.3 Examples of test rigs

Some examples may now be given to illustrate this process. As an example of one-dimensional testing for amplitude and phase error, Fig. 6.1a shows simulations of the propagation of a large-amplitude short wave using the formulations of Sections 2.7 and 4.12. The ratio of the amplitude of the wave to the mean water depth is such that cnoidal wave theory has to be used to generate the boundary conditions: a sinusoidal wave disperses under propagation in this case. It is seen that, although the amplitude remains well represented, a phase difference is observed between the wave train generated numerically from the Boussinesq equation (Section 2.7) and the wave train given by cnoidal wave theory. This raises the question of whether the phase shift is due to a disparity between the formulations of the Boussinesq equations and the cnoidal wave equations, or whether it could be due to numerical influences even though these had not manifested themselves in this way in tests with low-amplitude waves. The simple expedient of improving the resolution provides the result shown in Fig. 6.1b, which shows that the discrepancy has a numerical cause and that the performance envelope must be extended by a further dimension in order to account for numerical accuracy variations with ratios of wave height to mean water depth.

The importance of the correct posing of boundary conditions has been repeatedly emphasized in this work. A typical problem of practice is that measurements of water elevations made to investigate a proposed construction, for example, can only be usefully made before the construction is in place and would, in all probability, look very different if they were made after the construction were completed. One way out of this problem is to make the measurements so far away from the proposed construction that its subsequent influence upon them could be expected to be small, and then to model from the measurement area into the construction area numerically. In most cases this gives rise to grossly overdimensioned models and an inordinately expensive measuring program – for the problem has a simple modelling solution. The method of characteristics of Chapter 3 shows that any plane wave directional component crossing a boundary line can be divided further into an incident wave component and a reflected wave

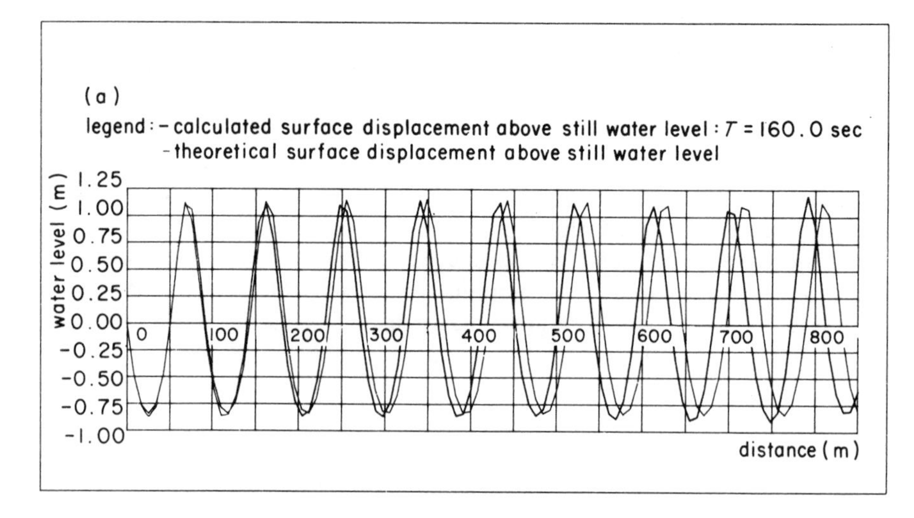

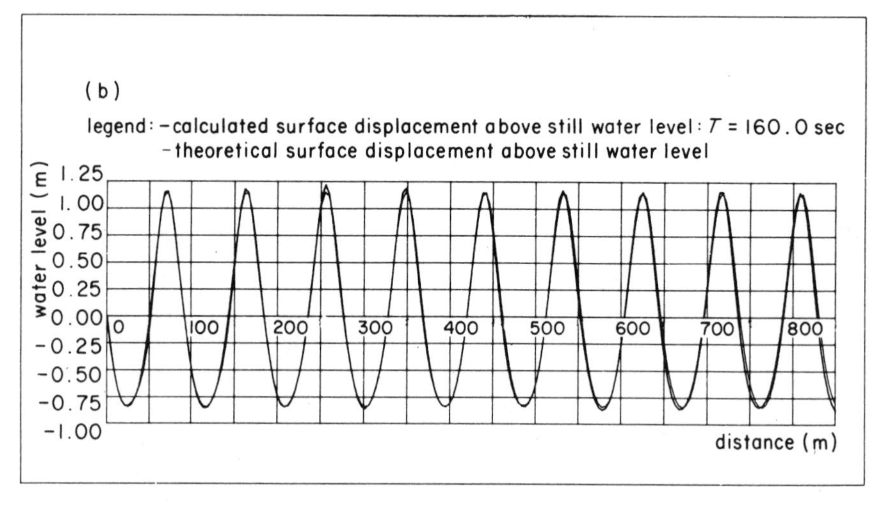

Fig. 6.1. (a) Propagation of a short-wave train with a large ratio of wave amplitude to water depth, showing a phase error that is not observed with small-amplitude waves. (b) The phase error disppears with improved resolution showing that the error is numerical but depends on the ratio of wave amplitude to water depth*

component. The wave component incident upon the construction will be the same both in the measurement situation and in the situation when the construction is completed, but the wave component reflected in the pre-construction measurement situation will differ from that obtaining in the post-construction situation. It is easy, in principle, to use the method of characteristics to separate

* All illustrations in this chapter are by courtesy of the Danish Hydraulic Institute unless otherwise stated.

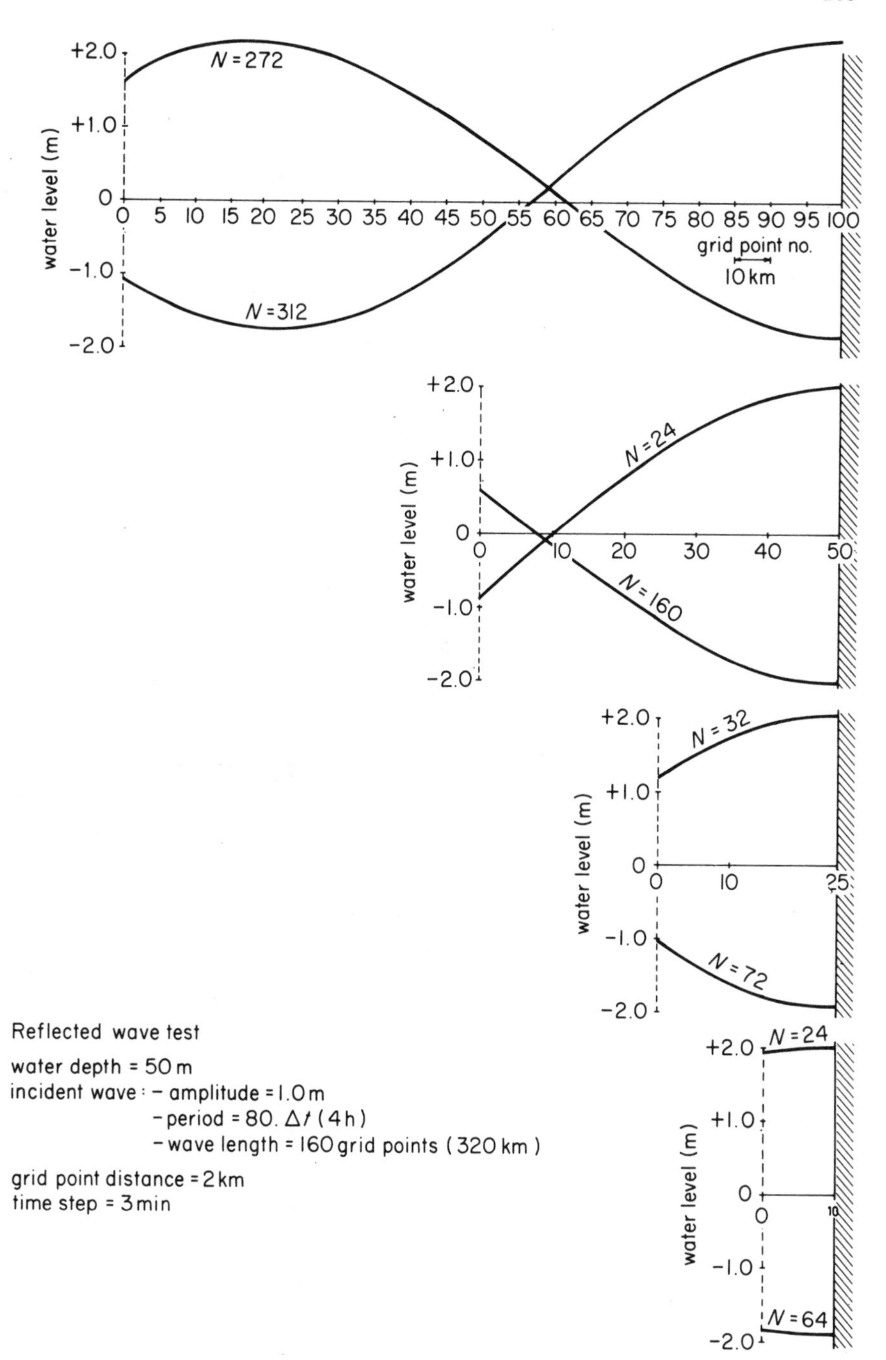

Fig. 6.2. Use of a module that synthesizes boundary data from a given ingoing wave train and the model-generated outgoing wave train so that the model can be cut off at any point without changing conditions within the model area

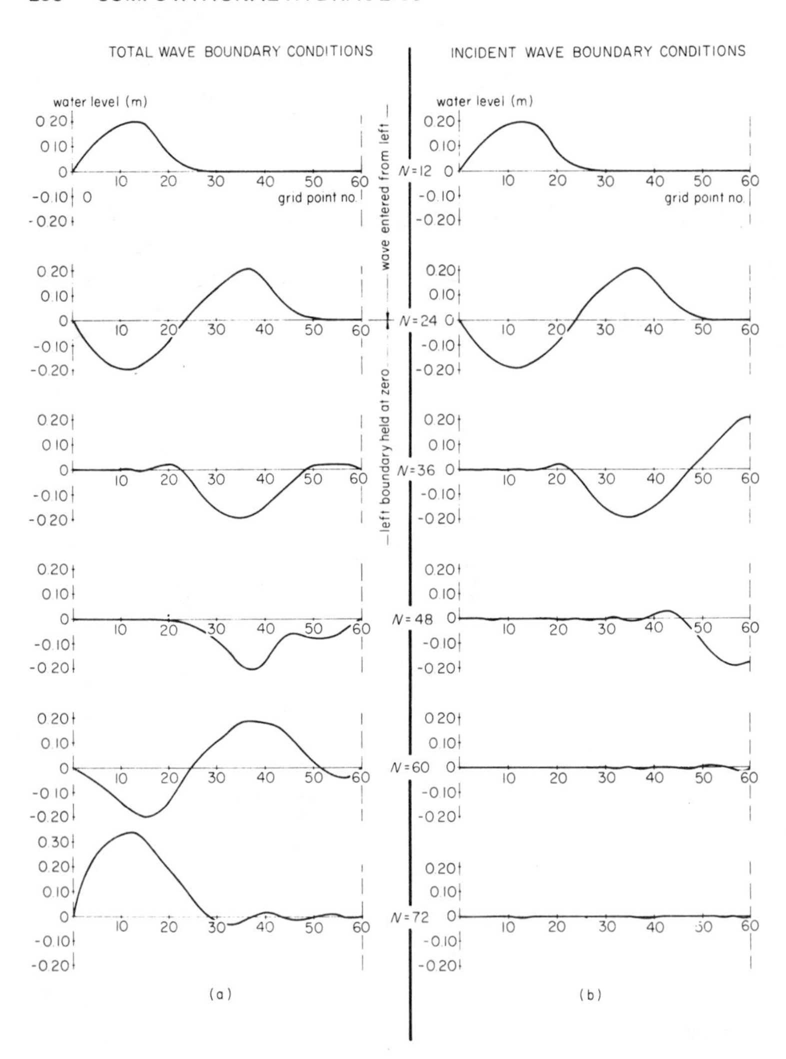

Fig. 6.3. The special boundary module may (a) totally reflect a wave or (b) allow it to pass completely out of the system, or indeed it can provide any degree of reflection between these limits

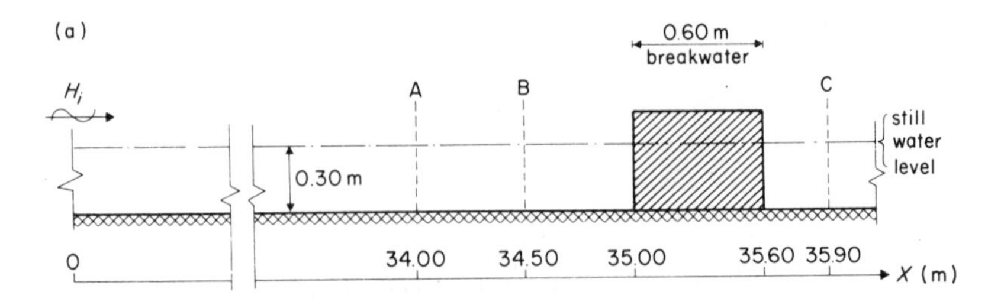

Fig. 6.4a. *(For caption see page 257)*

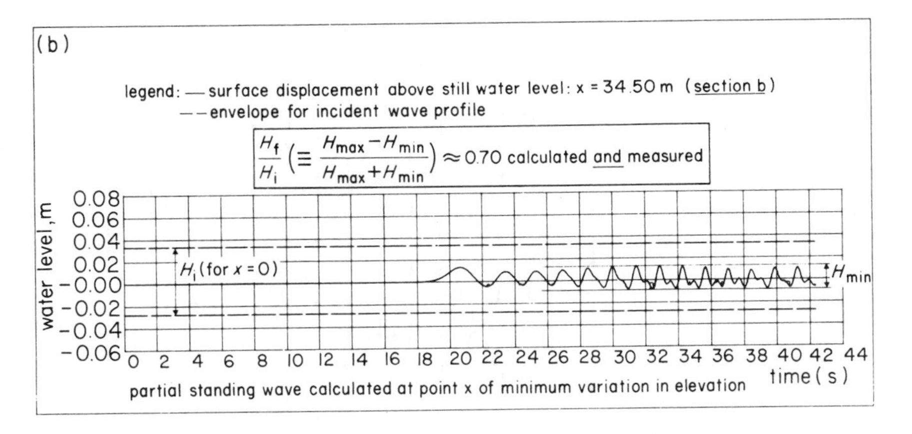

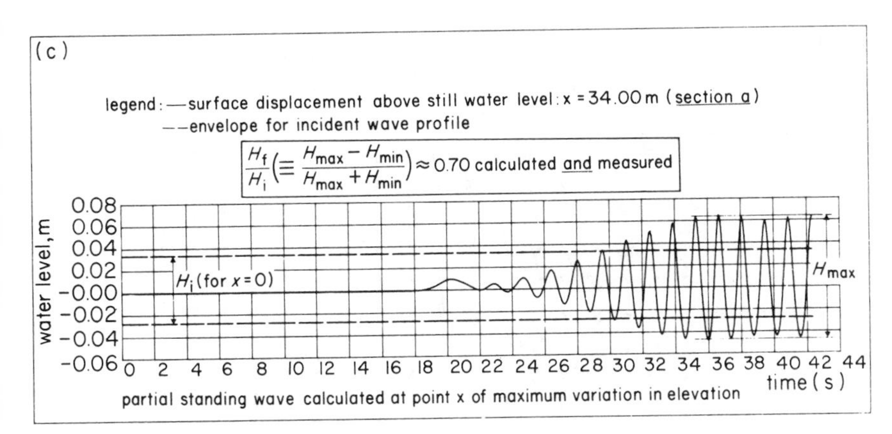

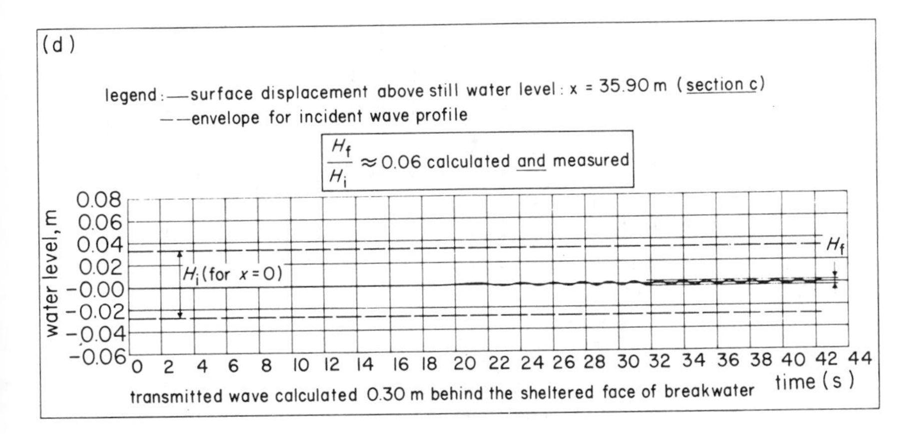

Fig. 6.4. Computation of reflection and transmission of short waves at a permeable breakwater with comparison with laboratory observations. Channel with breakwater; (a) vertical sketch; (b), (c), (d) computed time series (*For (a) see opposite.*)

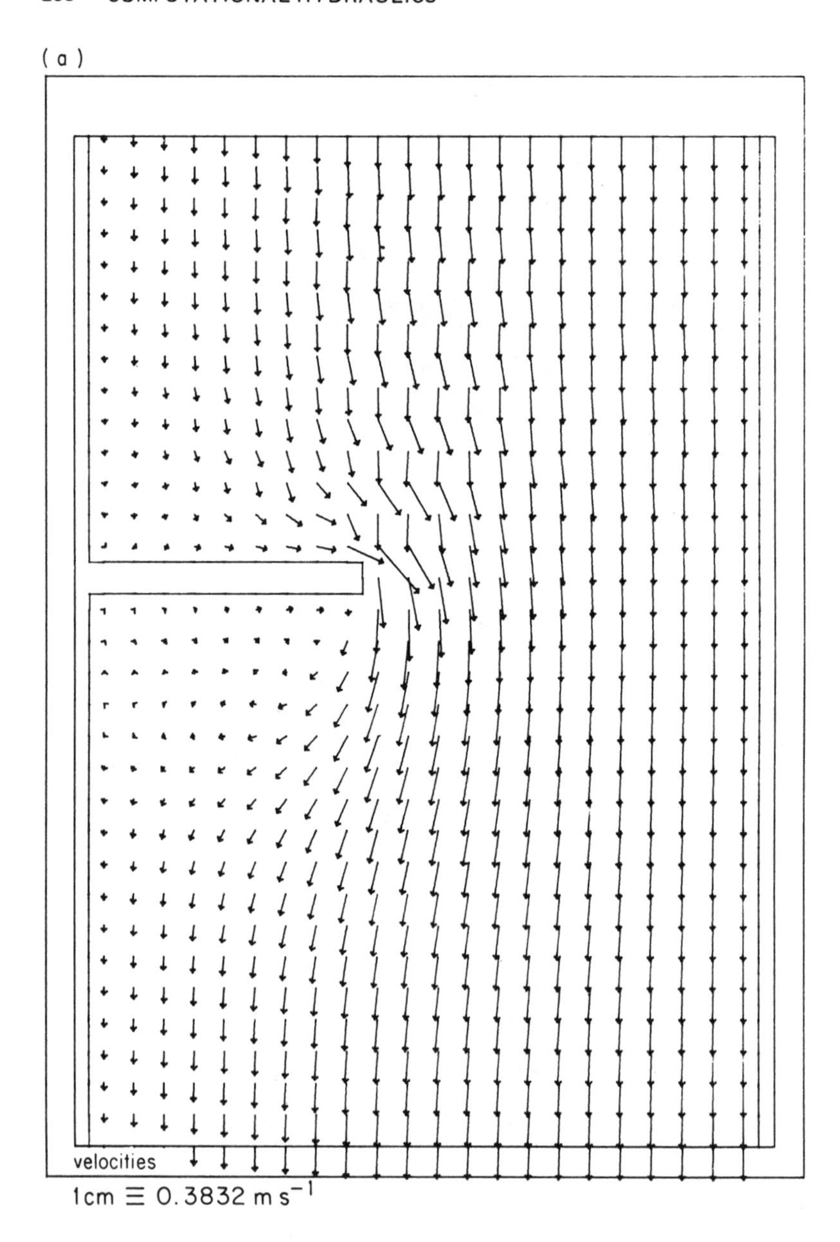

Fig. 6.5. (a) The characteristic 'zig-zagging' of velocity vectors in a second-order-accurate convective momentum approximation is accentuated in the rig shown (*see also* Fig. 4.52). (b) Increasing the accuracy of the approximation removes the zig-zagging (*For* (*b*) *see opposite.*)

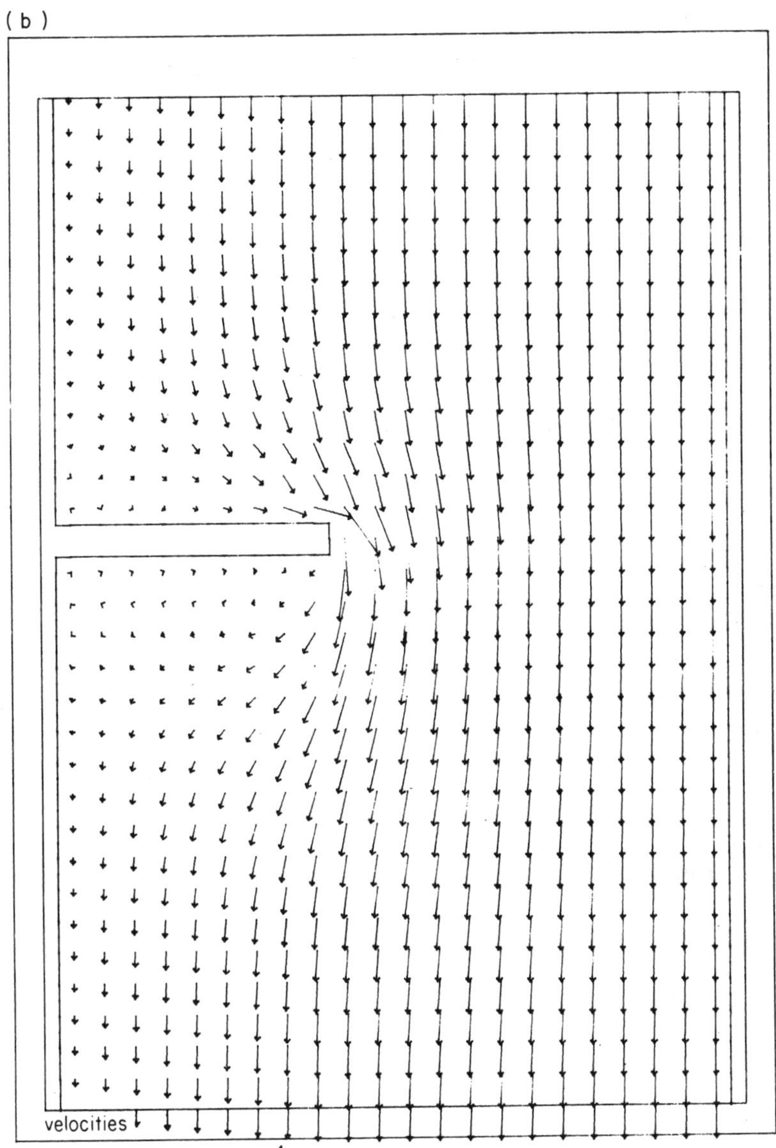

Fig. 6.5. (cont.)

out the two components from the field measurement using a model of the existing situation, and then to rework the model into its post-construction configuration and input the filtered incident wave component while letting the reflected wave component pass straight out through the boundary. The test rig results shown in Fig. 6.2 illustrate how effective this can be: the same incident wave has been used in all the cases shown but the 'radiation boundary module' synthesizes boundary data from this and the reflected wave such that the same conditions prevail within the model, no matter where the boundary is placed. Figure 6.3 shows further test-rig results for such a boundary module, in the one case (6.3a) for total reflection from a boundary and the other (6.3b) for zero reflection, in this last case the wave being allowed to pass out through the boundary. It is easy to generalize the module to provide partial reflections as well, but in practice these reflections are usually associated with reductions in flow areas (as in the case of permeable breakwaters or around ships and caissons) and they can better be directly related to this physical basis. Figure 6.4 shows a test on wave reflection and transmission at a permeable breakwater, in this case intermediate between a rig test and a field test in that comparisons are made with physical model results.

As an example of a development test rig, Fig. 6.5a shows a rig that accentuates convective-momentum-induced 'zig-zagging' in a second-order-accurate scheme while Fig. 6.5b shows how the 'zig-zagging' is eliminated by raising the accuracy of the convective momenta terms. Figure 6.6a shows that, even with this accuracy, a circulation can be induced in a steady-state, Manning-resisted flow even with very minor mass and momentum diffusion terms in the differential formulation, so that simply by the application of Bernoulli's law on closed streamlines with head loss, some numerical momentum diffusion (shear stress) must be present to drive the circulation. Figure 6.6b shows that introducing quite a heavy shear stress changes the flow field relatively little, thus giving a first indication of the order of magnitude of the numerically induced momentum transfer. Figure 6.7 shows the same test conducted with the convective momentum terms suppressed: the circulation quite disappears, indicating that the circulation is indeed driven by computational grid scale momentum transfers generated by the convective terms. Constructing a filter on this solution, at constant Courant number, shows that a fundamental sequence of such solutions tends to a limiting solution, as Δt, $\Delta x \to 0$, differing very little from that of Figure 6.6a, as hypothesized in Section 5.8 (*see also,* Flokstra, 1977).

The theory of information loss introduced in Chapter 5 suggests that boundary data continually 'refresh' the information in a computation (Abbott, 1974), so that, for example, in a region strongly influenced by boundary data one could expect to see very little loss of accuracy so that good agreement with measurements is to be expected in this region. On the other hand, as it was shown in Section 3.9 that a zero-velocity boundary condition is equivalent to a reflection of the whole flow system, a system enclosed within zero-velocity boundaries would receive no refreshment whatsoever from boundary conditions and would lose information indefinitely. This leads to the use of closed-flow systems for the

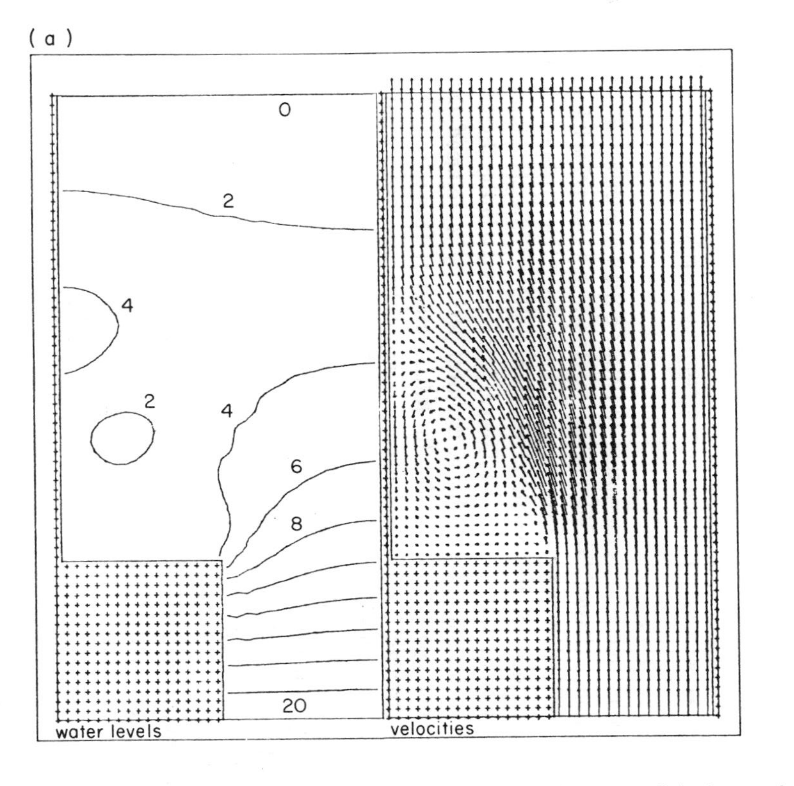

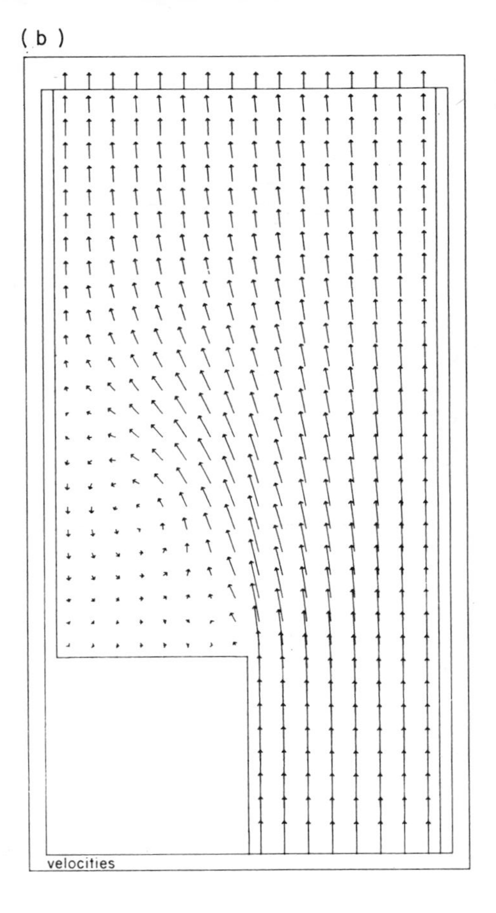

Fig. 6.6. A test rig to study circulation induced by resistance and the interaction of convective momenta, mass and momentum diffusion and truncation errors. Case (a) is without diffusion while in the case (b) a relatively strong diffusion has been introduced.

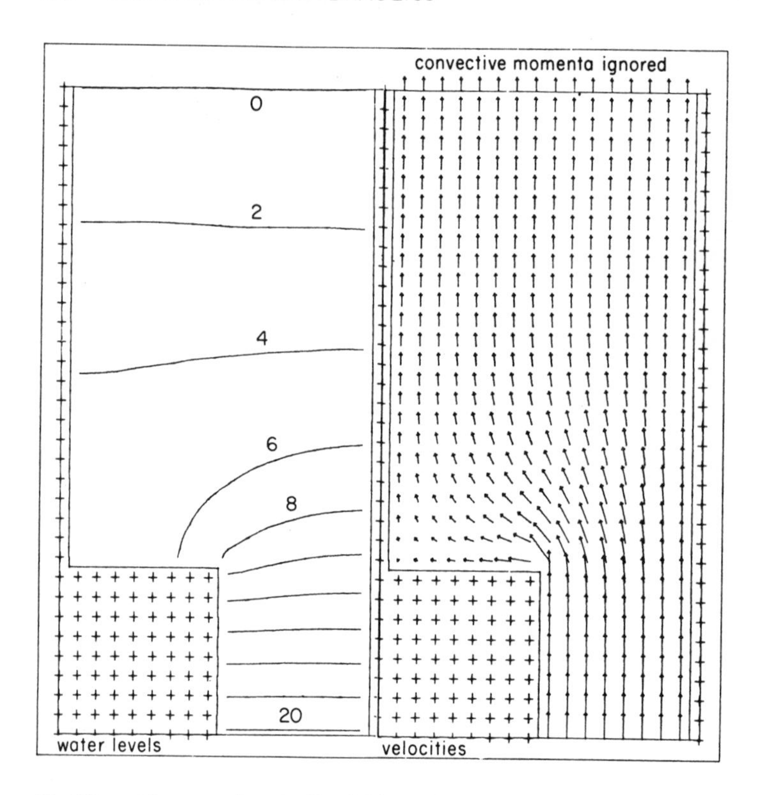

Fig. 6.7. The same rig as in Fig. 6.6 but with the convective momentum terms completely suppressed

strenuous testing of energy conservation and other critical properties of schemes. Figure 6.8 shows a classical 'lock-exchange' test in which a channel is initially divided by a vertical plate into two regions, the one being filled with heavier (salt) water and the other with lighter (fresh) water. At time $t = 0$ the plate is removed and the exchange of water bodies is followed. It is seen that, in the absence of dissipation and resistance, the exchange is almost perfectly reproduced, indicating near-zero energy falsification. With the introduction of dissipative interfaces of increasing strength, however, the main features of the exchange are quickly lost and the flow reduces to a mere oscillation. The thin layers extending each water body across the full domain of computation have the purpose of carrying the sweep algorithms across the whole domain without the need for specific 'flood-and-dry' procedures and their associated frame codes.

As a further example of the influence of dissipative interfaces, Fig. 6.9 illustrates a harbour oscillation calculated using two different schemes, both mass- and momentum-conserving, but the one conserving energy and the other dissipating energy. A radiation boundary module is, of course, used at the harbour

Fig. 6.8. (*opposite*). Numerical simulation of a 'lock-exchange test', which is a severe test for energy conservation in a two-layer fluid model

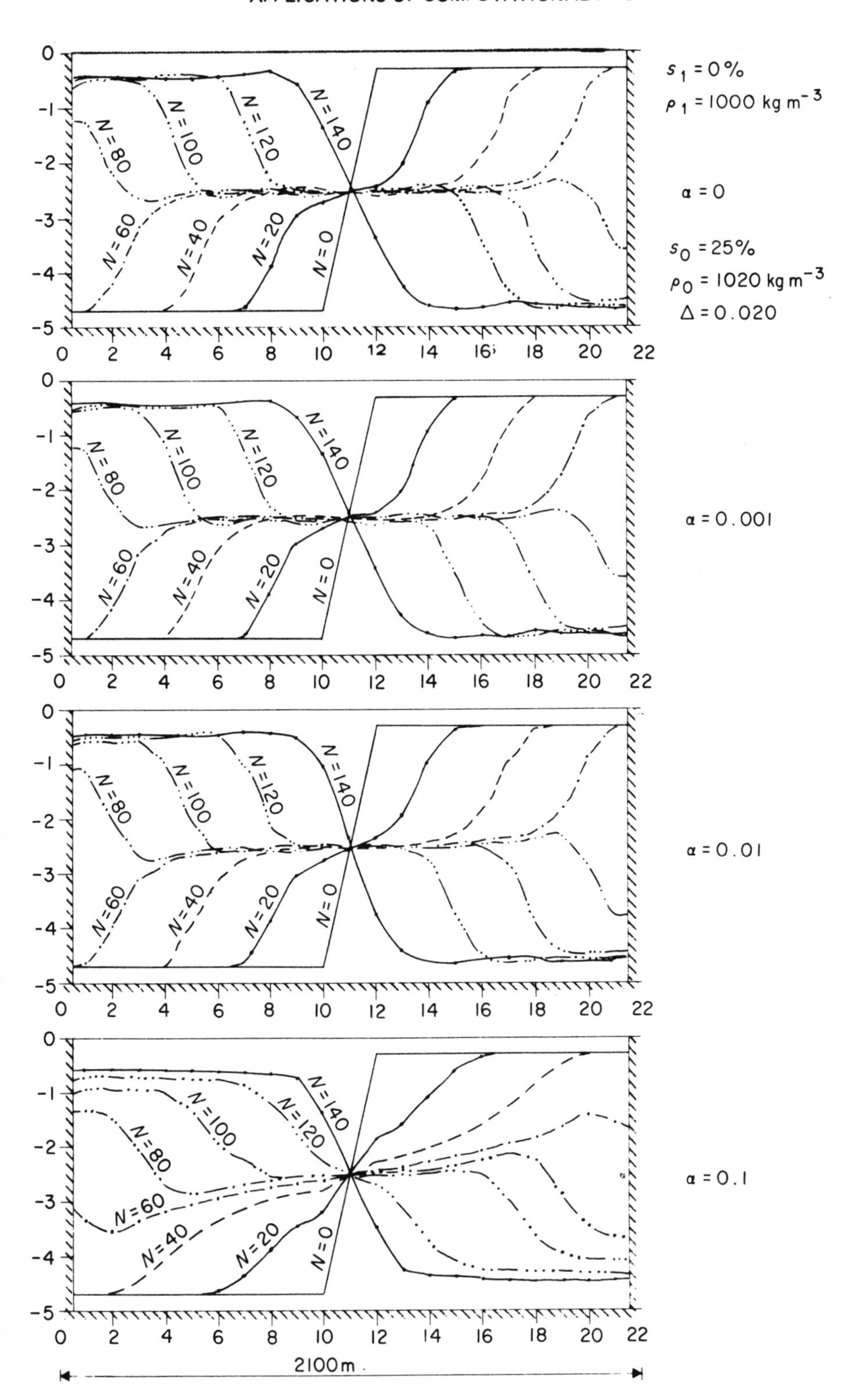
$s_1 = 0\%$
$\rho_1 = 1000\ \mathrm{kg\,m^{-3}}$
$\alpha = 0$
$s_0 = 25\%$
$\rho_0 = 1020\ \mathrm{kg\,m^{-3}}$
$\Delta = 0.020$
$\alpha = 0.001$
$\alpha = 0.01$
$\alpha = 0.1$
N=0
N=20
N=40
N=60
N=80
N=100
N=120
N=140
2100m

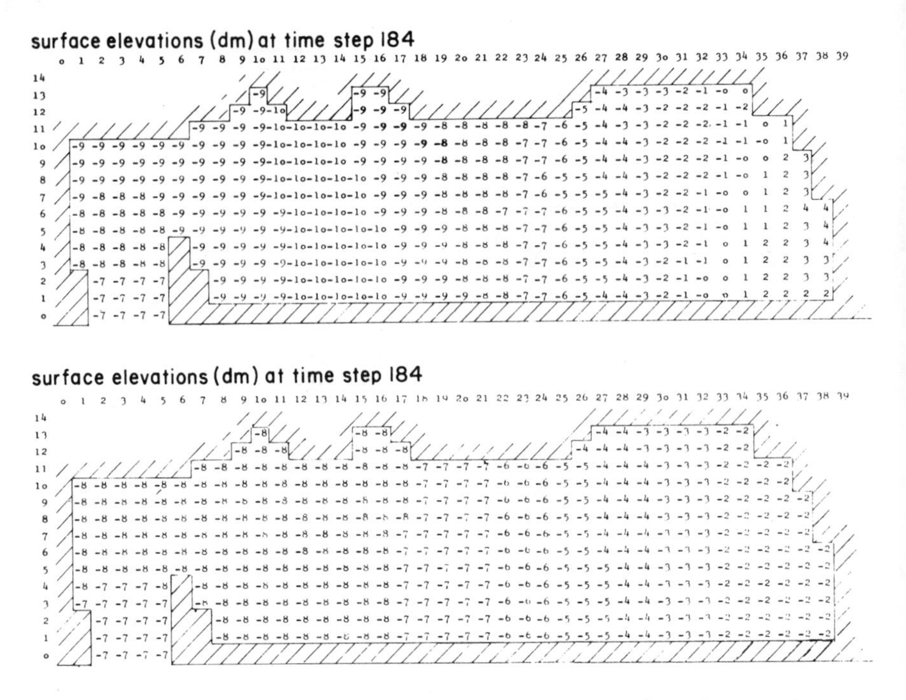

Fig. 6.9. Comparison of harbour seiching model results obtained (above) with a scheme that is numerically energy conserving and (below) with a numerically dissipative scheme

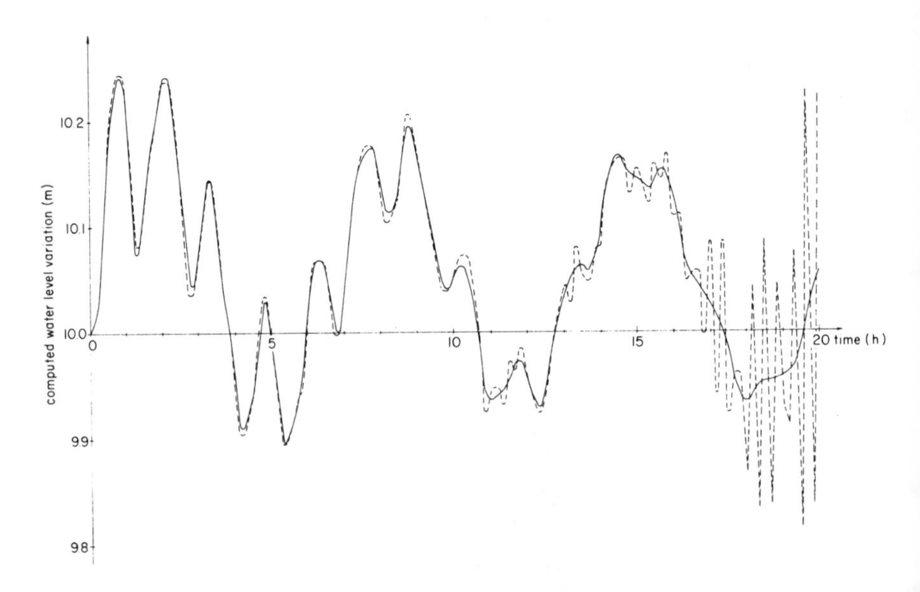

Fig. 6.10. Elimination of a long-term instability by use of a light dissipative interface. (- - -) $\alpha = 0$, $\Delta t = 10$ min; (—) $\alpha = 0.05$, $\Delta t = 10$ min

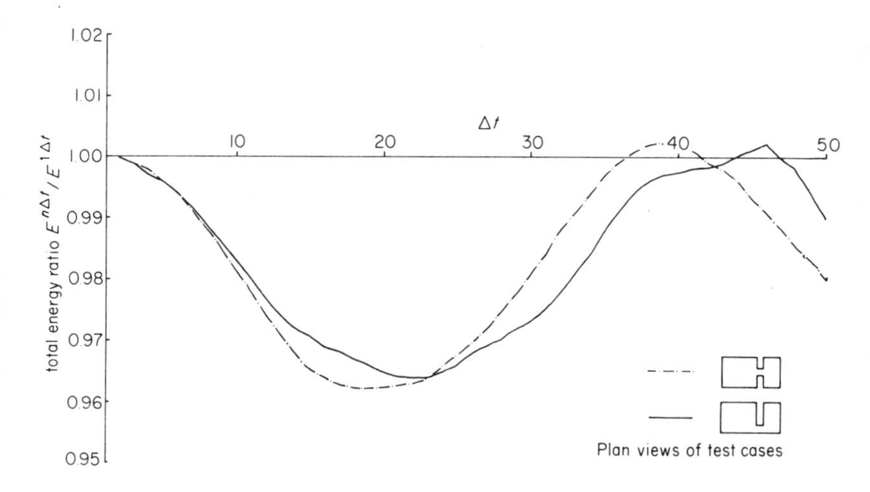

Fig. 6.11. Energy variation in a three-dimensional model in a closed fluid system. Ratio of total energy at $n\ \Delta t$ to total energy at 1 Δt for 50 Δt (from Hodgins, 1977)

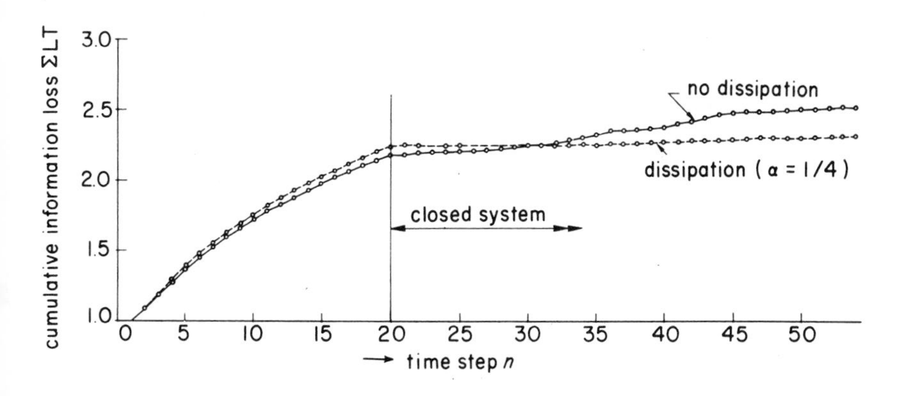

Fig. 6.12. Cumulative information loss in a channel flow test

entrance in both cases. Very serious errors are seen to be introduced in this case by the dissipative scheme. As seiching simulations are run at lower Courant numbers (usually $Cr = 1$ or $Cr = 2$ and rarely for $Cr > 4$), the phase errors play a smaller role in these applications, while these errors can be further reduced by using a higher-accuracy scheme of the type outlined in Section 4.12. By way of comparison with Fig. 6.9, Fig. 6.10 shows how a physically realistic (Kolmogorov–Ozmidov scale) dissipation can be used to eliminate a long-term instability in a finely balanced scheme, without spoiling the detail.

Much can be learnt about a scheme from a study of one of its 'integral invariants': in computational hydraulics these are essentially any norms based on variables not explicitly conserved in the difference scheme. For schemes based on mass and momentum conservation laws, the most convenient norm is the energy norm, or its information loss derivative. By way of examples, Fig. 6.11 shows the energy variation in a three-dimensional model while Fig. 6.12 shows the cumulative information loss in a simple channel flow test. The first, Fig. 6.11, illustrates the swinging of the energy norm due to components of higher wave number riding up over components of lower wave number, as outlined in Section 5.4. The second, Fig. 6.12, shows that although the cumulative information loss is initially greater for the scheme carrying a dissipative interface, after about $n = 30$ this scheme appears to have lost less information. This is because, in the closed system used here, the components of high wave number continue further to confuse the 'message' being transmitted by the elementary scheme, whereas in the scheme with a dissipative interface these components have been largely eliminated and can have little further confusing effect. The circumstances are, of course, different in an *open*-flow system.

The notion has been repeatedly propagated that schemes built upon mass and momentum conservation laws may be total-mass- and total-momentum-conserving. However, this does not prevent the solution of such a scheme from accumulating more or less mass and momentum in a particular region than are accumulated in the physical reality. Figure 6.13a, for example, shows how a nearly horizontal flow solution can 'put its nose down' when run at low numbers of points per wave length, and therefore with large truncation errors. (Hence the large amount of noise in Fig. 6.13.) It is corrected by centring the leading coefficients in the schemes, to give results shown in Fig. 6.13b.

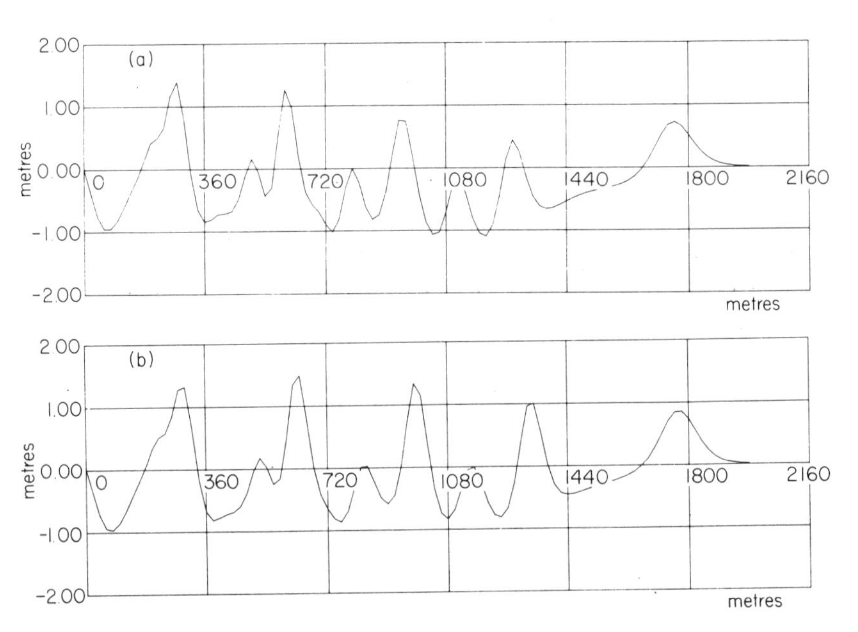

Fig. 6.13a. (a) A second-order, nearly horizontal flow scheme may develop a 'set down' when surface gradient coefficients are introduced explicitly. (b) This can be corrected by an improved centring of these coefficients

6.4 Examples of field tests

The field test involves those processes of *calibration* and *verification* that are associated with all modelling work. The process of calibrating a model is one of improving its realism by confronting it with results from the modelled nature, or *prototype*. Measurements on the prototype are used to pinpoint deficiencies in the physical realism of the model, and the model is changed accordingly to improve its description of the physical reality. It should be emphasized that 'calibration' does *not* mean the use of physically unrealistic parameters to force a poorly conceived model into satisfying prototype data. If there is a discrepancy between model results and calibration data then either there is something wrong in the physical realism of the model or there is something wrong in the physical realism of the data. The verification process is concerned with checking that the calibrated model continues to describe the prototype behaviour even when describing flow situations that are quite different from those utilized in the calibration process.

Turning now to examples, and following the order of Chapter 4, the first example is that of kinematic wave propagation, in this case of short storm waves. Figure 6.14a shows an area of the North Atlantic over which short waves are generated by the wind and simultaneously propagated. In the scheme, each frequency–direction wave energy spectrum component is translated with its group velocity using a high-accuracy kinematic scheme (Sections 3.7, 4.5). From a calculated surface wind, the rate of growth or decay of all components of the frequency–directional wave energy spectrum is then calculated in each point of the numerical grid. By integrating over directions one obtains the classical frequency spectrum and then by a further (rescaled) integration over frequencies one obtains the significant wave height. Figure 6.14b shows a comparison of calculated significant wave heights with measurements over three characteristic storms. The verification run r.m.s. error is about 0.7 m. For one time, there is a comparison of the calculated and measured frequency spectra – Fig. 6.14c. Figure 6.14d shows the corresponding calculated frequency–directional spectrum, while Fig. 6.14e shows schematized example inputs and outputs for this model. The full description of such a short-wave routing system (here the System 20 'Thor') belongs, of course, to a specialized monograph.

A further model of the routing type, shown in Fig. 6.15, uses methods developed in geometrical optics, to refract waves onto a coast, but in this case also accounting for bed resistance effects.

The next case considered in Chapter 4 was that of one-dimensional flow, building on the differential formulations of Chapter 2. One-dimensional schematizations are often characterized in practice by the rapidity of variation of flow sections that have to be accommodated in the schematization. Thus, either a very great number of grid points have to be used in order to provide a reasonable energy conservation in a mass- and momentum-based scheme, or this scheme itself must be of such an accuracy that its energy falsification is negligible. This

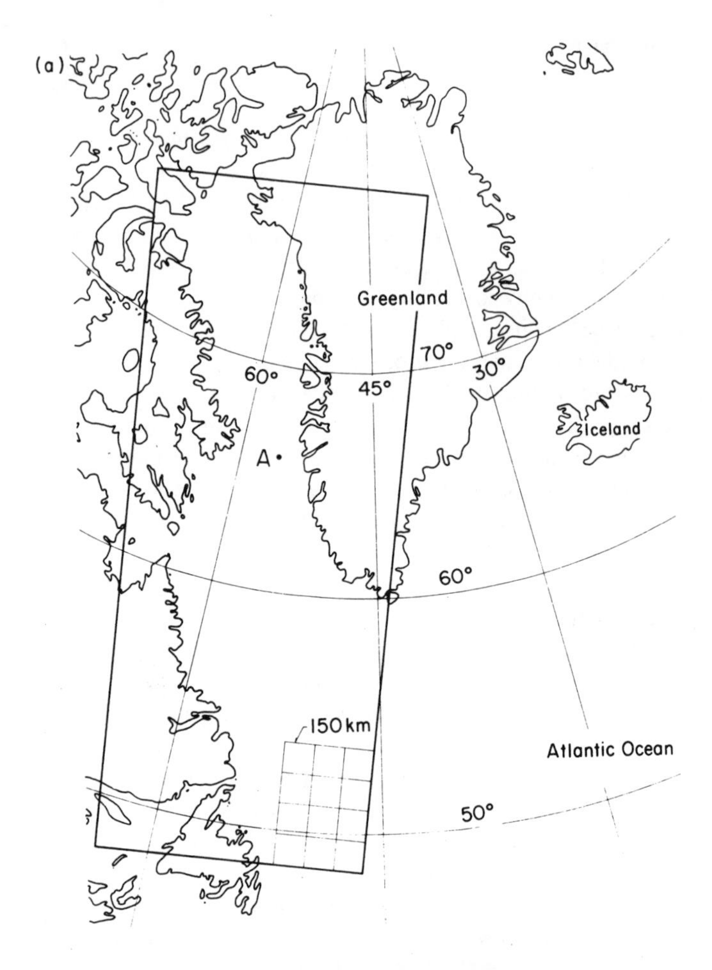

Fig. 6.14. A field test of a short-wave kinematic routing model allowing also the amplification and decay of the waves from surface wind, this last being in turn computed through geostrophic wind from a meteorological barometric pressure chart. (a) The model area with A marking the point at which control measurements have been made. (b) A comparison between calculated (- - -) and measured (——) wave heights for three characteristic storms. (c) Comparison of the calculated (- - -) and measured (——) frequency–amplitude spectrum for time 75/08/22 as indicated in (b). (d) The corresponding calculated frequency–directional spectrum. (e) Schematized inputs and outputs of this model (*See pages 268 to 271.*)

accuracy problem is frequently compounded in one-dimensional models by problems of *connectedness* of the model 'channels'. One-dimensional models may be *simply connected* or *multiply connected* (Fig. 6.16). In the first case (Fig. 6.16a) it is possible to travel between any two points in the model by one and only one route, whereas in a multiply connected model it is possible to travel between certain pairs of points, such as A and B in Fig. 6.16b, by more than one route. At the junctions the E and F-coefficients of the double-sweep algorithms have to

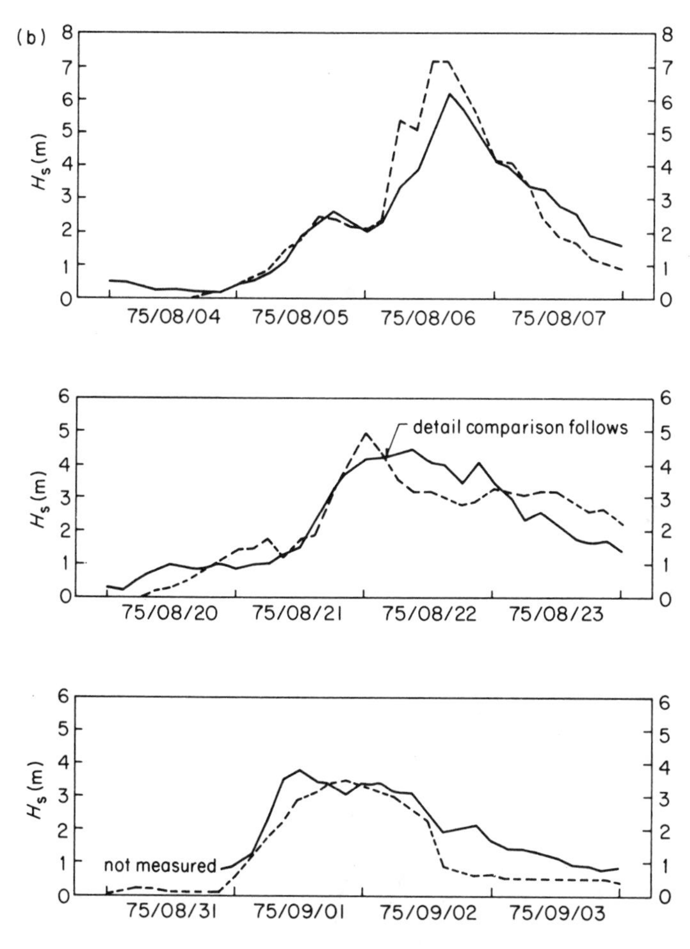

Fig. 6.14. (cont.)

be associated or dissociated, but whereas in a simply connected model only simple associations are required, in a multiply connected model a form of dissociation has also to be introduced. This last form necessitates the linking together of the various junction or node coefficients of the model through elimination procedures over the grid points between these nodes. If this elimination is not very well done, the algorithms become ill-conditioned and spurious circulations may slowly build up around the internal loops of the multiply connected regions, such as along BC in Fig. 6.16b. These may entirely destroy the stability of the model, in which case they will at least be detected, or, which is much more dangerous, they may remain so small as to go undetected while still spoiling the accuracy of the model. The full and proper description of these aspects of one-dimensional modelling belongs to a specialized text, of course, together with test-rig, field-test and prototype examples. The illustration

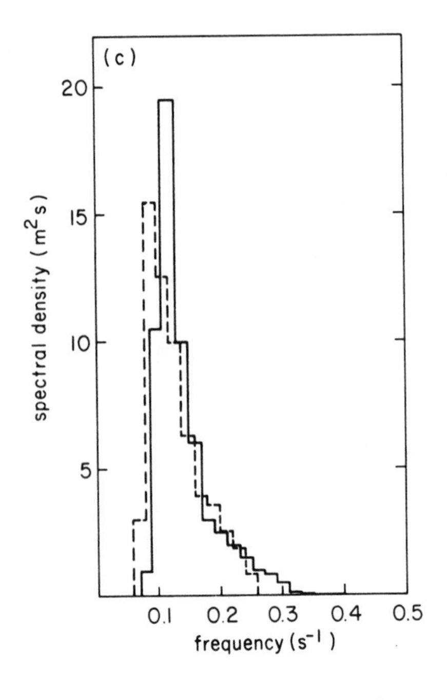
(c)
spectral density (m^2s)
20
15
10
5
0.1
0.2
0.3
0.4
0.5
frequency (s^{-1})

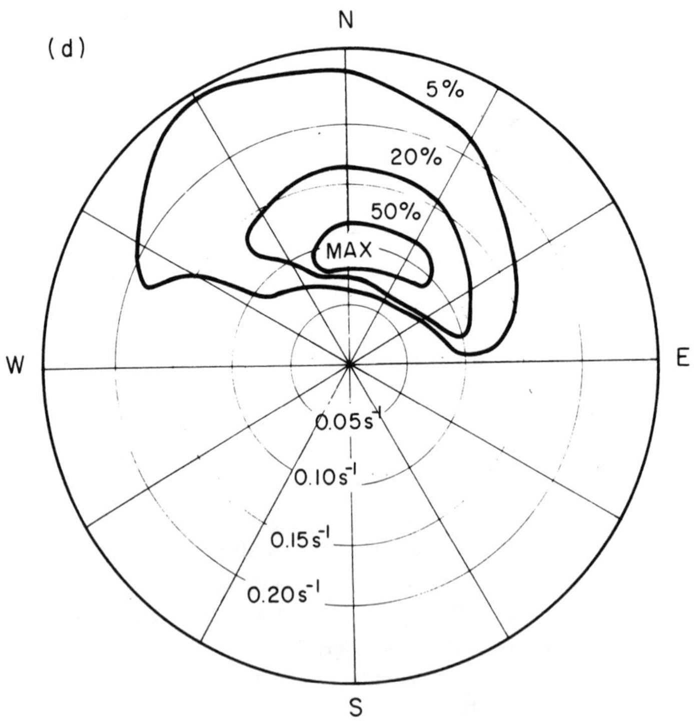
(d)
N
E
S
W
5%
20%
50%
MAX
0.05s^{-1}
0.10s^{-1}
0.15s^{-1}
0.20s^{-1}

Fig. 6.14. (cont.)

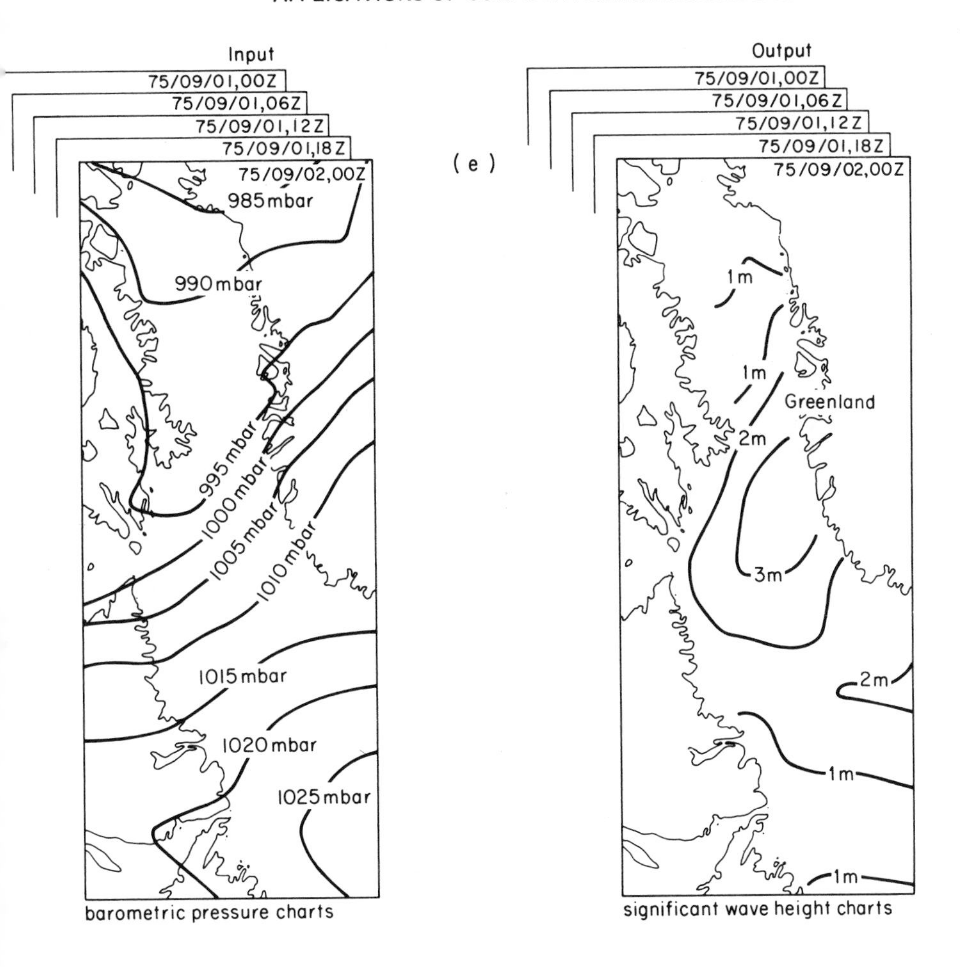

Fig. 6.14. (cont.)

used here is that of the Limfiord, a fiord in northern Denmark that is multiply connected and has exceptionally rapid changes in section, as can be seen from the satellite image of Fig. 6.17a. The associated numerical grid is shown in Fig. 6.17b. The Limfiord model was calibrated by applying observed water-level variations at its North Sea and Kattegat boundaries, applying an observed wind field across it (vectored within the model to provide components in the flow directions) and then modifying the schematizations until the results provided by the model agreed with those obtained in the field. The model was then verified by testing it under a second set of boundary and wind-field conditions. Figures 6.17c–e show the boundary conditions and wind conditions used in the verification while some corresponding computed and measured discharges and elevations are compared in Figs 6.17f and 6.17g.

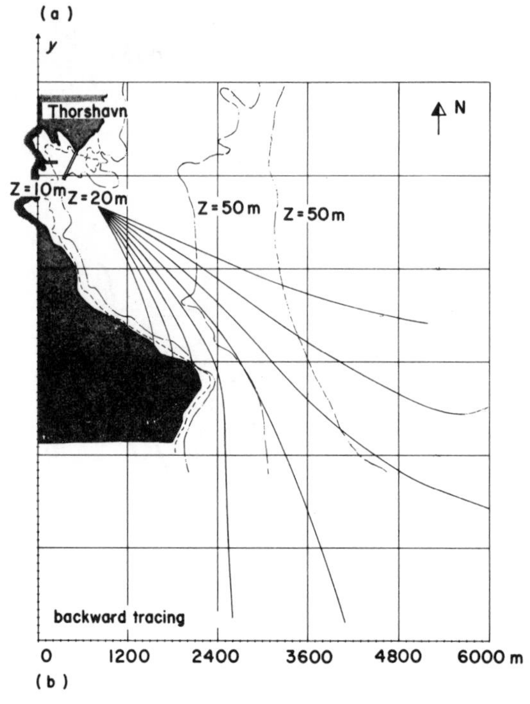

HARBOUR DESIGN—FAROE ISLANDS
grid spacing = 100 m
wave period = 12 s
nikuradse roughness = 0.05 m
initial wave height = 10 m
computed wave heights in meters

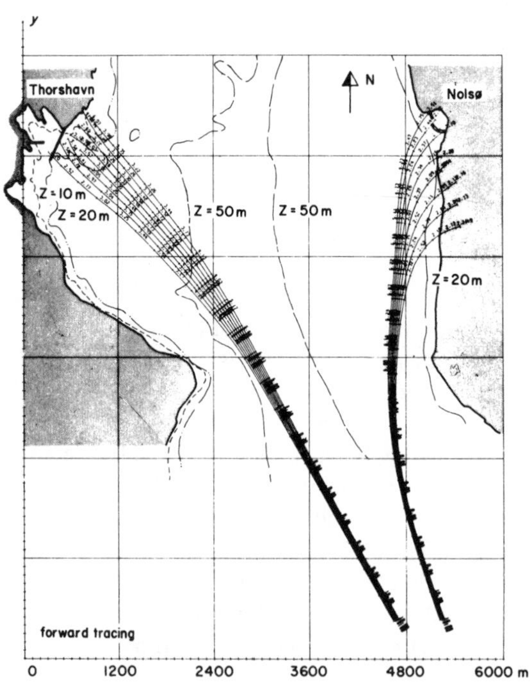

Fig. 6.15a and b. *(For caption see page 273)*

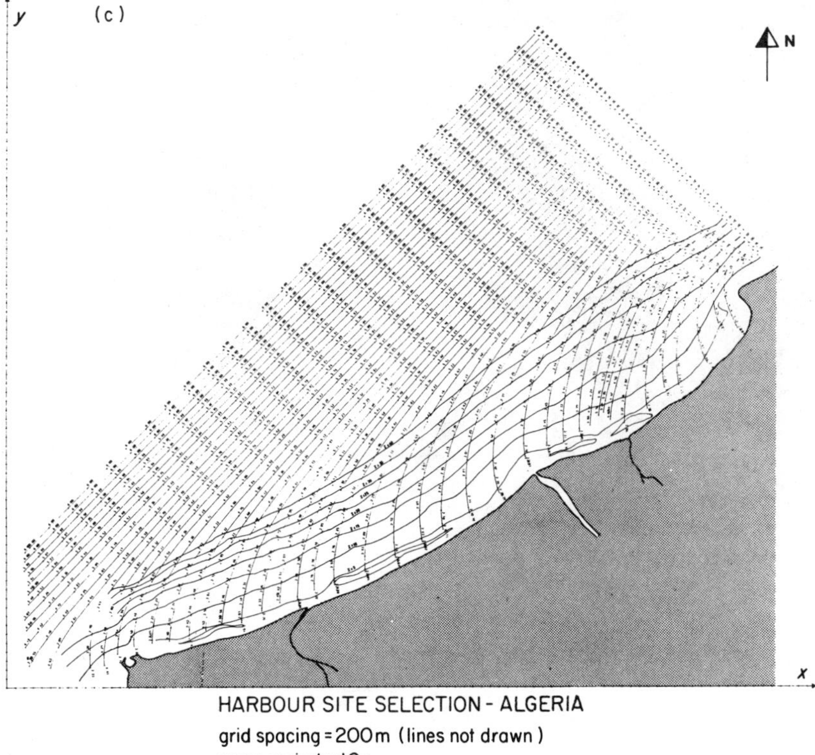

Fig. 6.15. A typical application of the wave refraction model, showing (a) the model used in inverse mode, to trace back wave orthogonals from an area of interest, thus determining the areas offshore at which (b) the model should be initiated from deep water in direct mode. A typical detailed application to a harbour site selection study is shown in (c) (from Skovgaard and Bertelsen, 1974) (*For (a) and (b) see page 272.*)

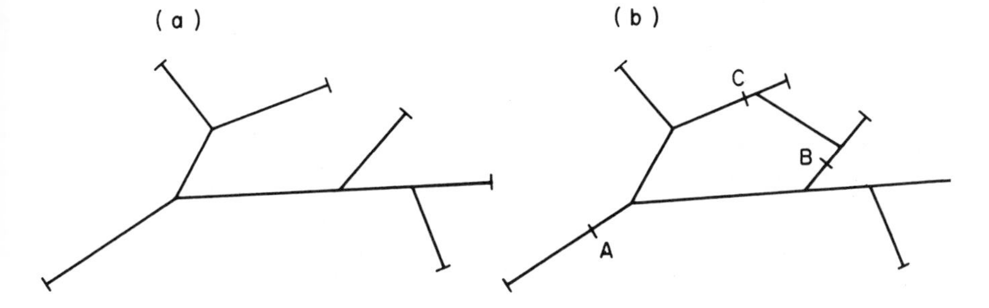

Fig. 6.16. Examples of (a) a simply connected and (b) a multiply connected one-dimensional model. In the latter case there exist points, such as A and B, that may be connected along more than one path

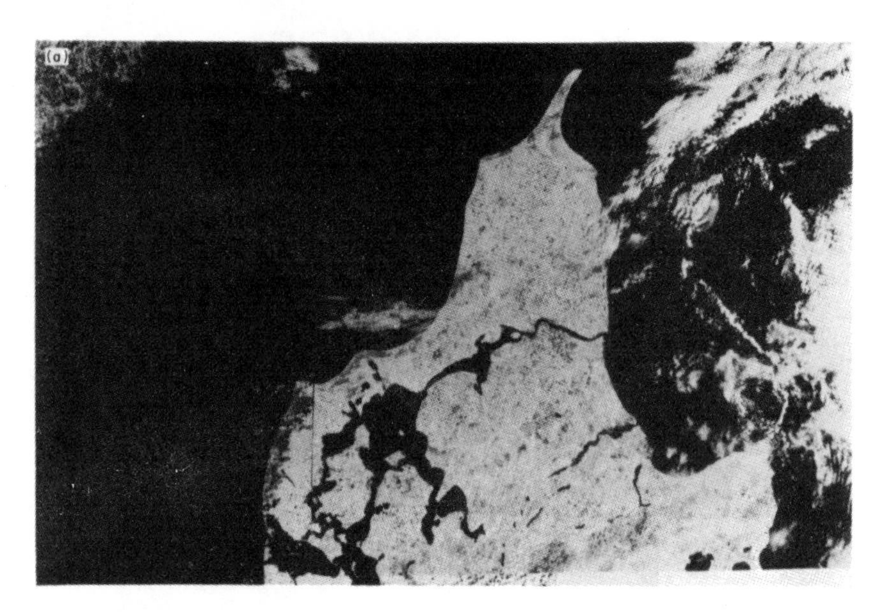

(b)

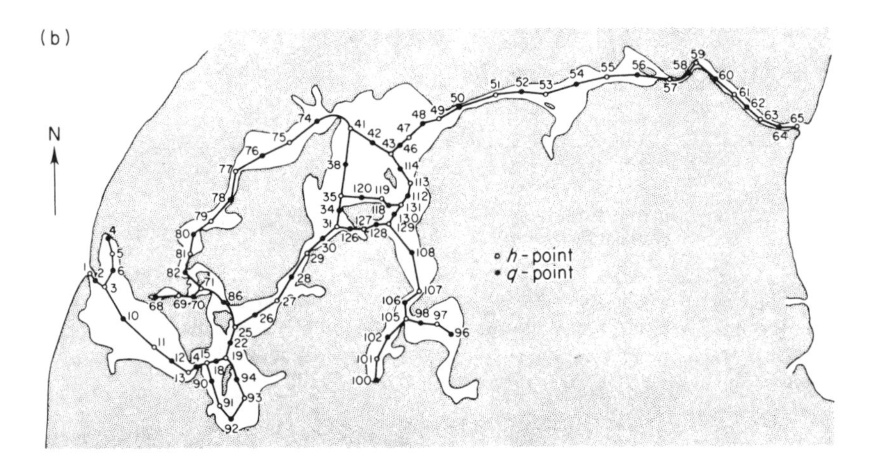

Fig. 6.17. Example of a multiply connected system: the Limfiord in northern Denmark. Its properties are clear from the satellite image (a) and these are carried over to the model grid (b). The boundary conditions (c) and wind conditions (d) and (e) are shown for a verification run of the Limfiord model. (f), (g) Comparisons between computed and observed discharges and elevations over this same verification period. (h), (i), (j) Typical water-quality results based upon the hydrodynamic stage results: (h) Comparison between measured and computed salinities in point 49, (i) Comparison between measured (○) and computed (——) concentrations in point 26, (j) Predictions of conditions obtaining in point 53 following the removal of 90% of phosphorus from domestic and industrial sources. (Before ——, After - - -) (LANDSAT image, Danish Hydraulic Institute and Danish Water Quality Institute Systems) *(See pages 274 to 279.)*

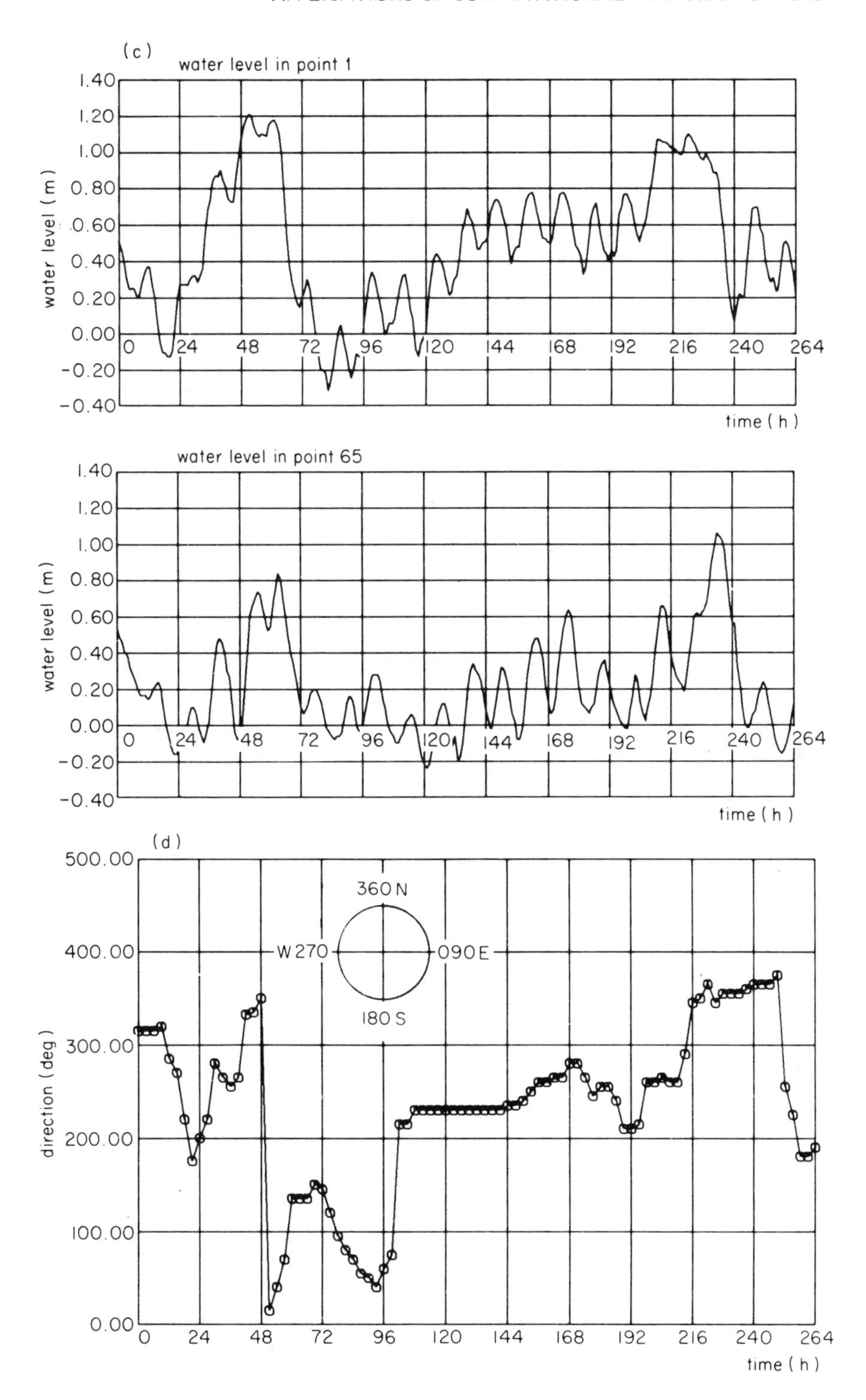
(c)
water level in point 1
water level (m)
time (h)
water level in point 65
water level (m)
time (h)
(d)
360 N
W 270
090 E
180 S
direction (deg)
time (h)

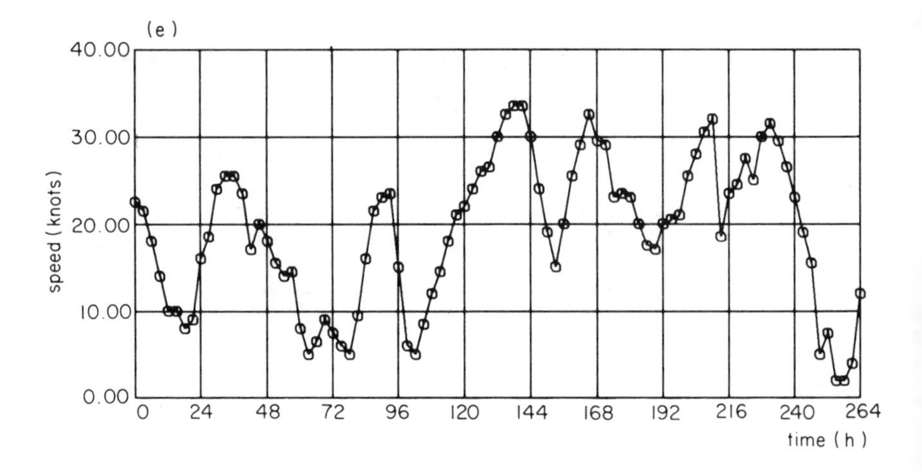
(e)
speed (knots)
40.00
30.00
20.00
10.00
0.00
0
24
48
72
96
120
144
168
192
216
240
264
time (h)

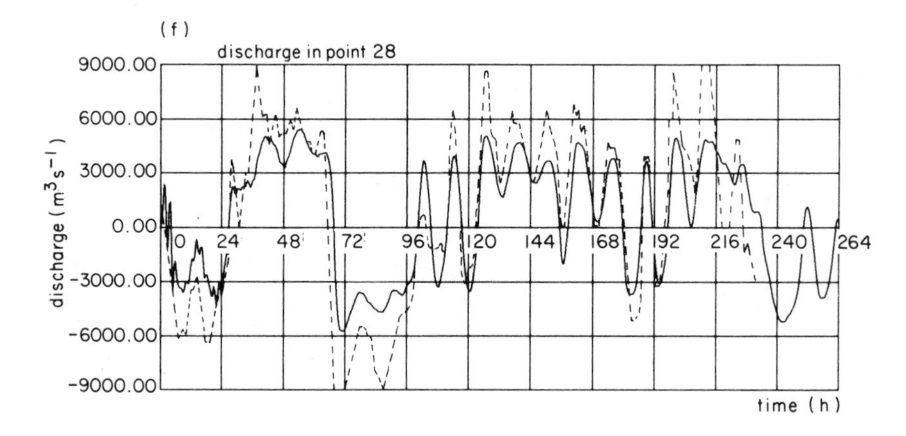
(f)
discharge in point 28
discharge ($m^3 s^{-1}$)
9000.00
6000.00
3000.00
0.00
-3000.00
-6000.00
-9000.00
0
24
48
72
96
120
144
168
192
216
240
264
time (h)

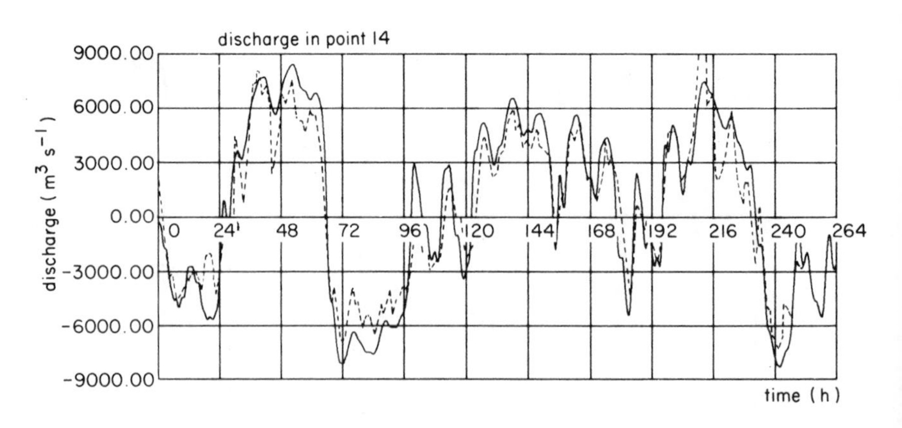
discharge in point 14
discharge ($m^3 s^{-1}$)
9000.00
6000.00
3000.00
0.00
-3000.00
-6000.00
-9000.00
0
24
48
72
96
120
144
168
192
216
240
264
time (h)

Fig. 6.17. (cont.)

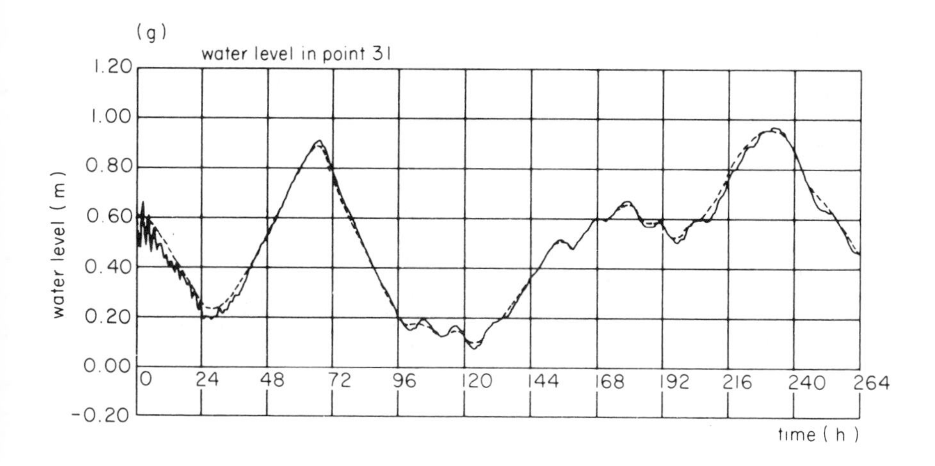
(g)
water level in point 31
water level (m)
1.20
1.00
0.80
0.60
0.40
0.20
0.00
-0.20
0
24
48
72
96
120
144
168
192
216
240
264
time (h)

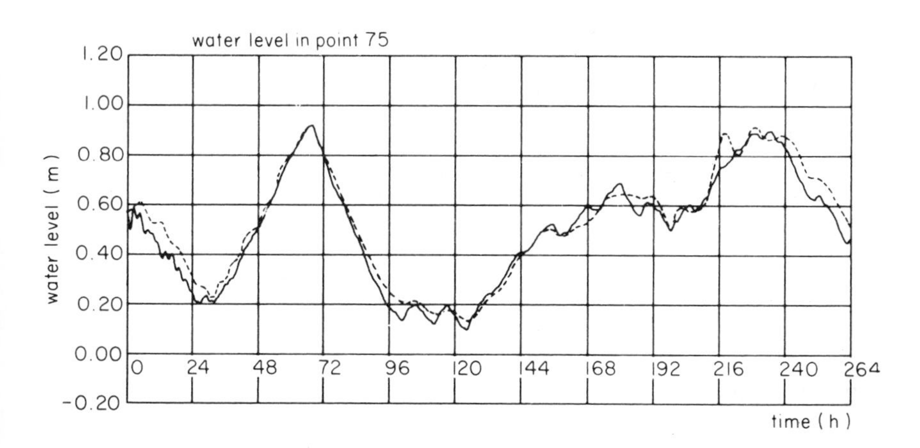
water level in point 75
water level (m)
1.20
1.00
0.80
0.60
0.40
0.20
0.00
-0.20
0
24
48
72
96
120
144
168
192
216
240
264
time (h)

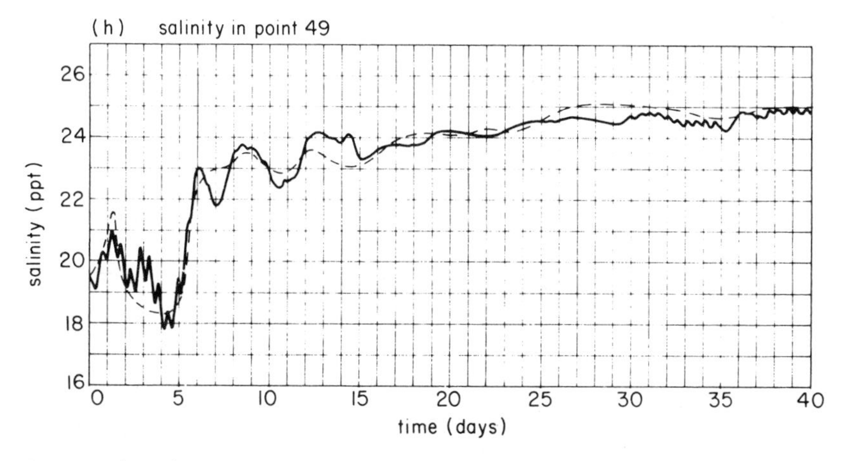
(h) salinity in point 49
salinity (ppt)
26
24
22
20
18
16
0
5
10
15
20
25
30
35
40
time (days)

Fig. 6.17. (cont.)

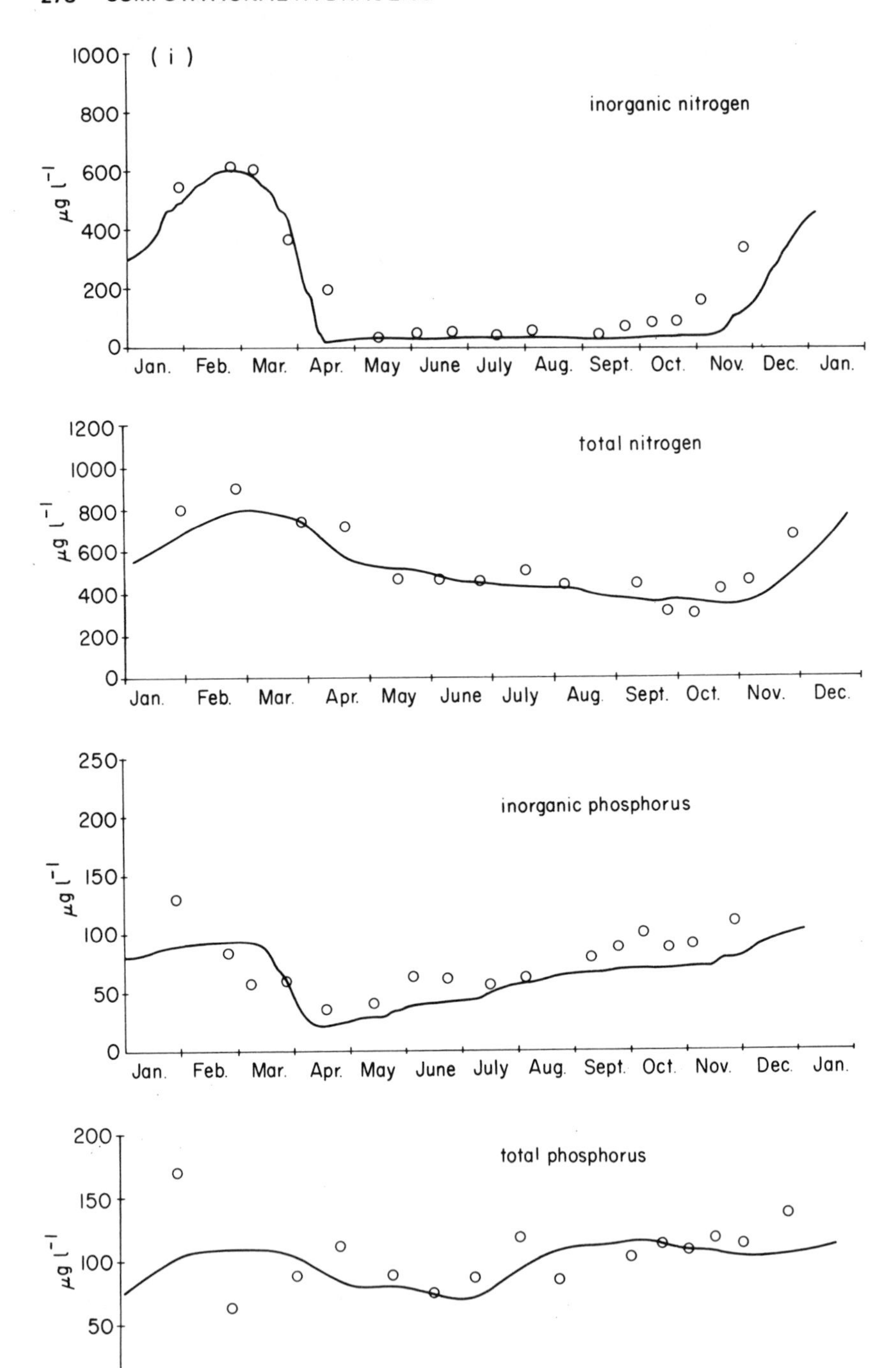
(i)
inorganic nitrogen
μg l⁻¹
1000
800
600
400
200
0
Jan. Feb. Mar. Apr. May June July Aug. Sept. Oct. Nov. Dec. Jan.
total nitrogen
μg l⁻¹
1200
1000
800
600
400
200
0
Jan. Feb. Mar. Apr. May June July Aug. Sept. Oct. Nov. Dec.
inorganic phosphorus
μg l⁻¹
250
200
150
100
50
0
Jan. Feb. Mar. Apr. May June July Aug. Sept. Oct. Nov. Dec. Jan.
total phosphorus
μg l⁻¹
200
150
100
50
0
Jan. Feb. Mar. Apr. May June July Aug. Sept. Oct. Nov. Dec.

Fig. 6.17. (cont.)

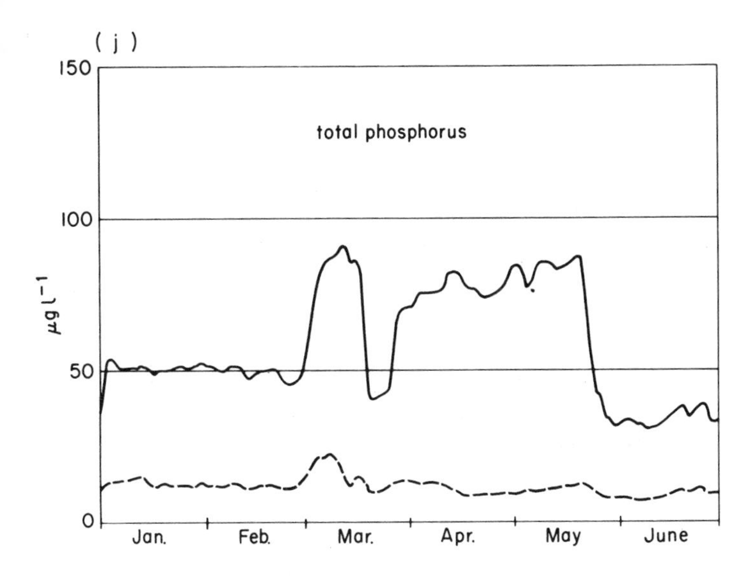
(j)
150
100
50
0
total phosphorus
μg l⁻¹
Jan.
Feb.
Mar.
Apr.
May
June

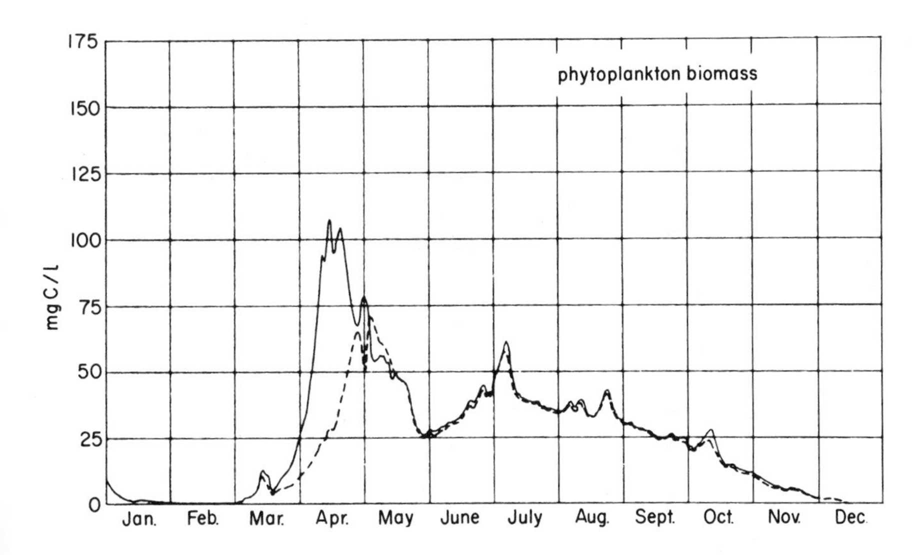
175
150
125
100
75
50
25
0
phytoplankton biomass
mg C/l
Jan.
Feb.
Mar.
Apr.
May
June
July
Aug.
Sept.
Oct.
Nov.
Dec.

Fig. 6.17. (cont.)

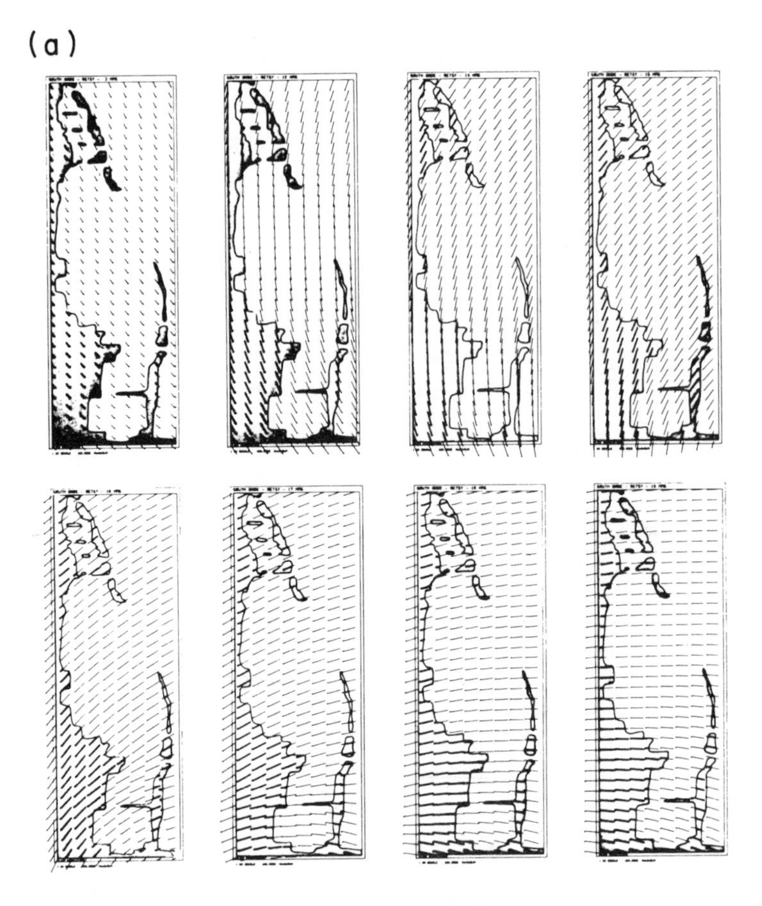

Fig. 6.18. Simulation of 'a hurricane-induced surge'. The wind field of (a) is generated from standard hurricane descriptive parameters while being routed across the model area to provide the corresponding water elevation (b) and velocity (c) fields. Both the flooding and drying processes appear here (*See pages 280 to 282.*)

The hydrodynamic stage of the Limfiord model used as an example here was in fact only a 'platform' for further transport dispersion and water quality stages, also of considerable complexity. Also, in this case the calibration and verification not only provided a field test for the modelling system (System 11, 'Siva') but also provided a foundation for production runs. Taking advantage of the unconditional stability of the implicit scheme, some of these used time steps of up to 21 600 s (6 h) while continuous simulations of up to 6 prototype months were made at low cost, using $\Delta t = 3600$ s. A description of higher stages and production applications belongs, again, to more specialized monographs, and only a few representative results from the higher stages are presented in Fig. 6.17h to 6.17j.

Turning now to models that are two-dimensional in plan, the range of applications and associated field tests is so great that only a short selection of examples

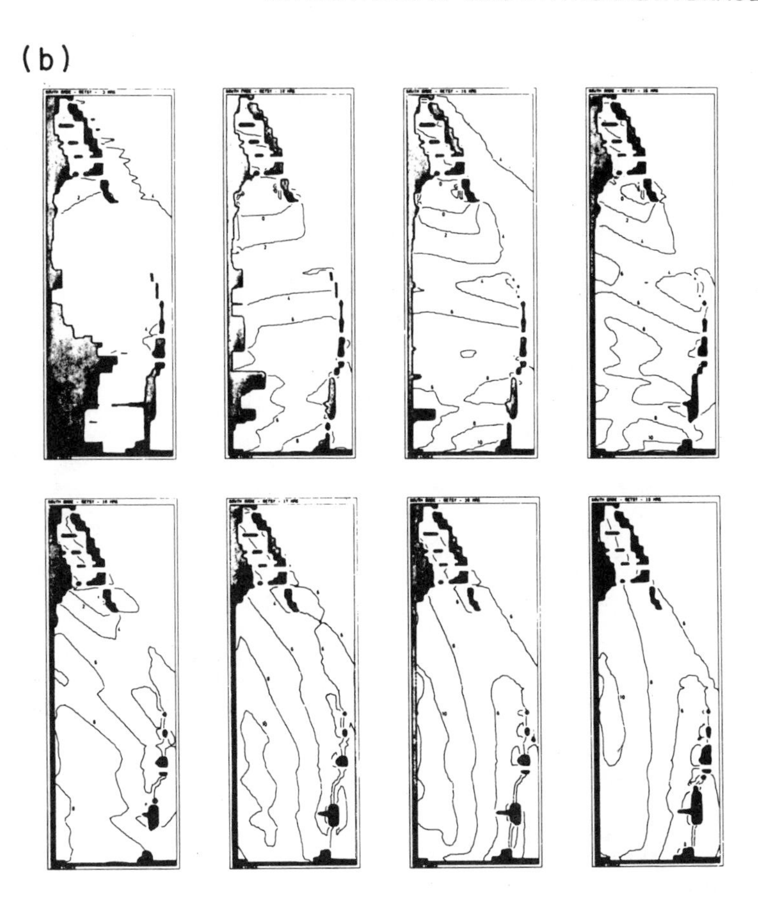

Fig. 6.18. (cont.)

can be made. In order to illustrate how an area can be flooded and dried in a model under the guidance of its grid code, Fig. 6.18 illustrates the flooding of a coastal area in Florida caused by a hurricane-induced surge. In this study carried out with an early production System 21 'Jupiter' (Mark 5-2) model, a special procedure constructs the time-varying barometric pressure fields and associated wind fields for any given hurricane parameters, these fields then being applied, after corrections to wind speed and direction for sea-level conditions, to force the hydrodynamic scheme. The model was used to determine surge levels at the embankments of a large nuclear power plant built on low-lying land, not shown in Fig. 6.18.

Results from a very different study for a nuclear power plant, in Texas, are illustrated in Fig. 6.19, which shows, after a check against laboratory results (Fig. 6.19a), the water elevations engendered in the first seconds following the failure of a reservoir embankment just above two 500-MW reactor blocks. In this case the whole area is covered by a thin layer of water at all times so as to carry the double-sweep algorithms without recourse to a grid-code structure.

(c)

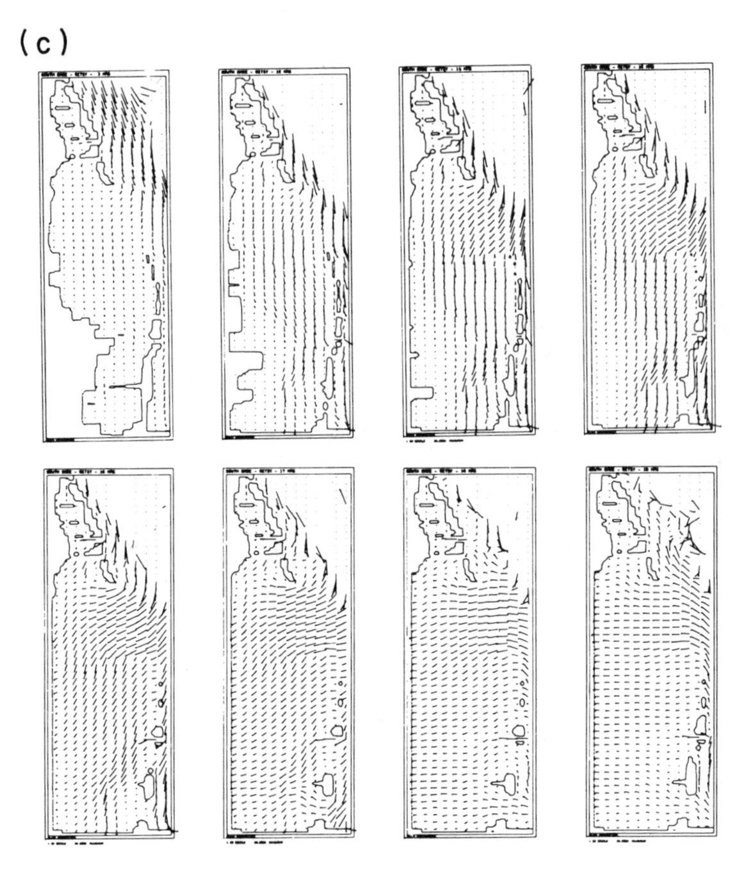

Fig. 6.18. (cont.)

This method of treating flooding – adopted essentially because extensive testing, including tests against physical experiments, had shown it did not inhibit the propagation of the front – is essentially that used in the lock exchange test of Fig. 6.8 and as illustrated again later. Some further typical results of tests on this plant are shown in Fig. 6.20. The results were all obtained with a production Mark 5-2 version of the 'Jupiter' system. Nuclear power plants present considerable waste-heat problems, often necessitating computations of flow fields from a hydrodynamic stage to provide velocity and dispersion coefficient fields for a combined transport–dispersion and heat-balance stage. A typical temperature field output is shown for a projected Danish plant in Fig. 6.21. Extensive calibrations and verifications have been made of models of this type, working with salinities, but as these necessitated very expensive field investigations and subsequent data processing they could be carried out only on a contract basis.

The next examples are all taken from field tests of a later System 21 'Jupiter' variant, the Mark 6-3 (Abbott, 1976, 2). This system constructs and runs models

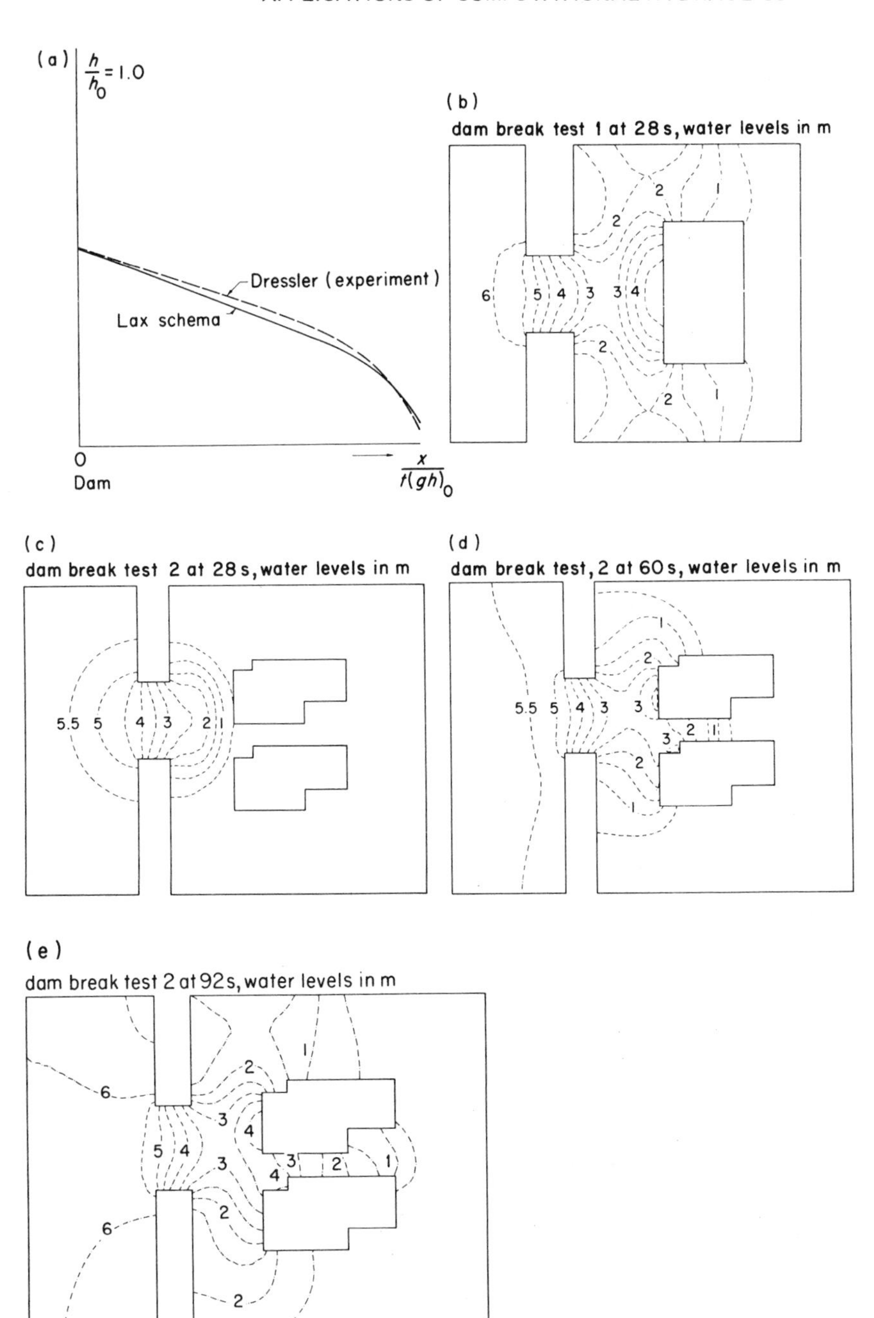

Fig. 6.19. (a) A comparison of computational and experimental dam break followed ((b), (c), (d), (e)) by a typical simulation of water elevations obtaining just after the failure of an embankment

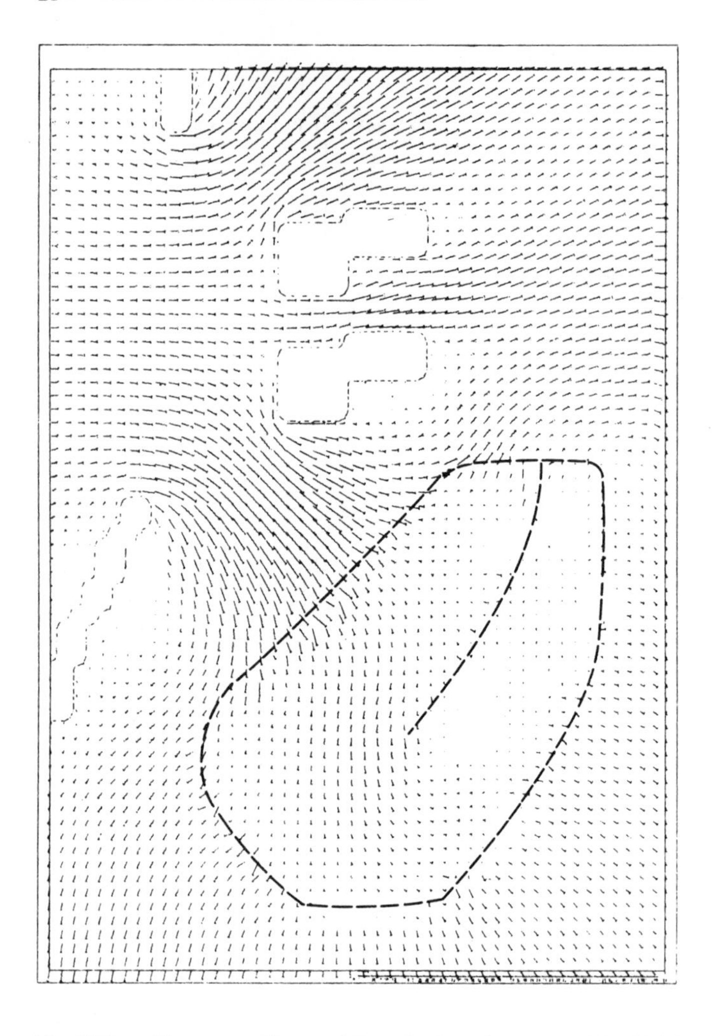

Fig. 6.20. The more widespread flow developed some considerable time after the embankment failure of Fig. 6.19, after the flood wave has successively overtopped the three dikes of an emergency cooling basin and is settling down into a steady flow

with several different scales of resolution solved simultaneously at each time step. A typical model area, the North Sea, is shown in Fig. 6.22a while Figs 6.22b and 6.22c show the bathymetry resolved by the three grid sizes, of 18 km^2, 6 km^2 and 2 km^2, corresponding to the computational domain SD_0 and the subdomains SD_1 and SD_2. All these domains are coupled together through a four-stage implicit difference scheme with two alternating-direction components, so that outward sweeps have to be dissociated as they pass from SD_i to SD_j with $i < j$ and associated again as they pass from SD_i to SD_j with $i > j$ while return sweeps have to reflect these dissociation and association processes. This type of model has to be particularly extensively tested within a very sophisticated organizational

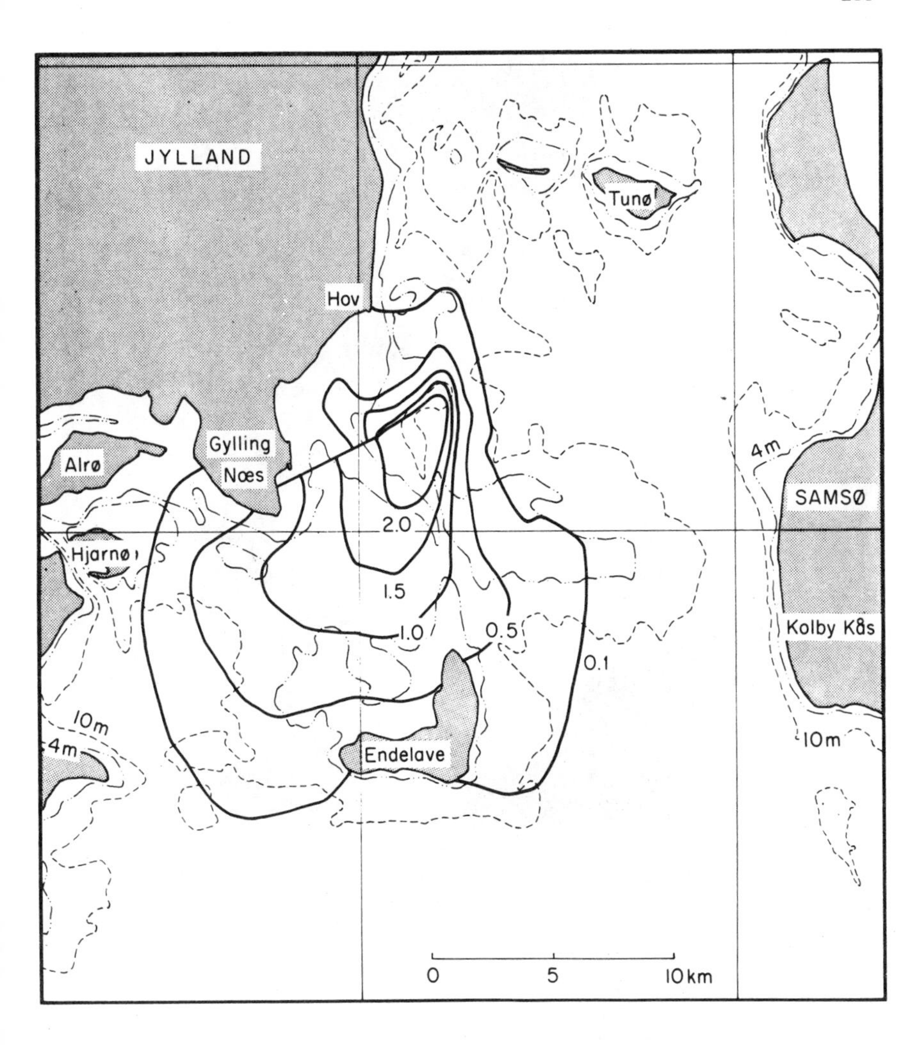

Fig. 6.21. The instantaneous far-field temperature distribution around the cooling water outlet of a projected nuclear power plant. These fields are computed at 10-minute intervals for various tidal and meteorolpgical conditions using velocities and dispersion coefficients given by a two-dimensional hydrodynamic modelling stage

frame if it is to function consistently through floodings and dryings, if pseudo-reflections and -refractions are to be avoided at the changes of scale, and accuracy and stability are to be uniformly maintained.

The North Sea model of Fig. 6.22 is fitted with special boundary modules that synthesize tidal elevations with 10 tidal components, corresponding to German tidal atlas parameters. The resulting elevation field at the first (pre-calibration) run is shown in Fig. 6.22d while Fig. 6.22e compares the corresponding resulting velocity field with that given for the same instant in the British

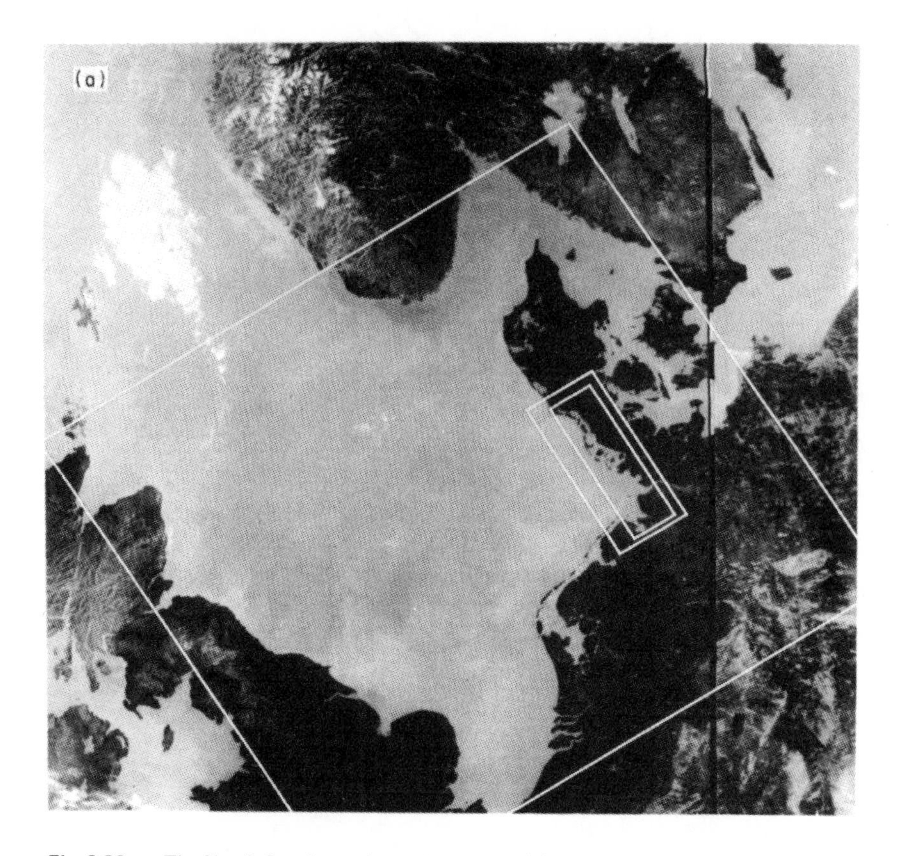

Fig. 6.22. The North Sea shown in satellite image (a) is modelled using a computation proceeding in the coarse grid domain (SD_0) of (b) simultaneously with computations in the finer grid domains (SD_1 and SD_2) of (c). A typical elevation field is shown in (d), while the computed velocity field is compared with the *Admiralty Tidal Atlas* velocity field in (e). (In this last field the velocity vectors are not drawn to scale.) (f) Comparisons between computations (——) and tidal predictions (– – –) at Tynemouth, Aberdeen and Heligoland, all obtained at the very first run of the model. (g) The cut-off of the tide disappears in a finer-scale model, embedded in a coarser grid so as to use boundary data from outside of Heligoland. (——) Computed; (---) recorded. (h) A further illustration of this effect, showing a detail from the fine grid model (g) to illustrate how the stream follows the main sea arms in this area. The divergence between tidal atlas predicted and numerically computed low waters observed at Heligoland can also be partially corrected by using another schematization of the Wadden Sea and estuarine areas behind the Heligoland. In this first run, the deeper channels have been averaged out in the bathymetry, so that the tidal wave propagates in the model at a considerably lower speed than its bulk propagates in the deeper channels of the prototype. By providing suitable locally one-dimensional channels in the model schematization, higher celerities and hence other standing waves can be obtained. Such adjustments form part of the calibration process, sometimes called a 'tuning process'. It is a rule that calibration adjustments must always have a physical justification and considerable care must be exercised in order not to introduce 'calibration adjustments' on the basis of errors in scheme-numerics and field-data. (a) Courtesy NCAA/University of Dundee (*See pages 286 to 293.*)

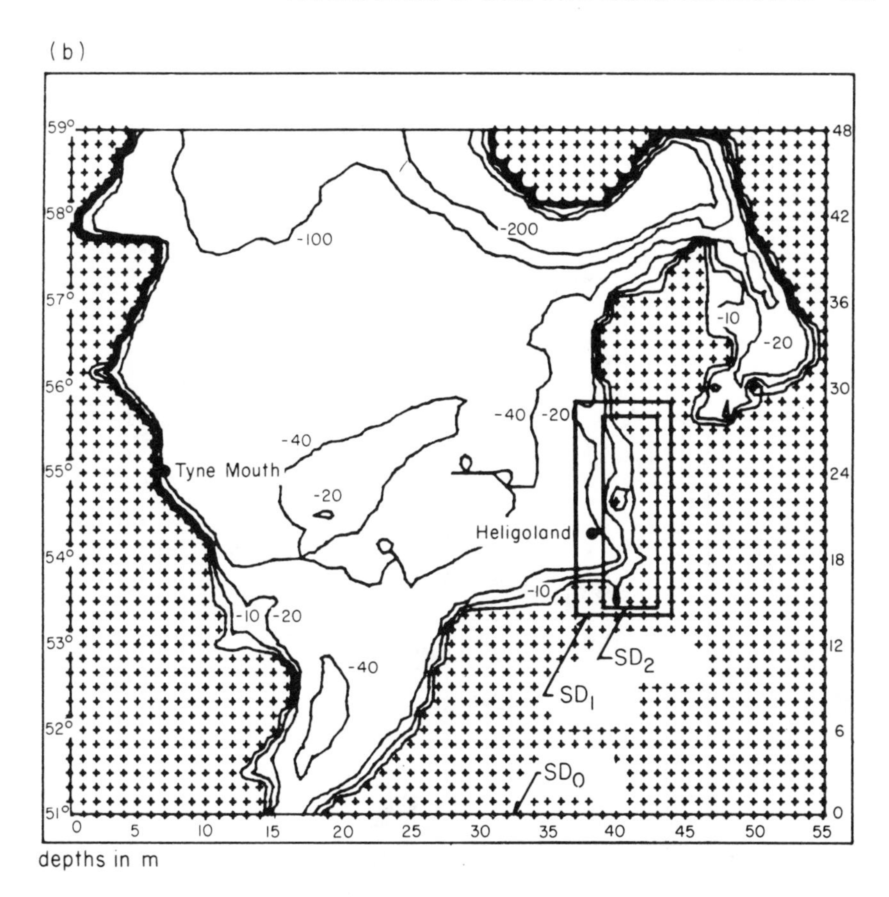

The North Sea model is nested through three sub-domains: SD_0, SD_1 and SD_2

Grid	Mesh size	No. of grid points
SD_0	18 km	49 * 56' = 2.744
SD_1	6 km	46 * 22 = 1.012
SD_2	2 km	124 * 37 = 4.588
Total number of grid points		= 8.344

Fig. 6.22. (cont.)

(c)

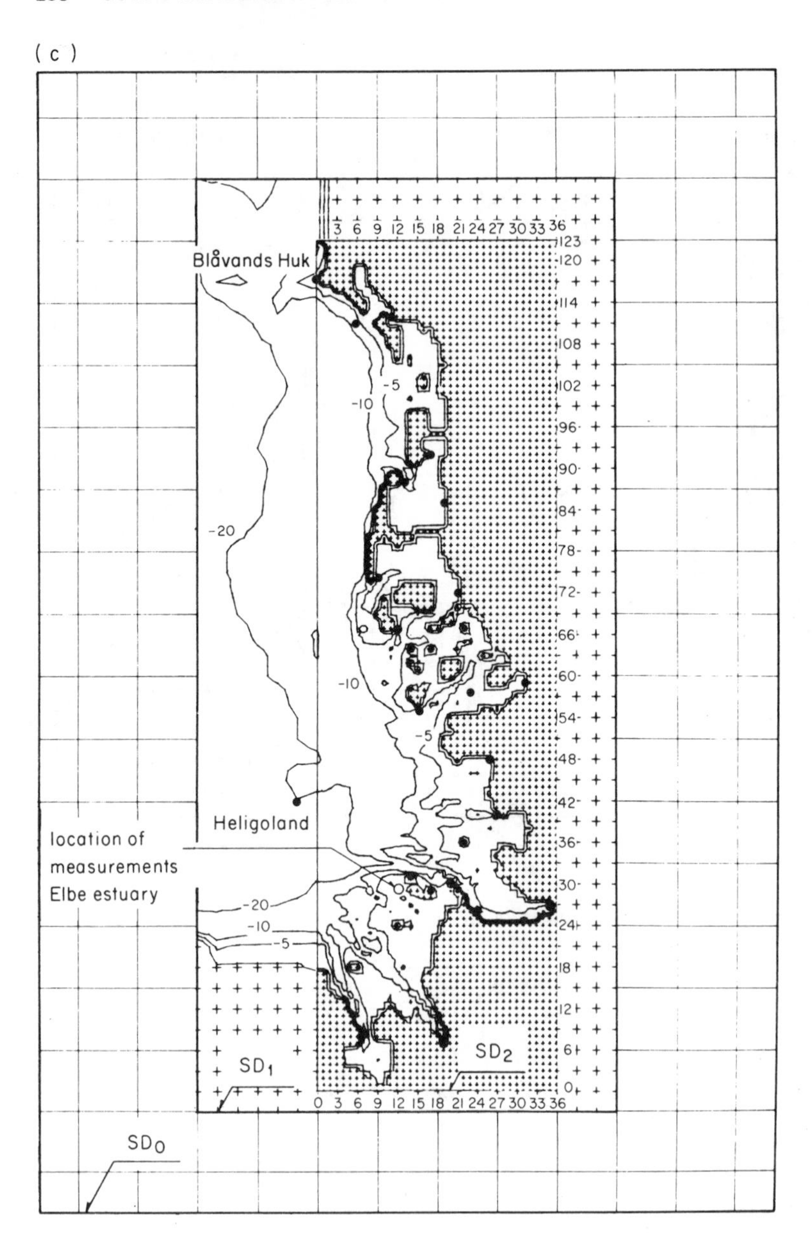

Fig. 6.22. (cont.)

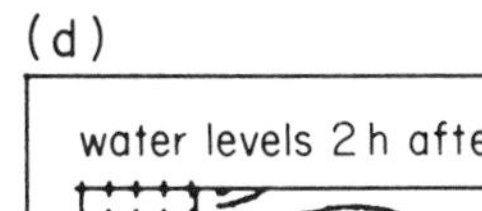

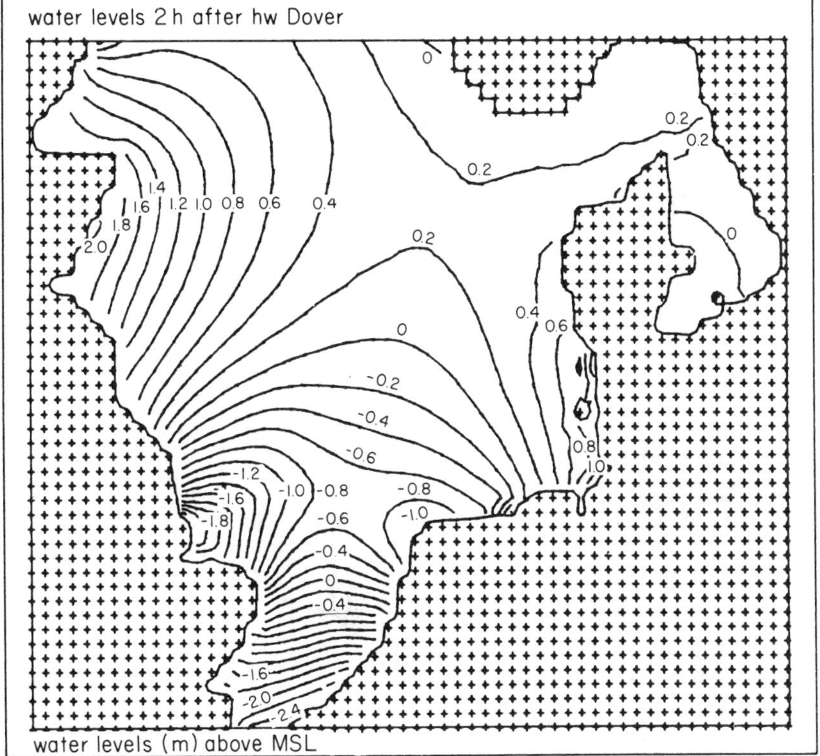

Fig. 6.22. (cont.)

tidal atlas. Figure 6.22f shows comparisons between computed results and tidal-atlas predictions for Tynemouth and Aberdeen, UK, both situated close to deep water, and for Heligoland, West Germany, situated offshore from the German Bight, an area of shallows penetrated by deep sea arms. The cut-off at the bottom of the elevations at Heligoland shows, following the methods of Section 3.6, that the reflected wave is not being properly phased against the incident wave in this first version of the model, corresponding to an averaging out of water depths in the German Bight. Figure 6.22g shows comparisons between computed and recorded elevations and velocities in this area following some calibration on a still further refinement of the resolution, to a 667 m grid. Figure 6.22h shows how this finer grid allows the resolution of flows over flats and in the deeper sea arms.

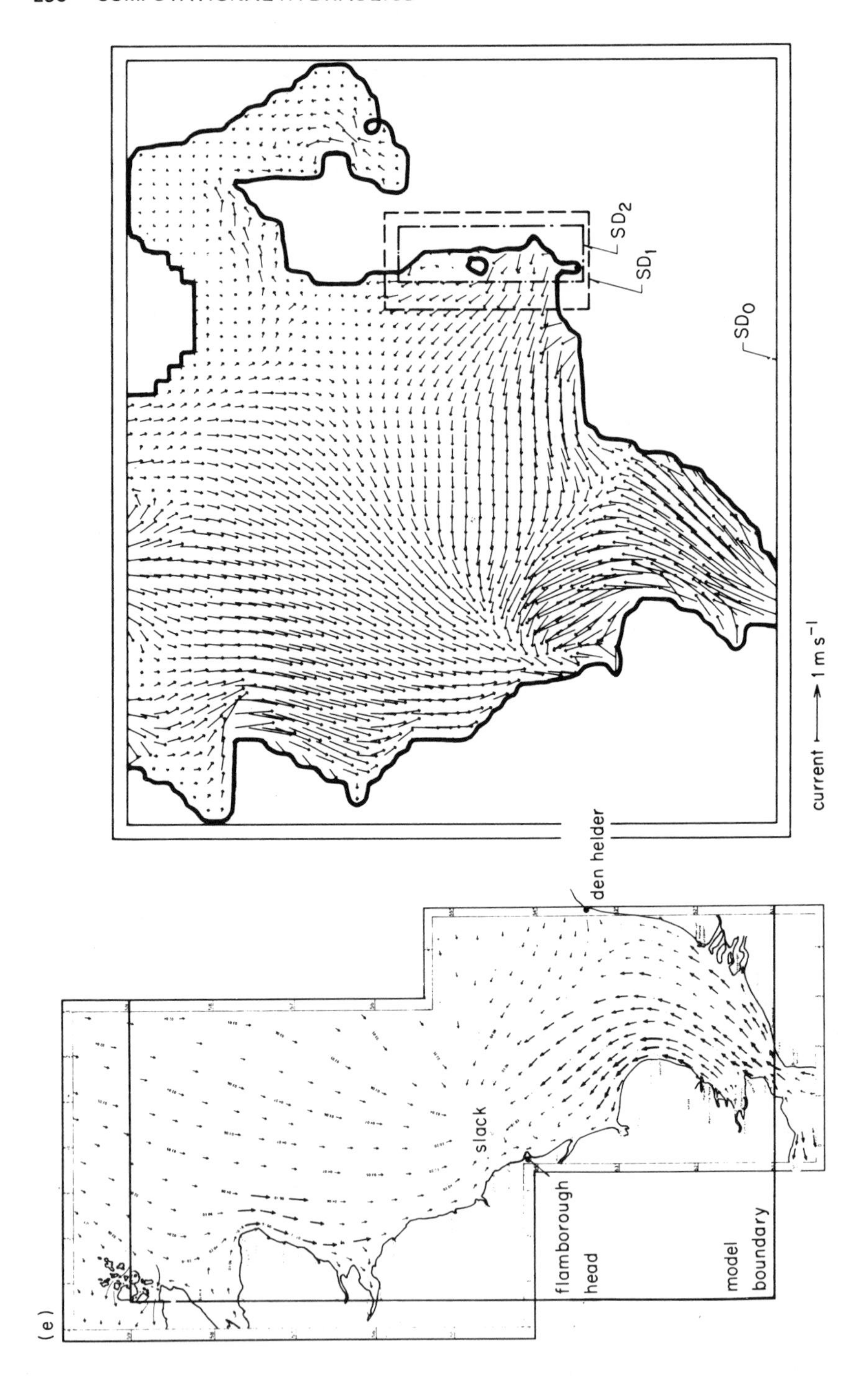
(e)
flamborough
head
model
boundary
slack
den helder
current ⟶ 1 m s⁻¹
SD0
SD1
SD2

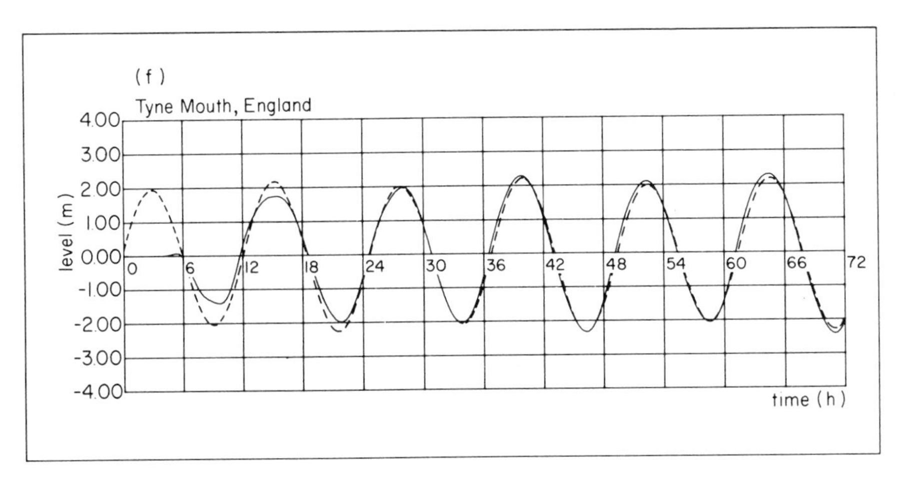
(f)
Tyne Mouth, England
4.00
3.00
2.00
1.00
0.00
-1.00
-2.00
-3.00
-4.00
level (m)
0
6
12
18
24
30
36
42
48
54
60
66
72
time (h)

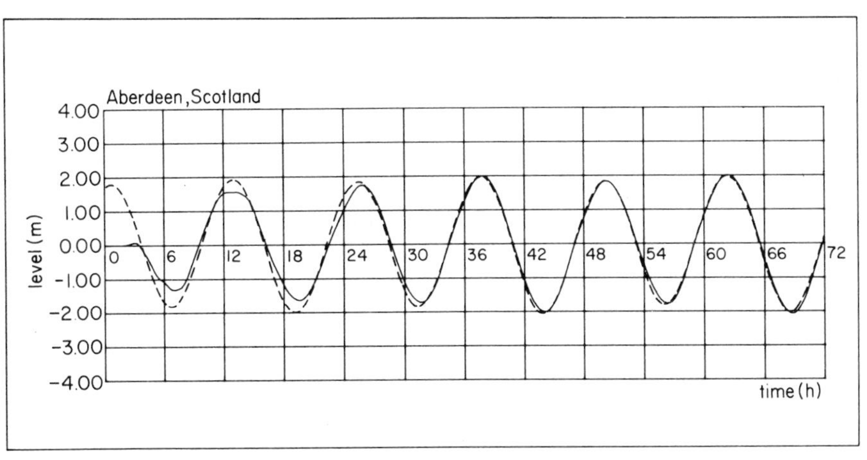
Aberdeen, Scotland
4.00
3.00
2.00
1.00
0.00
-1.00
-2.00
-3.00
-4.00
level (m)
0
6
12
18
24
30
36
42
48
54
60
66
72
time (h)

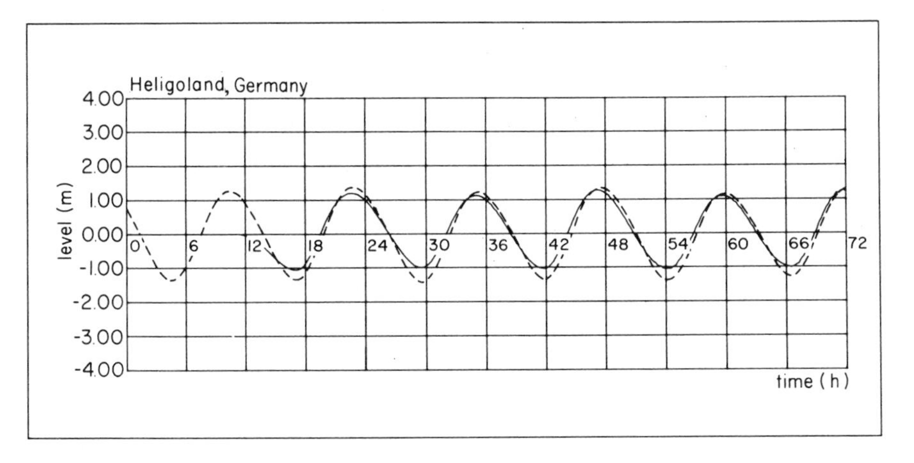
Heligoland, Germany
4.00
3.00
2.00
1.00
0.00
-1.00
-2.00
-3.00
-4.00
level (m)
0
6
12
18
24
30
36
42
48
54
60
66
72
time (h)

Fig. 6.22. (cont.)

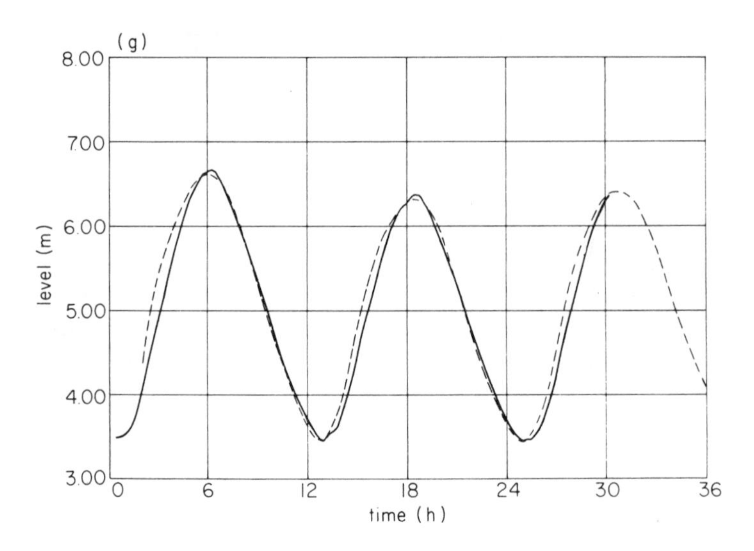

Fig. 6.22. (cont.)

An extension of this field test to include a storm surge simulation is illustrated in Fig. 6.23. Figure 6.23a shows a meteorological barometric pressure field on a 50-nautical-mile grid provided at 10 800-s (3-h) intervals, while Fig. 6.23b shows the corresponding computed geostrophic wind field interpolated onto the hydrodynamic grid of SD_0 at 900-s (15-min) time steps. A further space interpolation is, of course, used for SD_1 and SD_2. Figures 6.23c and 6.23d show the first-run, precalibration elevation field and velocity field computed using the Mark 6-3 for the same instant while Fig. 6.23e shows the first-run combined tide and storm surge elevations at Heligoland, as compared with observations.

The problems arising in two-dimensional modelling from the convective momentum terms have been introduced in Section 4.11, and the various numerical solutions have to be verified by field testing. All the two-dimensional example results given above were from System 21 models, all of which solve the full equations of nearly horizontal flow (Equations (2.4.1), (2.4.5), (2.4.6)), with additional terms (barometric pressure, surface wind stress, evaporation, etc.) as required. However, in order to test the convective momentum terms to their limits, field tests were made on the Bay of Fundy, Canada, where velocities of up to 7 knots occur under normal tidal conditions. Figures 6.24a and 6.24b show the model domains and subdomains, the special boundary module enabling the model to be truncated wherever required, outside regions of multiple reflections from subsequent constructions. A typical computed flow field, showing the exceptional velocities that occur in this area, is given in Fig. 6.24c.

The above examples were all taken from situations in which the water was vertically homogeneous. Figure 6.25, however, shows results from a region of strongly stratified flows, the Sound between Denmark and Sweden in which the highly saline waters of the North Sea meet the brackish waters of the Baltic. The

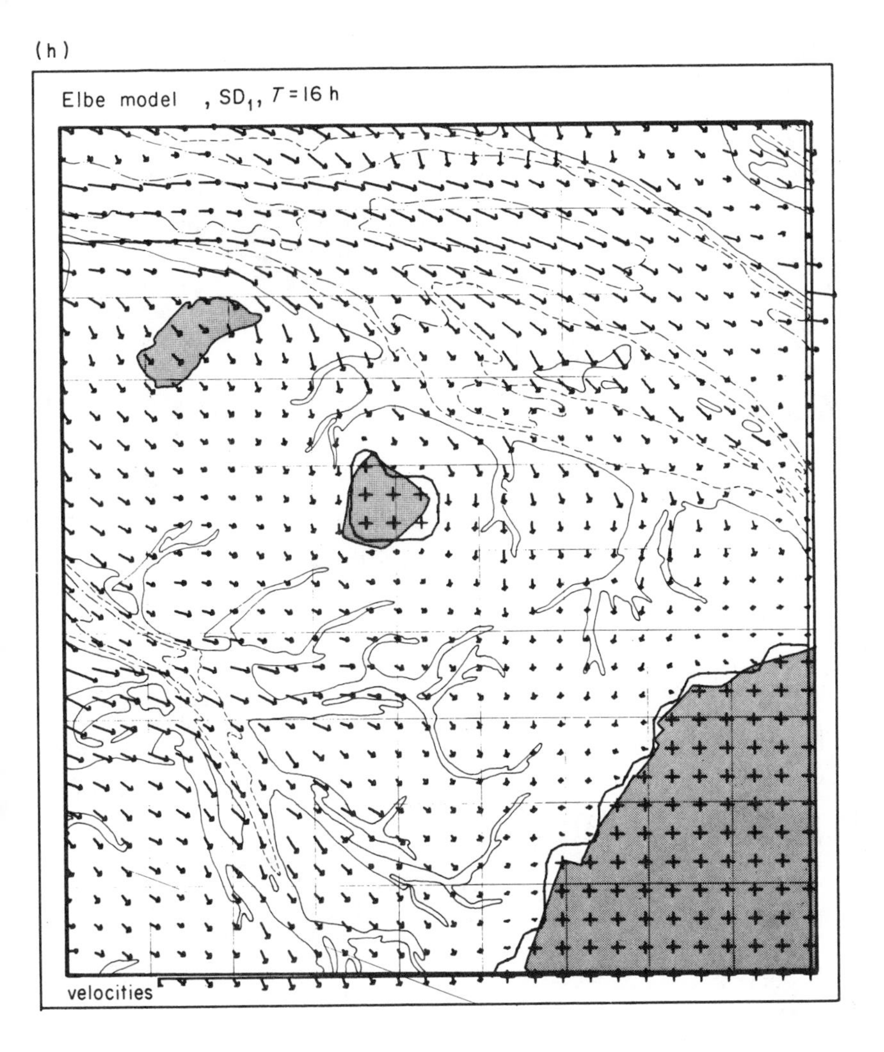

Fig. 6.22. (cont.)

model region is indicated in the satellite view, Fig. 6.25a, while Fig. 6.25b shows the bathymetry. Figure 6.25c shows velocities in the upper brackish waters and in the more restricted lower saline waters at one time while Fig. 6.25d shows these velocity fields again for a later time as the lower layer has intruded and spread under the influence of the wind-driven upper layer. Figure 6.25e shows comparison of computed and recorded velocities in the two layers during such a simulation. These results were obtained using a pre-production System 22, 'Neptune', modelling system.

A test of results given by a short-wave model against classical diffraction theory is shown in Fig. 6.26, together with some perspective plots.

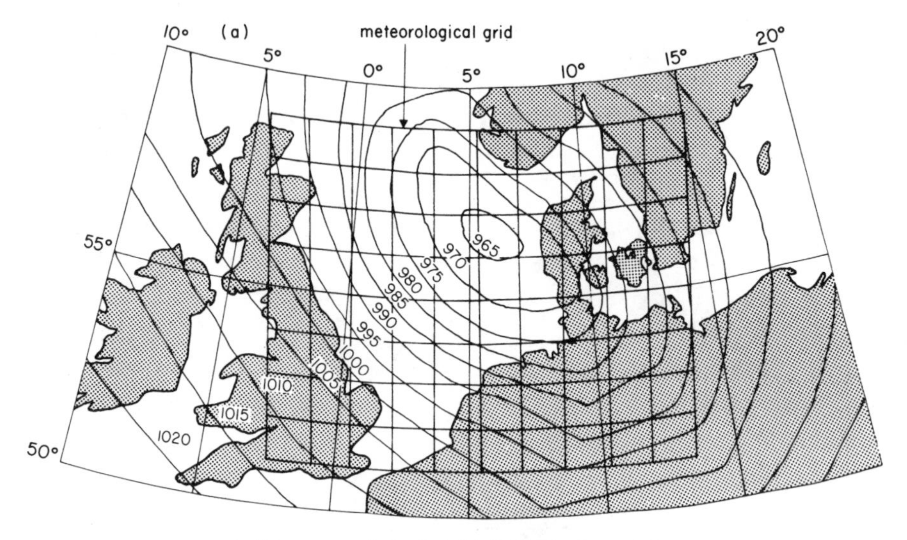

surface pressure (mbar) at 03:00 GMT on 3 Jan, 1976

(b)

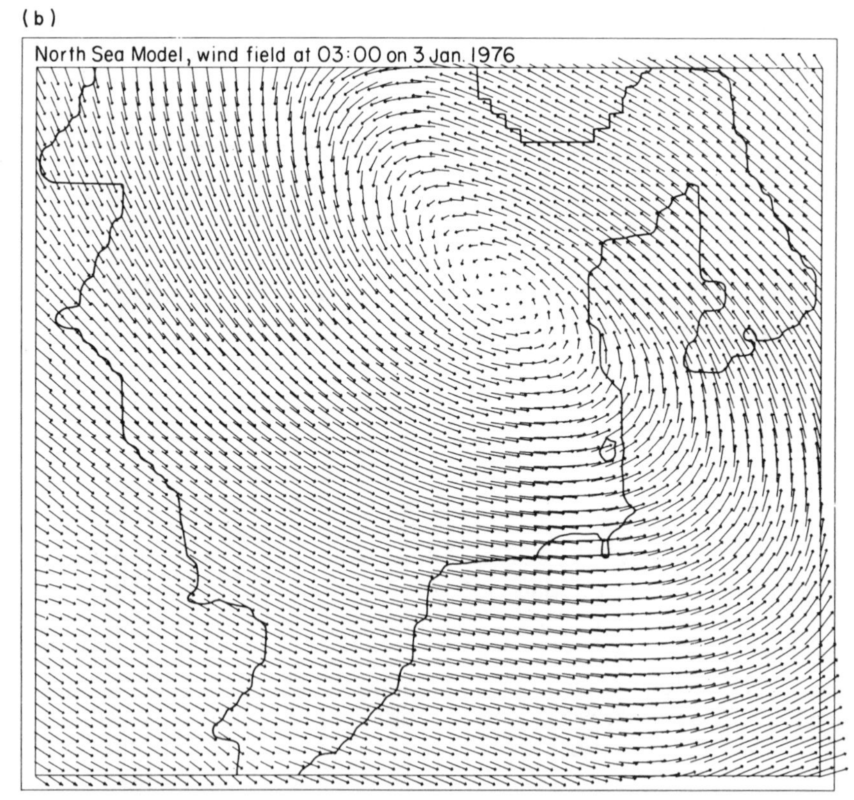

wind speed: ——→ 50 m s^{-1}

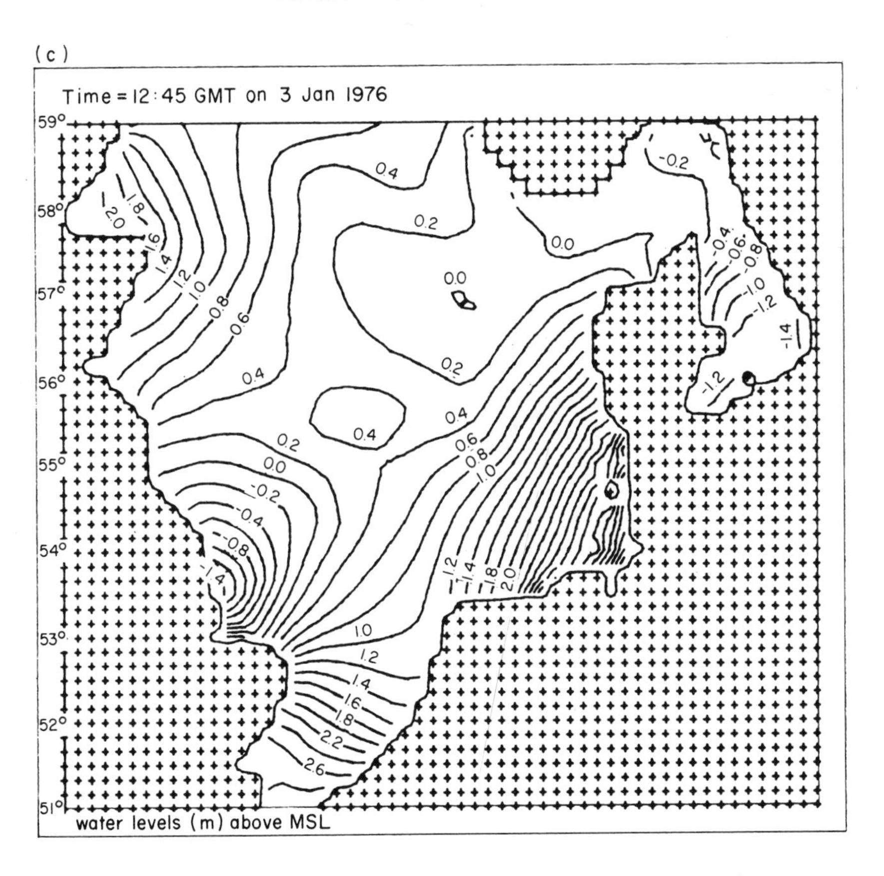

Fig. 6.23. The meteorological barometric pressure field (a) is transformed to a geostrophic wind field (b) and this is corrected to sea level to provide a surface stress. (c) The resulting first-run elevation field and (d) the velocity field produced by this barometric pressure and wind stress and the tide. (e) The corresponding computed combined tide and storm surge at Heligoland, computed at this first run (——), is compared with observations (– – –). (f) Two perspective views of the water surface of the North Sea under these storm conditions (see pages *294 to 297*)

To conclude these illustrations of the applications of computational hydraulics, Fig. 6.27 extends the test-rig results of Fig. 6.1 and the test simulation of Fig. 6.2b, and the field results of Fig. 6.4 using laboratory data to a field test on a complete harbour, with physical model tests for calibration and verification. The harbour concerned, Hantholm in northern Jutland, is shown in prototype in Fig. 6.27a and, skipping over the physical model, as a measured and a numerical model bathymetry in Fig. 6.27b. Figure 6.27c shows the spreading of short (12-s) periodic waves during the starting of the model from cold, together with a fully developed periodic wave situation. Figure 6.27e shows a perspective view and 6.27f and 6.27g illustrate some results obtained with irregular waves. These results were obtained using a prototype (up to 6.27c) and production versions

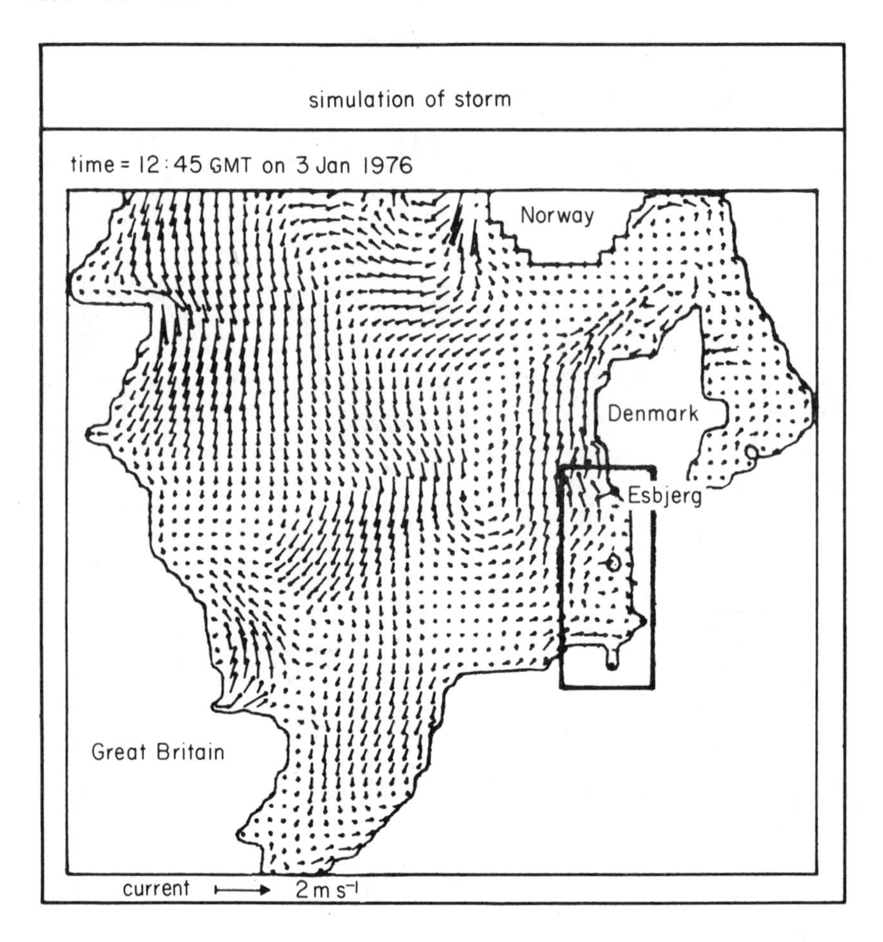
simulation of storm
time = 12:45 GMT on 3 Jan 1976
Norway
Denmark
Esbjerg
Great Britain
current
2 m s⁻¹

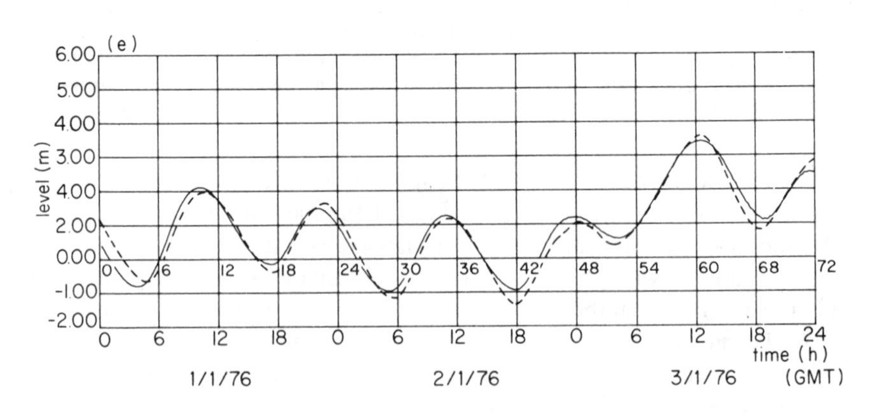
(e)
6.00
5.00
4.00
3.00
4.00
2.00
0.00
-1.00
-2.00
level (m)
0
6
12
18
24
30
36
42
48
54
60
68
72
0 6 12 18 0 6 12 18 0 6 12 18 24
time (h)
1/1/76
2/1/76
3/1/76
(GMT)

Fig. 6.23. (cont.)

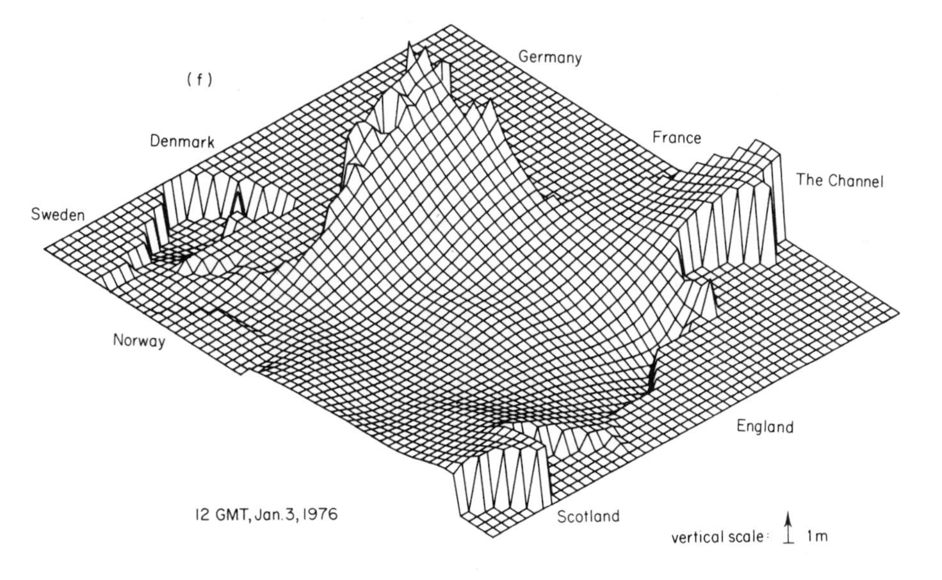
(f)
Germany
Denmark
France
The Channel
Sweden
Norway
England
12 GMT, Jan. 3, 1976
Scotland
vertical scale: 1m

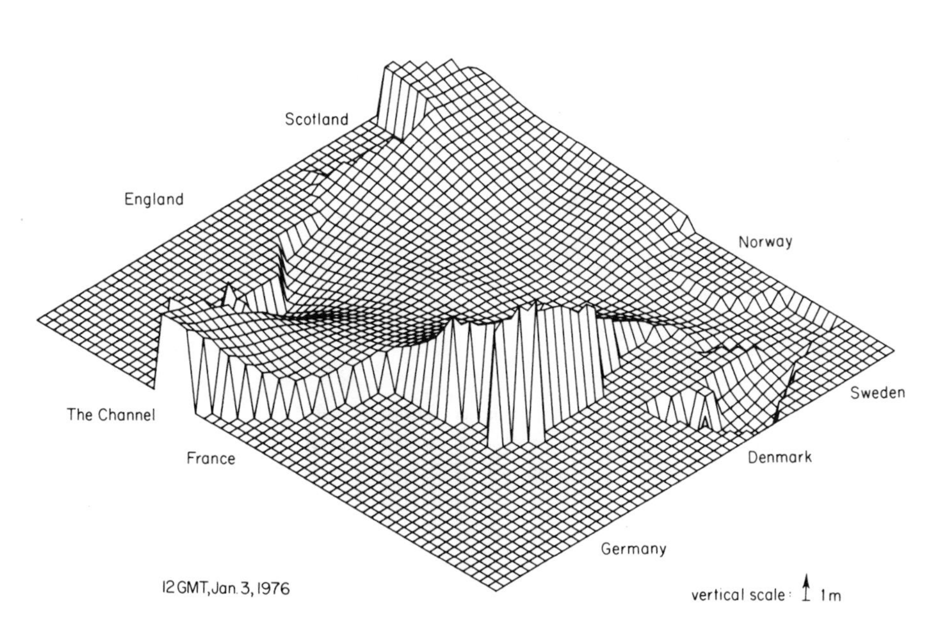
Scotland
England
Norway
The Channel
Sweden
France
Denmark
Germany
12 GMT, Jan. 3, 1976
vertical scale: 1m

Fig. 6.23. (cont.)

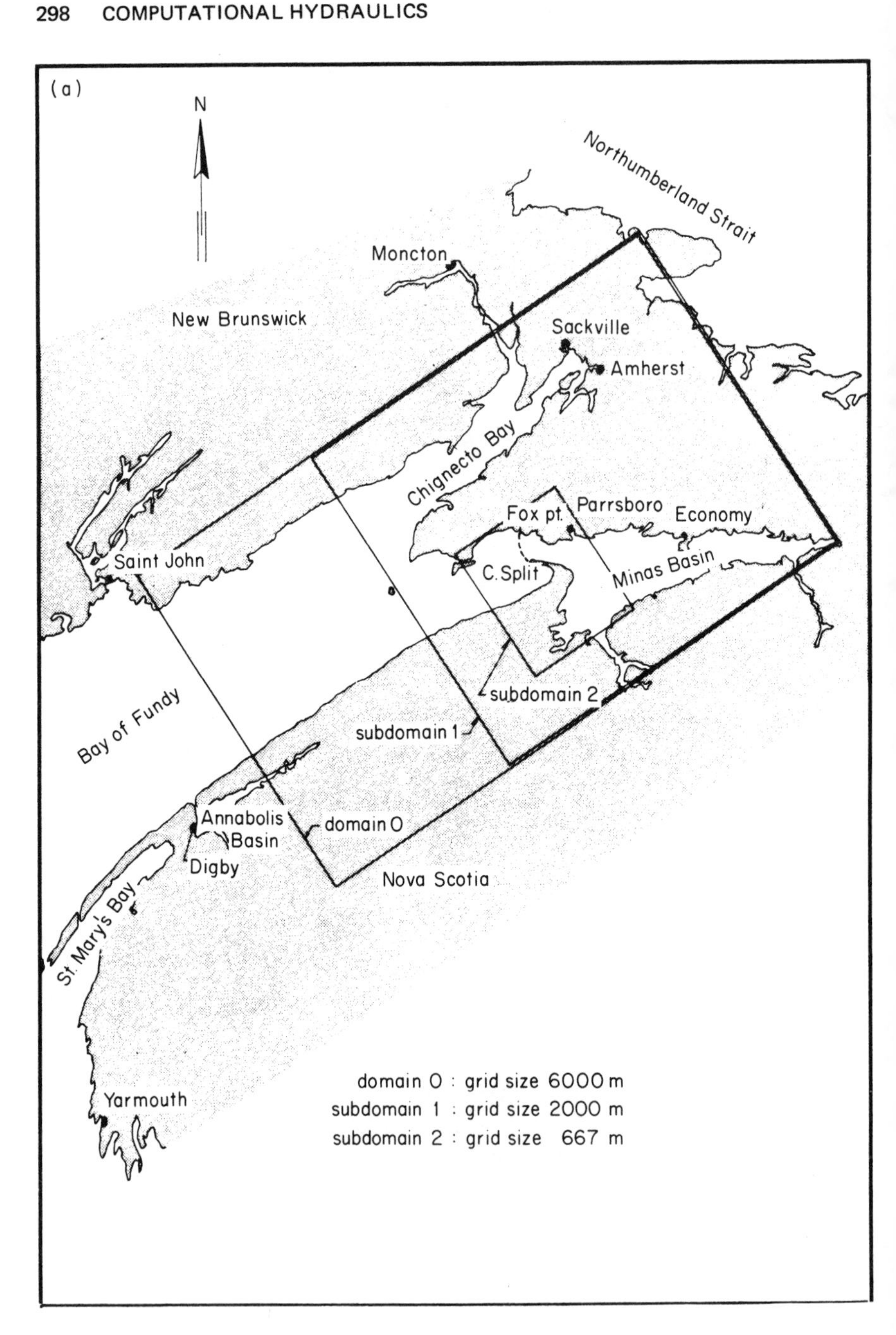

Fig. 6.24. (a) The model area and (b) the grid used for a high-velocity field test on the Bay of Fundy. (c) A typical velocity field in the finest grid. (*See pages 298 to 300*)

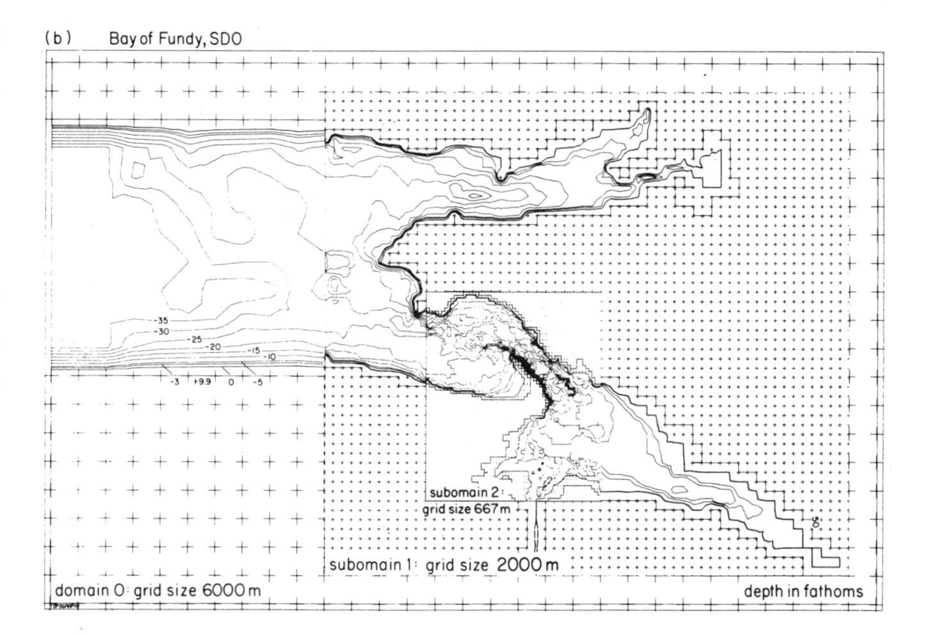

Fig. 6.24. (cont.)

of the System 21, 'Jupiter', Mark 8 modelling system that solves the full two-dimensional Boussinesq equations (2.7.9)–(2.7.11) using methods similar to those which have been outlined in Section 4.12. The four-stage alternating-direction schemes are here corrected to at least third-order accuracy so that a grid size of 10 m and a time step of 1 s can be used, providing acceptable running costs. It should be added that periodic waves have been used in most of Fig. 6.27 only for clarity, and in practice field-measured irregular wave trains have to be used as input in order to obtain physically useful harbour disturbance test results. These wave trains are synthesized from a special boundary module just like the periodic waves of Fig. 6.25 and they similarly refract and diffract in the model. However, in Fourier decomposition, only the components down to a period of about 4 s would be well represented in the model of Fig. 6.25. Fortunately, this is about the limit of practical interest for many if not most applications.

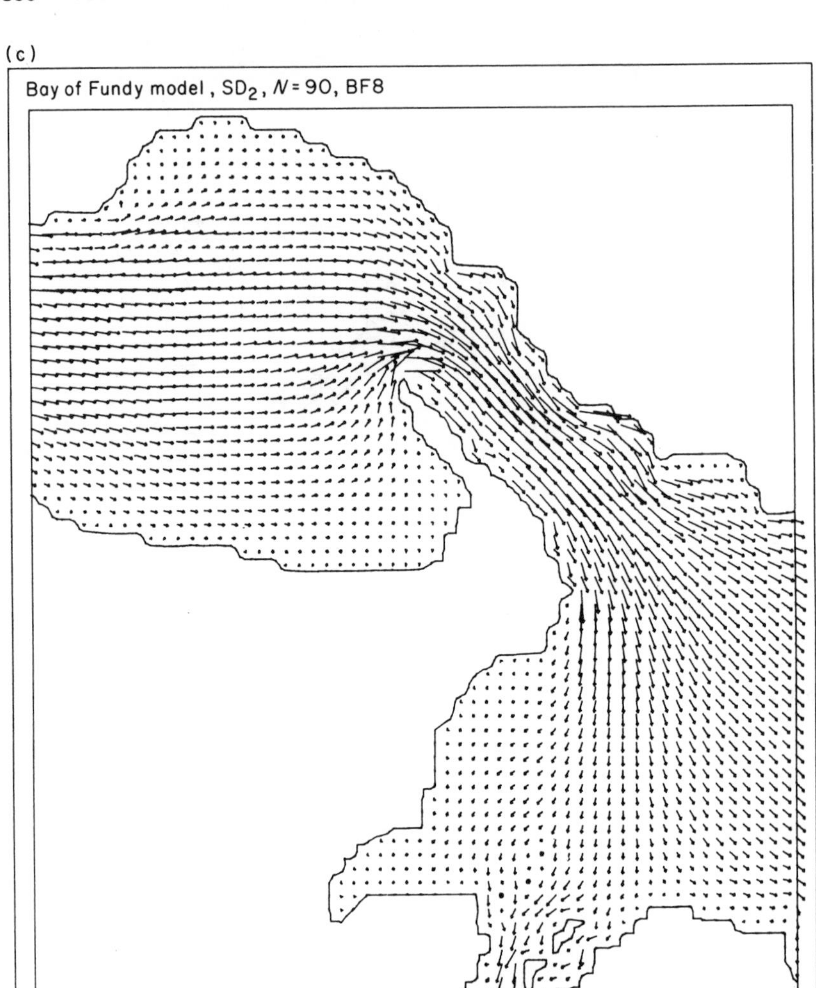

Fig. 6.24. (cont.)

Fig. 6.25. (a) Satellite image of the often strongly stratified Sound between Denmark and Sweden showing the model area. (b) The model bathymetry. (c), (d) Flux density fields in the upper and lower fluid layers of the model at different times during one simulation. (e) A comparison between computed and recorded flux densities in the two layers at a station near the centre of the model
(—) computed upper layer; (- - -) measured lower layer; (—) computed lower layer; (- - -) measured lower layer. The measured results correspond to recordings at 7, 10, 15, 20 and 25 m, with the rotorstop exception shown. (LANDSAT image) (*See pages 301 to 305.*)

(b)

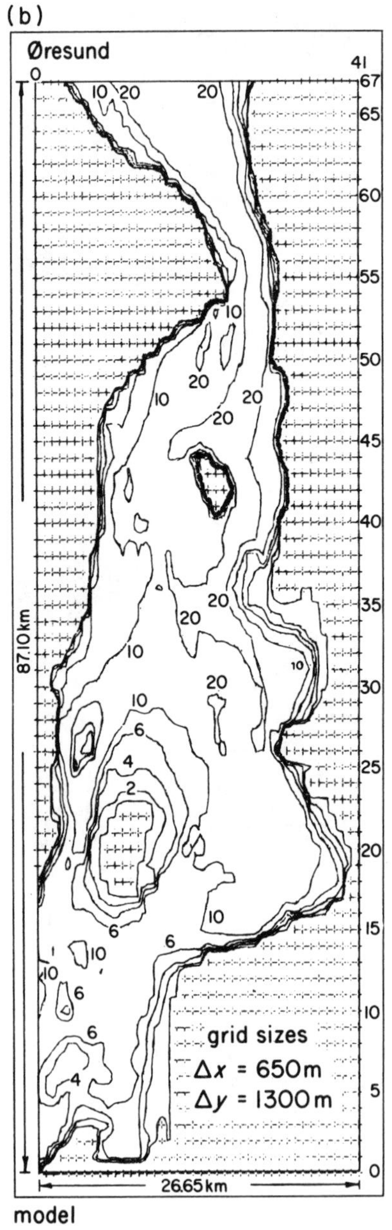

Höganäs
Gilleleje
Hornbæk
Helsingør
Helsingbo
Ven
Vedbaek
Landskron
København
Barsebacl
Salt- holm
Dragør
Malmö
Klagshamn
Skanör

model
depth (m) below DNN

chart

Fig. 6.25. (cont.)

(c)

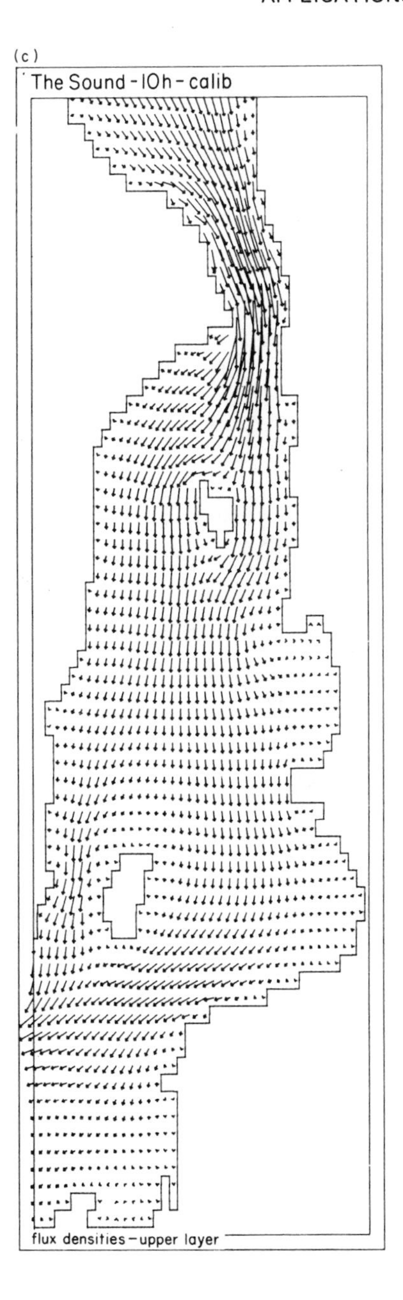

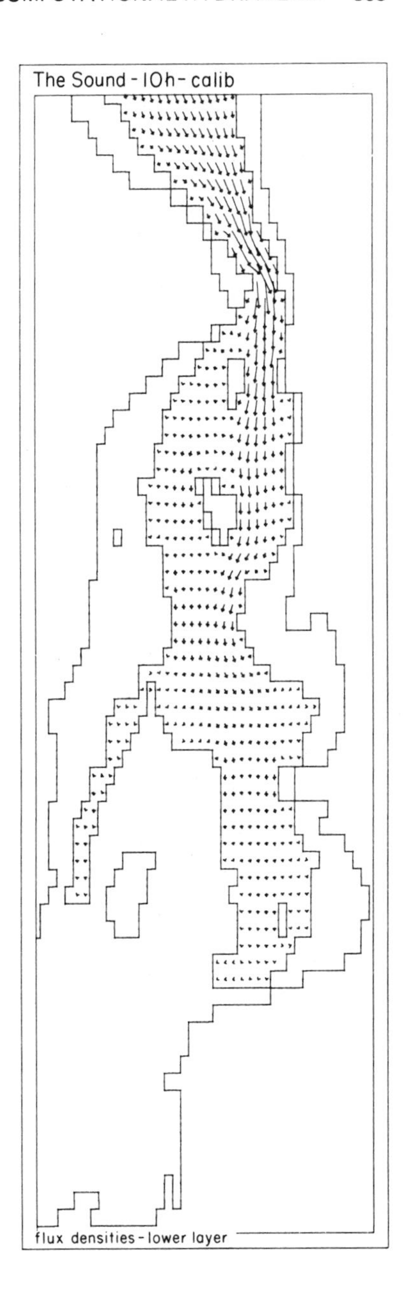

Fig. 6.25. (cont.)

(d)

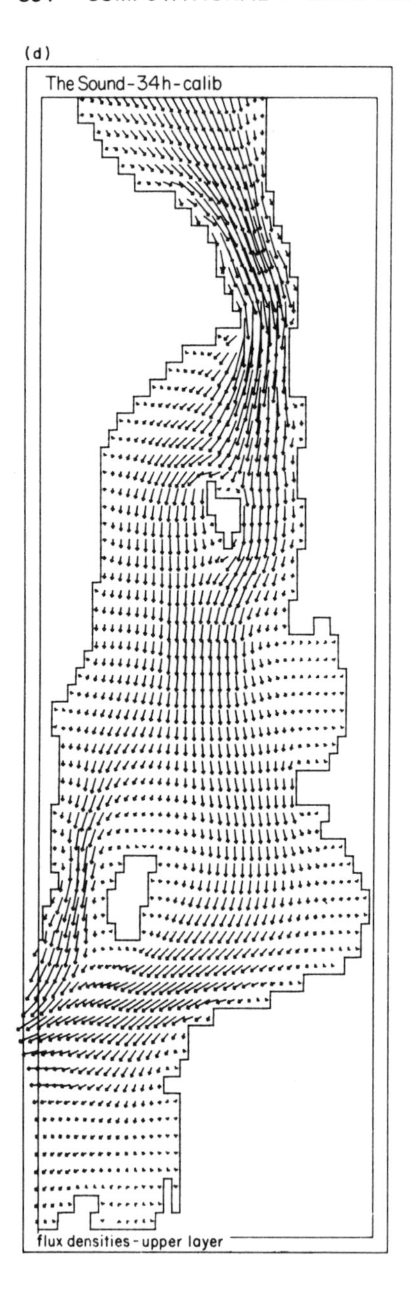

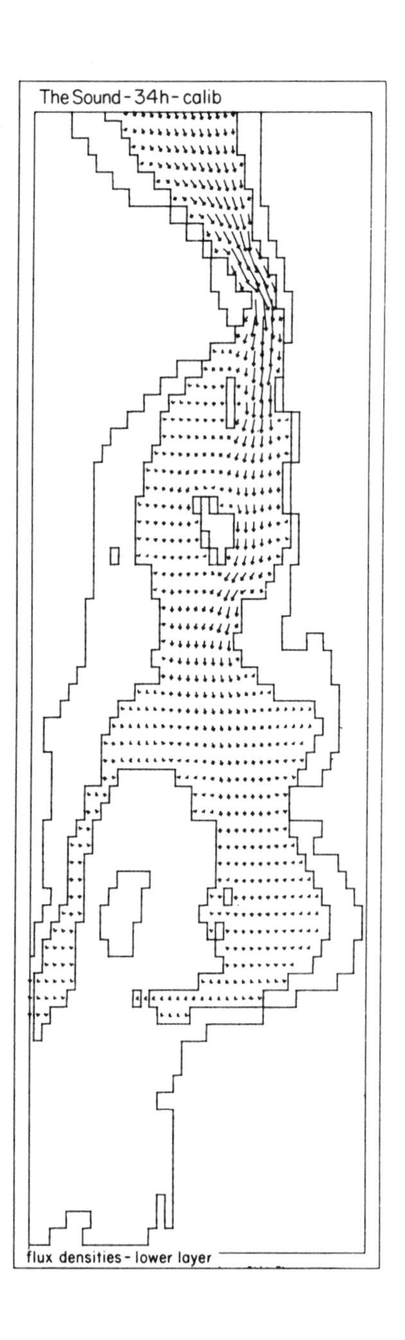

Fig. 6.25. (cont.)

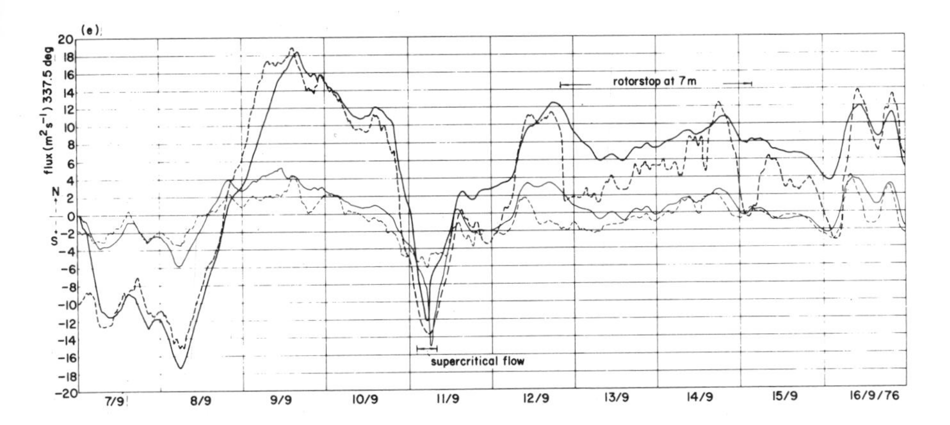

Fig. 6.25. (cont.)

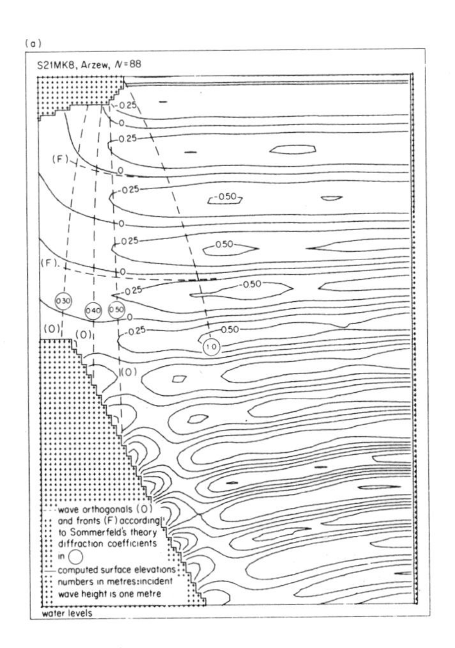

Fig. 6.26. Comparison of results obtained for the short-wave model with classical diffraction theory. (a) In the plan plot, the 0.5 lines of the classical theory are seen to demarcate correctly the limits of ±0.25 m elevations while the 1 lines correctly demarcate the limits of the 0.5 m elevations. (b) Perspective plots provide a more pictoral representation. *(See pages 305 to 307)*

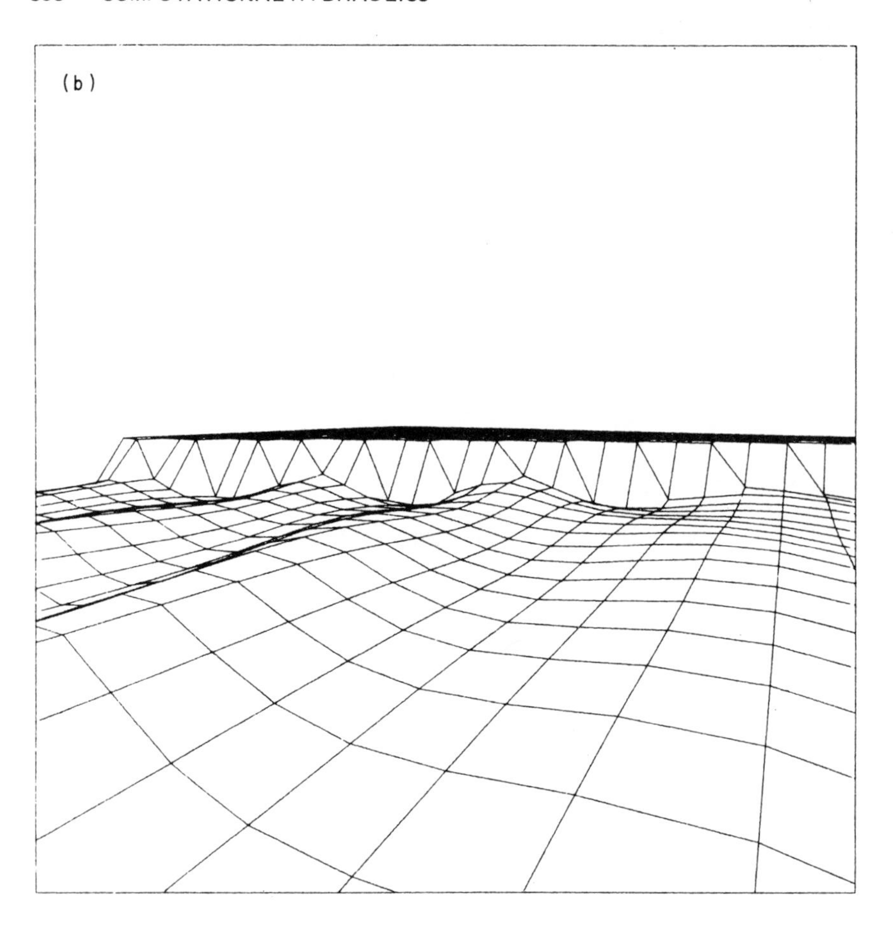

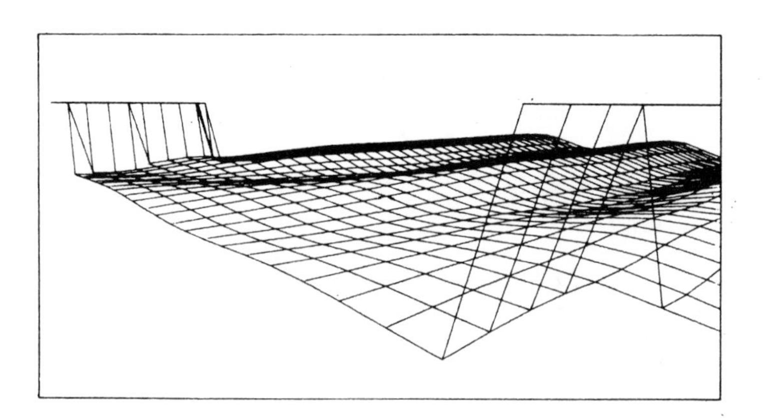

Fig. 6.26. (cont.)

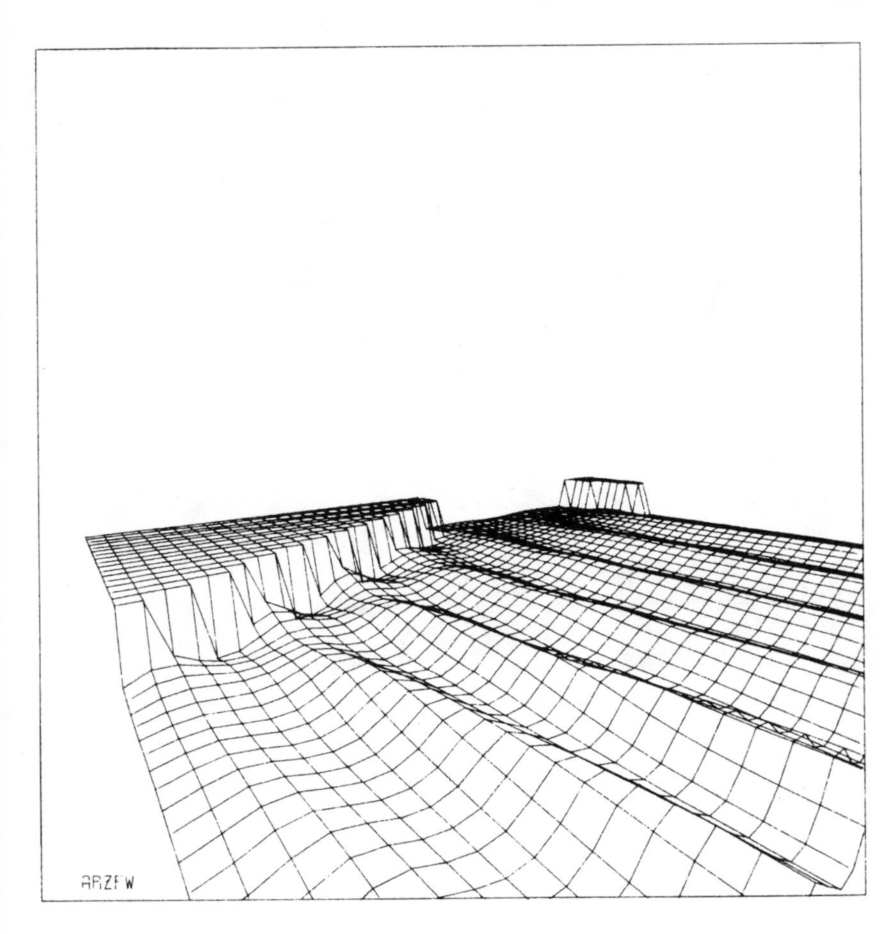

Fig. 6.26. (cont.)

(a)

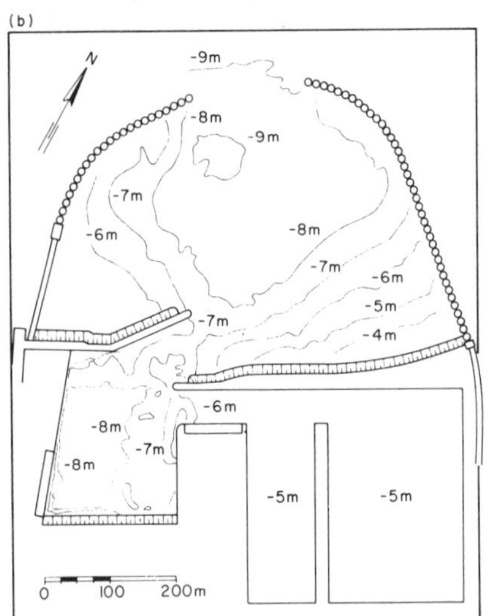
(b)
N
-9m
-8m
-9m
-7m
-6m
-8m
-7m
-6m
-5m
-7m
-4m
-6m
-8m
-7m
-8m
-5m
-5m
0 100 200m

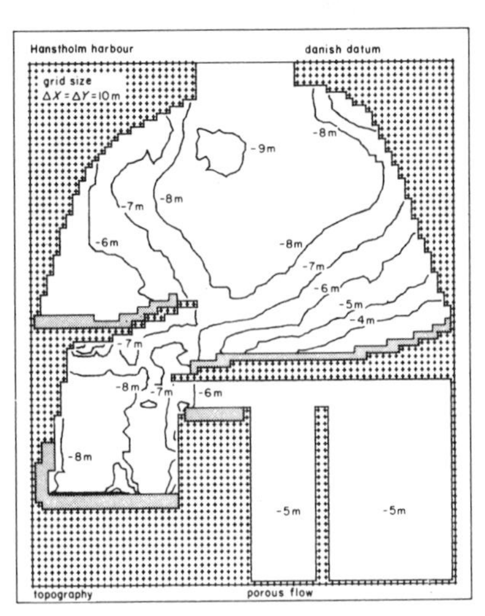
Hanstholm harbour
danish datum
grid size
ΔX = ΔY = 10m
-9m
-8m
-7m
-8m
-6m
-8m
-7m
-6m
-5m
-4m
-7m
-8m
-7m
-6m
-8m
-5m
-5m
topography
porous flow

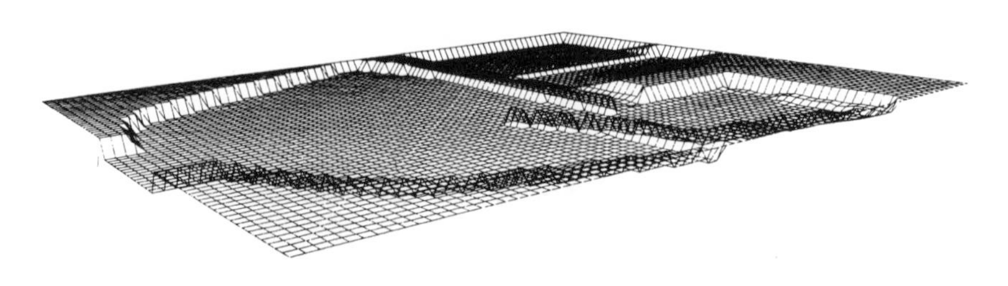

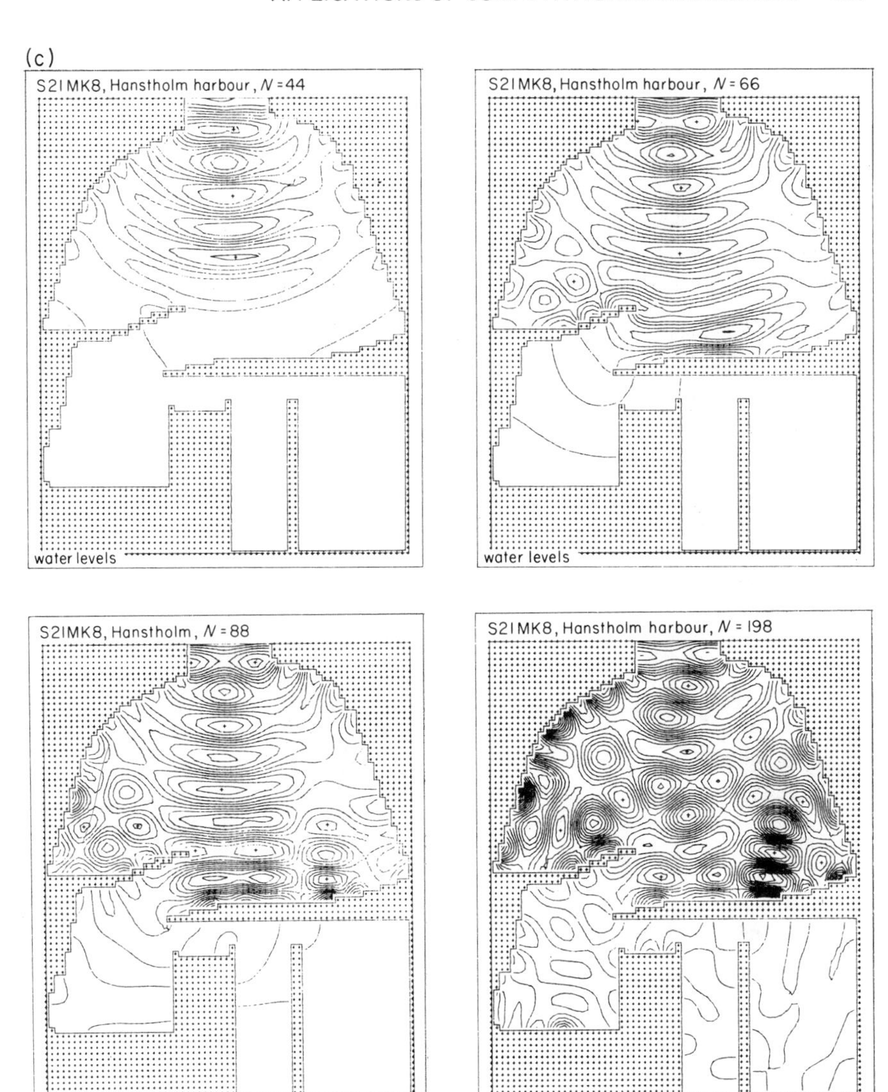

Fig. 6.27. Short-wave field test on Hanstholm harbour in northern Jutland. (a) An aerial view. (b) The bathymetries of the area as measured and as resolved in the model. (c) Periodic short waves in the harbour during a starting situation and (d) during an established oscillation. (e) A perspective view. (f) Irregular wave patterns. (g) Measuring stations. (h) A comparison between computed and physical model test r.m.s. wave heights for irregular waves (*See pages 308 to 312*)

(d)

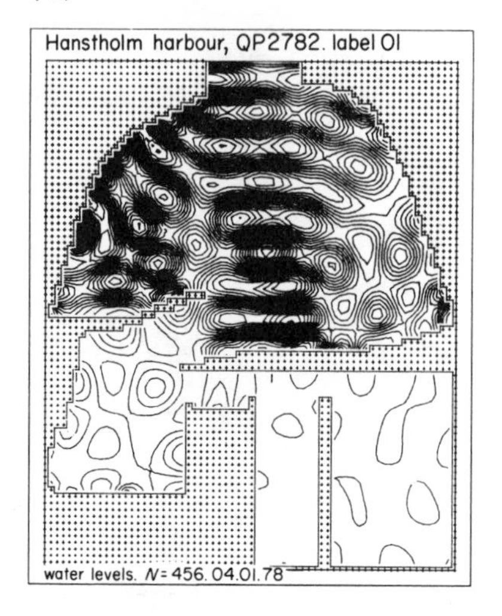

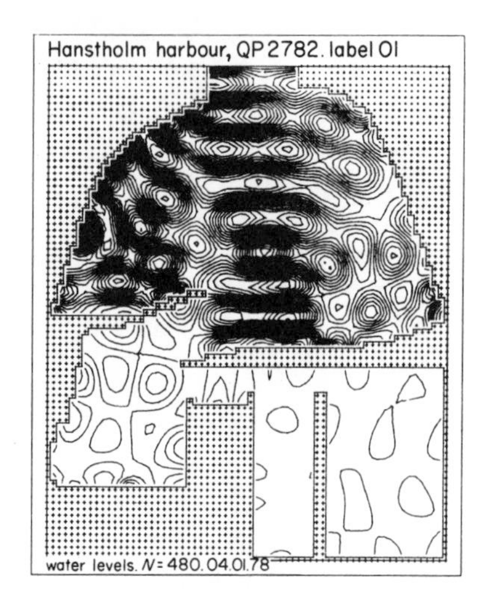

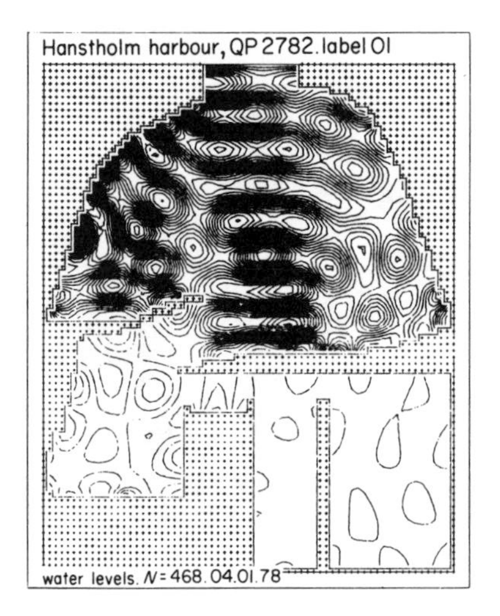

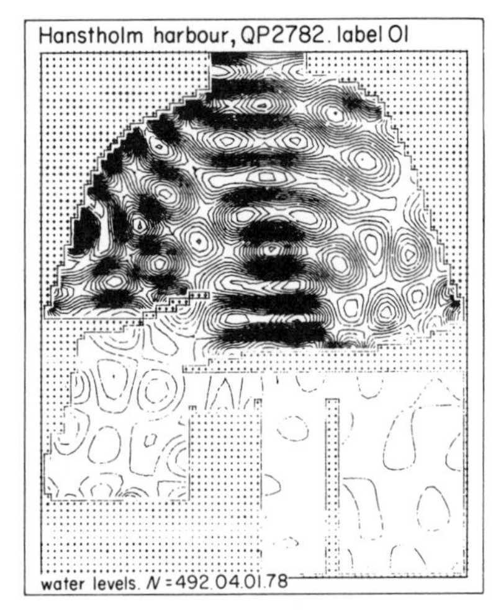

Fig. 6.27. (cont.)

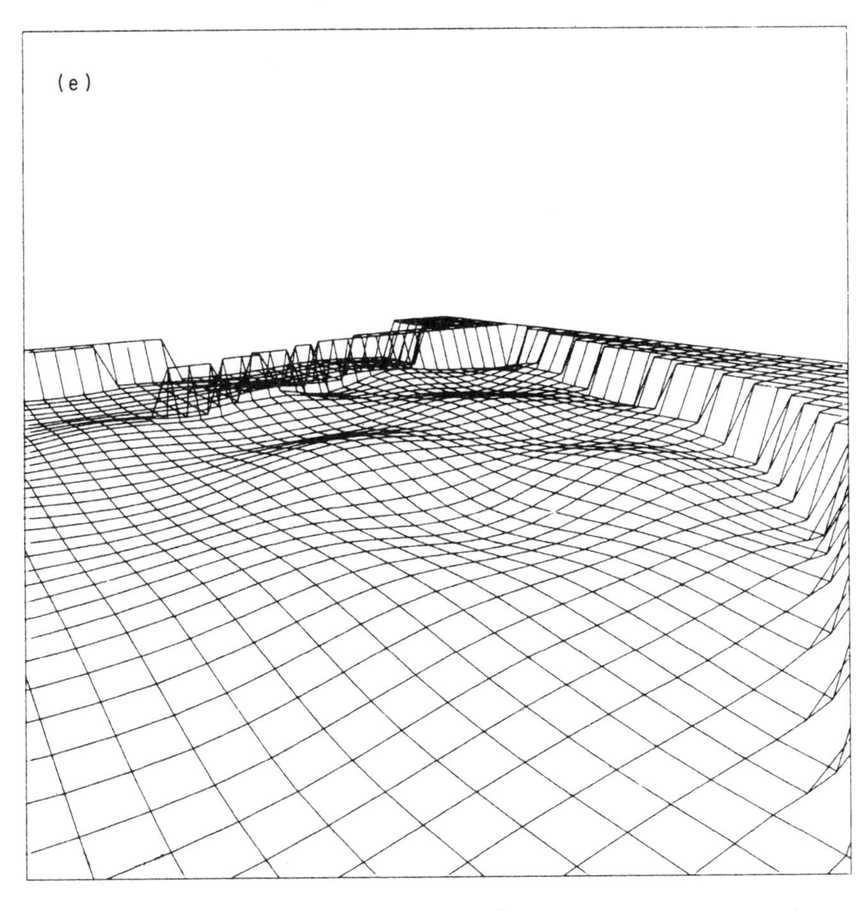
(e)

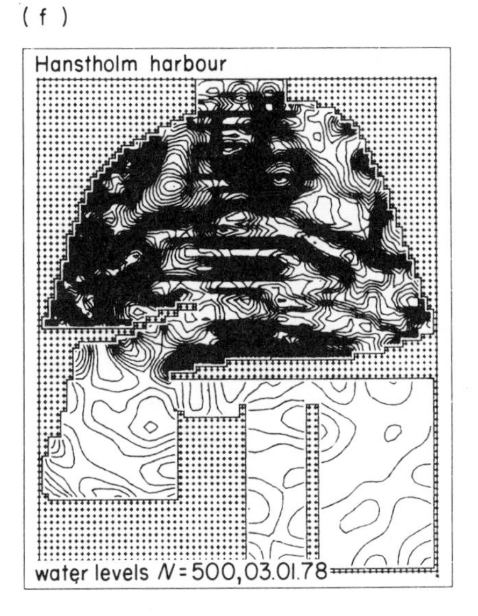
(f)
Hanstholm harbour
water levels N = 500, 03.01.78

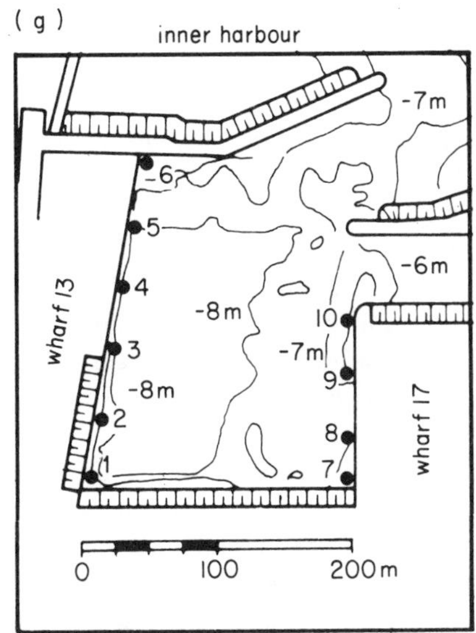
(g)
inner harbour
-7m
-6
5
4
wharf 13
3
-8m
2
1
-8m
-6m
10
-7m
9
8
7
wharf 17
0
100
200m

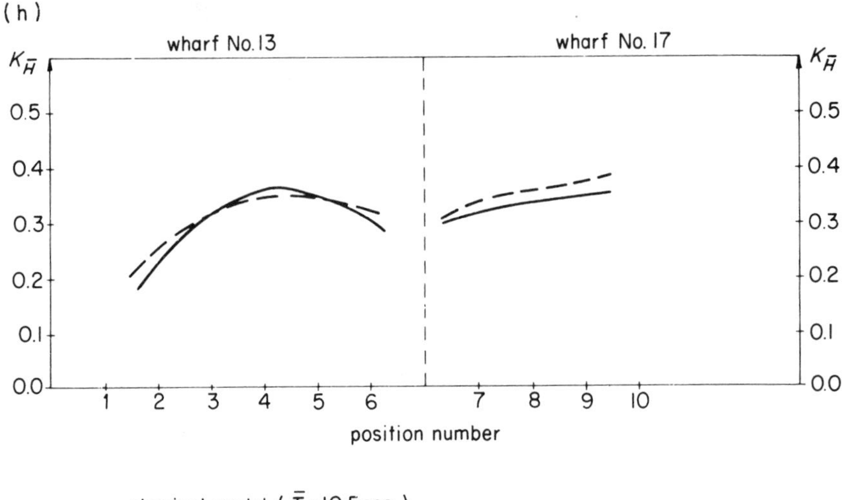

Fig. 6.27. (cont.)

6.5 Production models

The testing outlined above provides production modelling systems that are used to obtain solutions to important engineering problems. The models generated by these systems may be used to decide on the heights of dykes; on expected flood levels for insurance and other risk calculations; on the layout of harbours and offshore terminals; on the positioning and dimensioning of intake and discharge structures; and they obviously have very many other applications besides. Accordingly, these models frequently play a strategic role in the engineering process and their accuracy and reliability often underpin the entire engineering operation. However, any model is only as good as the data used to calibrate and verify it and it is thus essential in practice that good field investigations are made, necessarily backed up by reliable instrumentation and data-processing facilities, and that these investigations are properly integrated with the modelling work. The field teams then have to be aware of the new data requirements of numerical models, while the modelling teams must similarly be aware of the opportunities and limitations and corresponding economics of field investigation, instrumentation and data interpretation techniques. When this engineering complex works properly, computational hydraulics provides methods of hitherto unprecedented power and flexibility, so that dramatic economies can be made in all manner of engineering works, both at their execution stage and during their subsequent operation. If, however, the field investigation is not carried out well, if the instruments malfunction, if the data interpretation is incorrect, if there is insufficient redundancy in the data for their inadequacies to appear in the modelling, or if the modelling itself is not properly carried out, the design of works will proceed on incorrect premises and the works themselves will not function properly. The results may range from a mild inconvenience to a disaster on the greatest scale.

Thus, although the very scale and nature of modern construction methods necessitate that their design uses all possible aids – and in hydraulic, coastal and offshore engineering, computational hydraulics is one of the most significant of these aids – the whole modern complex of design aids is a delicate one and it is even potentially dangerous. The elaboration of a sound computational hydraulics is one contribution to making the modern engineering process somewhat less dangerous. And even this most ethical of objectives has its own aesthetic.

Fig. 6.28. Numerical modelling for engineering practice depends upon well planned and thoroughly executed field studies, using instruments that are accurate and reliable by virtue of their sound construction, calibration and maintenance. Instruments must be accurately positioned, with anchoring, buoying and signalling arrangements appropriate to expected wave and current conditions, while their recordings must be retrieved and the instruments reset in good time. It is only when the numerical model is properly complemented by sound field investigations that it becomes a true instrument of engineering practice. The photograph shows one of the vessels used in the field studies that provided the field data of figure 6.14.

References

Abbott, M. B. (1961). On the spreading of one fluid over another, *La Houille Blanche*, **16**, 5 and 6, pp. 622–635 and 827–846.

Abbott, M. B. (1966). *An Introduction to the Method of Characteristics*, Thames & Hudson, London, and American Elsevier, New York.

Abbott, M. B. (1974). Continuous flows, discontinuous flows and numerical analysis, *J. Hyd. Res.*, **12**, pp. 417–467.

Abbott, M. B. (1976, 1). Computational hydraulics, a short pathology, *J. Hyd. Res.*, **14**, pp. 271–285.

Abbott, M. B. (1976, 2). The application of design systems to problems of unsteady flow in open channels, *Int. Symp. on Unsteady Flow in Open Channels*, BHRA Fluid Engg, Cranfield, UK.

Abbott, M.B. (1979). Commercial and scientific aspects of mathematical modelling *Advances in Engineering Software*, **1**, pp. 147–152.

Abbott, M. B., Damsgaard, A. and Rodenhuis, G. S. (1973). System 21, Jupiter, a design system for two-dimensional nearly horizontal flows, *J. Hyd. Res.*, **11**, pp. 1–28.

Abbott, M. B. and Grubert, J. P. (1973). Towards a design system for stratified flows, *IAHR Int. Symp. on Stratified Flows, Novosibirsk*, pp. 113–124, ASCE, New York.

Abbott, M. B. and Ionescu, F. (1967). On the numerical computation of nearly-horizontal flows, *J. Hyd. Res.*, **5**, pp. 97–117.

Abbott, M. B., Marshall, G. and Ohno, T. (1969). *On Weak Solutions of the Equations of Nearly Horizontal Flow*, Report Series 3, International Institute for Hydraulic and Environmental Engineering, Delft.

Abbott, M. B., Pardo-Castro, E. and Tas, P. (1967). On the optimum recording of a type of hydrological data, *Bull. IASH*, **12**, pp. 44–58.

Abbott, M. B., Petersen, H. M. and Skovgaard, O. (1978). Numerical modelling of short waves in shallow water, to be published in *J. Hyd. Res.*, **3**.

Abbott, M. B. and Rasmussen, C. H. (1977). On the numerical modelling of rapid contractions and expansions in models that are two-dimensional in plan, *Proceedings 17th Congress IAHR, Baden-Baden*, **2**, pp. 229–238.

Abbott, M. B. and Rodenhuis, G. S. (1972). A numerical simulation of the undular hydraulic jump, *J. Hyd. Res.*, **10**, pp. 239–257, discussion in *J. Hyd. Res.*, **12**, pp. 141–152 (1974).

Abbott, M. B. and Torbe, I. (1963). On flows and fronts in a stratified fluid, *Proc. Roy. Soc.* A, **273**, pp. 12–40.

Abbott, M. B. and Vium, M. P. (1977). Computational hydraulics, an alternative view, *J. Hyd. Res.*, **15**, pp. 97–123.

Aitken, A. C. (1939). *Determinants and Matrices*, Oliver & Boyd, Edinburgh. (Second edition: 1958).

Anderson, D. and Fattahi, B. (1974). A comparison of numerical solutions of the advective equation, *J. Atm. Sci.*, **31**, pp. 1500–1506.

Aramanovich, I. G., Guter, R. S., Lyusternik, L. A., Raukhvarger, I. L., Skanavi, M. I. and Yanpol'skii, A. R. (1965). *Mathematical Analysis, Differentiation and Integration* (Trans. H. Moss), (2nd edn) Pergamon, Oxford.

Barnett, A. G. (1976). Numerical stability in unsteady open channel flow computations, *Int. Symp. on Unsteady Flow in Open Channels*, BHRA Fluid Engg, Cranfield, UK.

Benjamin, T. B. and Lighthill, M. J. (1954). On cnoidal waves and bores, *Proc. Roy. Soc.* A, **224**, pp. 448–460.

Bourbaki, N. (1960). *Eléments d'histoire des mathématiques*, Hermann, Paris.

Boussinesq, J. (1872). Théorie des ondes et des remous etc. *J. Math. Pure et Appliqué*, 2ème Ser., **17**, pp. 55–108.

Brillouin, L., (1956). *Science and Information Theory*, Academic Press, New York.

Bruce, G. H., Peacemann, D. W., Rachford, H. H. and Rice, J. D. (1953). Calculation of unsteady gas flow through porous media, Petroleum, *Trans. AIME*, **198**, p. 79.

Callen, H. B. (1960). *Thermodynamics*, John Wiley, New York.

Cantor, G. (1895–1897). *Contributions to the Founding of the Theory of Transfinite Numbers* (trans. P. E. B. Jourdain). (Reprinted by Dover, New York.)

Cunge, J. A. (1969). On the subject of flood propagation computation method (Muskingum method), *J. Hyd. Res.*, 7, pp. 205–230.

Cunge, J. A. (with J. A. Liggett) (1975). Numerical methods of solution of the unsteady flow equations, in *Unsteady Flow in Open Channels* (ed. K. Mahmood and V. Yevjevich), Chap. 4, Water Research Publications, Fort Collins, Colorado, USA.

Cunge, J. A. and Wegner, M. (1964). Intégration numérique des équations d'écoulement de Barré de St Venant par un schéma implicite de differences finies, *La Houille Blanche*, **1**, pp. 33–39.

Courant, R. and Friedrichs, K. O. (1948). *Supersonic Flow and Shock Waves*, Interscience, New York.

Courant, R., Friedrichs, K. O. and Lewy, H. (1928). Über die partiellen differenzengleichungen der mathematischen Physik, *Math. Ann.*, **100**, p. 32.

Craya, A., (1945/1946). Calcul graphique des regimes variables dan les canaux, *La Houille Blanche*, 1, Nov 1945–Jan 1946, pp. 19–38 and 2, Mar 1946, pp. 117–130.

Daubert, A. (1964). Quelques aspects de la propagation des crues, *La Houille Blanche*, **3**, pp. 341–346.

Daubert, A. and Graffe, O. (1967). Quelques aspects des écoulements presque horizontaux à deux dimensions en plan et non permanents: application aux estuaires, *La Houille Blanche*, **8**, pp. 847–860.

Dedekind, R. (1872). *Essays on the Theory of Numbers* (trans. W. W. Beman), Dover, New York. (Reprinted: 1901, 1963)

Dooge, J. C. I. (1973). *Linear Theory of Hydrologic Systems*, Tech. Bull. 1468, US Department of Agriculture, Washington, DC.

Dronkers, J. J. (1964). *Tidal Computations in Rivers and Coastal Waters*, North-Holland, Amsterdam.

Engelund, F. A. (1965, 1). *Laerbog: Hydraulik*, Private Ingeniorfond, DTH, Copenhagen. (In Danish)

Engelund, F. A. (1965, 2). A note on Vedernikov's criterion, *La Houille Blanche*, **8**, pp. 801–802.

Evans, E. P. (1977). The behaviour of a mathematical model of open channel flow, *Proceedings 17th Congress IAHR, Baden-Baden*, **2**, pp. 173–180.

Falconer, R. A. (1976). Mathematical model of jet-forced circulation in reservoirs and harbours, *Ph.D. Thesis*, University of London.

Fast, J. D. (1962). *Entropy, The Significance of the Concept of Entropy and its Applications in Science and Technology*, Philips Technical Library, Eindhoven.

Favre, H. (1935). *Ondes de Translation*, Dunod, Paris.

Flokstra, C. The closure problem for depth-averaged two-dimensional flow. *Proceedings 17th Congress IAHR, Baden-Baden,* **2**, pp. 247–256.

Friedrichs, K. O. (1948). On the derivation of the shallow water theory, *Comm. Pure and Applied Math.*, **1**, pp. 81–85.

Galileo Galilei (1590, 1600). *On Motion and Mechanics, Comprising 'De Mota' and 'Le Meccaniche'* (trans. I. G. Drabkin and S. Drake), University of Wisconsin Press, Madison, Wisconsin. (Reprinted: 1960)

Galileo Galilei (*ca.* 1600). *Dialogues Concerning Two New Sciences* (trans. H. Crew and A. de Silvio), Macmillan, London (1914). (Reprinted by Dover, New York: 1954)

Goldstein, H. (1957). *Classical Mechanics*, Addison-Wesiey, Reading, Massachusetts.

Gelfond, A. O. (1961). *Solution of Equations in Integers* (trans. L. F. Boron), Dover, New York.

de Groot, S. R. (1951). *Thermodynamics of Irreversible Processes*, North-Holland, Amsterdam.

Grijsen, J. G. and Vreugdenhil, C. B. (1976). Numerical representation of flood waves in rivers, *I.A.H.R. Symp. on Unsteady Flow in Open Channels, Newcastle upon Tyne*.

Gustafsson, B. (1971). An alternating direction implicit method for solving the shallow water equations, *J. Comp. Phys.*, **7**, pp. 239–254.

Hadamard, J. (1923). *Lectures on Cauchy's Problem in Linear Partial Differential Equations*, Dover, New York. (Reprinted: 1952)

Hamming, R. W. (1977). *Digital Filters*, Prentice Hall, Englewood Cliffs.
Hankins, T. H. (1967). The reception of Newton's second law of motion in the eighteenth century, *J. History of Ideas*, pp. 43–65.
Hansen, B. (1965). *A Theory of Plasticity for Ideal Frictionless Materials*, Teknisk Forlag, Copenhagen.
Hansen, W. (1956). Theorie zur Errechnung des Wasserstandes und Strömungen in Randmeeren nebst Anwendungen, *Tellus*, **3**, pp. 287–300.
Harlow, F. H. (1971). Contour dynamics for numerical fluid flow calculations, *J. Comp. Phys.*, **8**, pp. 214–229.
Hartree, L. R. (1952). *Numerical Analysis*, Oxford University Press, London.
Hasse, L. (1974). On the surface to geostrophic wind relationship at sea and the stability dependence of the resistance law, *Beitr. Phys. Atm.*, **47**, pp. 45–55.
Henderson, J. C. de C.,(1960). Topological aspects of structural linear analysis, *Aircraft Engineering*, **31**, pp. 137–141.
Holley, F. M. and Cunge, J. A. (1975). Time dependent mass dispersion in natural streams, *Symposium on Modelling Techniques*, American Society of Civil Engineers, New York.
Hill, M. N. (General Editor) (1962). *The Sea*, Vol. 1, Interscience, New York.
Hydén, H. (1974). Språngskiktsrörelser och turbulens i densitetsskiktade vatten, Thesis, Inst. Vattenbyggnad, Royal Technical University, Stockholm. (In Swedish)
Iwasa, I. (1966). *Lecture Note on Mathematical Analysis of Steady Behaviour of One-Dimensional Free Surface Shear Flows*, Hydraulics Laboratory, Department of Civil Engineering, Kyoto.
Jeans, J. (1940). *An Introduction to the Kinetic Theory of Gases*, Cambridge University Press.
Jollife, A. N. (1968). *The Investigation of Systems of Regular Channels of the General Type Occurring in Catchments*, Report Series No. 4, International Institute for Hydraulic and Environmental Engineering, Delft.
Jonsson, I. G. (1978). Energy flux and wave action in gravity waves propagating on a current, to be published in *J. Hyd. Res.*
Kamke, E. (1950). *Theory of Sets* (trans. F. Bagemihl), Dover, New York.
Khinchin, A. I. (1957). *Mathematical Foundations of Information Theory* (trans. R. A. Silverman and M. D. Friedman), Dover, New York.
Kolmogorov, A. N. and Fomin, S. V. (1957). *Elements of the Theory of Functions and Functional Analysis*, Vol. 1: *Metric and Normed Spaces* (trans. L. F. Boron), Graylock, Rochester, New York.
Kolmogorov, A. N. and Fomin, S. V. (1961). *Elements of the Theory of Functions and Functional Analysis*, Vol. 2: *Measure, The Lebesgue Integral, Hilbert Space* (trans. H. Kamel and H. Kumm), Graylock, Rochester, New York.
Kopal, Z. (1955). *Numerical Analysis*, Chapman & Hall, London.
Korteweg, D. J. and Vries, G. de (1895). On the change of form of long waves advancing in a rectangular channel, and on a new type of long stabilising wave, *Phil. Mag.*, **5**, p. 41.
Lamb, H. (1932), *Hydrodynamics,* 6th Edn., Cambridge University Press.

Landau, L. D. and Lifshitz, E. M. (1959). *Fluid Mechanics* (trans. J. S. Sykes and W. H. Reid), Pergamon, Oxford.

Landau, L. D. and Lifshitz, E. M. (1960), *Mechanics* (trans. J. S. Sykes and W. H. Reid), Pergamon, Oxford.

Launder, B. E. and Spalding, D.B. (1972), *Mathematical Models of Turbulence*, Academic Press, London and New York.

Lax, P. D. (1954). Weak solutions of non-linear hyperbolic equations and their numerical computation, *Comm. Pure and Applied Math.*, 7, pp. 159–193.

Lax, P. D. and Wendroff, B. (1960). Systems of Conservation Laws, *Comm. Pure and Applied Math.*, **13**, pp. 217–237.

Ledermann, W. (1957). *Introduction to the Theory of Finite Groups*, Oliver & Boyd, Edinburgh.

Leendertse, J. J. (1967). *Aspects of a Computational Model for Long Water Wave Propagation*, Rand Memorandum, RH-5299-PR, Santa Monica, California.

Lemoine, R. (1948). Sur les ondes positives de translation dans les canaux et sur le ressaut ondulé de faible amplitude, *La Houille Blanche*, **3**, pp. 183–185.

Leonard, A. (1974). Energy cascade in large-eddy simulations of turbulent fluid flows, *Adv. in Geophys.*, A **18**, pp. 237–248.

Leslie, D. C. and Quarini, G. L. (1979). The application of turbulence theory to the formulation of subgrid modelling procedures. *J.F.M.* **91** pp. 65–91.

Liggett, J. A. and Cunge, J. A. (1975). Numerical methods of solution of the unsteady flow equation, in *Unsteady Flow in Open Channels*, **1**, Edited by Mahmood K. and V. Yevjevich, Water Resources Publ., Fort Collins, Colorado.

Lighthill, M. J. and Whitham, G. B. (1955). On kinematic waves: I. Flood movement in long rivers; II. Theory of traffic flow on long crowded roads, *Proc. Roy. Soc.* A, **229**, pp. 281–345.

Liusternik, L. and Sobolev, V. (1961). *Elements of Functional Analysis* (trans. A. E. Labarre, H. Izbicki and H. W. Crawley), Unger, New York.

Long, R. R. (1964). The initial-value problem for long waves of finite amplitude, *J. Fluid Mech.*, **20**, pp. 161–176.

Longuet-Higgins, M. S. and Stewart, R. W. (1960). Changes in the form of short gravity waves on long waves and tidal currents, *J. Fluid Mech.*, **8**, pp. 565–583.

Longuet-Higgins, M. S. and Stewart, R. W. (1962). Radiation stress and mass transport in gravity waves, with applications to 'Surf Beats', *J. Fluid Mech.*, **13**, pp. 481–504.

Lundgren, H. (1962). *The Concept of the Wave Thrust*, Basic Research Progress Report 3, Coastal Engineering Laboratory, Technical University of Denmark.

Mahieux, F. (1972). *Le Calcul de la Rentabilité de la Recherche, Sélection des Projets*, Eyrolles, Paris.

Malvern, L. E. (1969). *Introduction to the Mechanics of a Continuous Medium*, Prentice Hall, Englewood Cliffs, New Jersey.

Manin, Yu. I. (1977). *A Course in Mathematical Logic*, Springer, New York.

Marchand, J.-P. (1962). *Distributions, An Outline*, North-Holland, Amsterdam.

Maxwell, J. Clerk (1877). *Matter and Motion*. Reprinted by Dover, New York.

Mikusinski, P. B. (1959). *Operational Mathematics*, Pergamon, Oxford and Panstwowe Wydawnictwo Naukowe, Warsaw.

Milne-Thomson, L. M. (1955). *Theoretical Hydrodynamics*, 3rd edn., Macmillan, London.

Molenkamp, C. R. (1968). Accuracy of finite difference methods applied to the advection equation, *J. App. Met.*, **7**, pp. 160–167.

von Neumann, J. (1949/1963). Recent theories of turbulence. *John Von Neumann, Collected Works*, (ed. A. H. Taub), **6**, Pergamon, Oxford.

von Neumann, J. and Richtmyer, R. D. (1950). A method for the numerical calculations of hydrodynamic shocks, *J. Appl. Phys.*, **21**, p. 232.

Newton, I. (1687). *Mathematical Principles of Natural Philosophy* (trans. A. Motte, ed. F. Cajuri), Berkeley University Press, California (1947).

Nihoul, J. C. J. (Editor) (1975). *Modelling of Marine Systems*, Elsevier, Amsterdam.

Norrie, D. H. and de Vries, G. (1973). *The Finite Element Method, Fundamentals and Applications*, Academic Press, New York.

O'Kane, J. P. J. (1971). A kinematic reference frame for estuaries of one dimension, in *Int. Symp. Math. Models in Hydrology*, **2**, Int. Assoc. Hydrol. Sci., Warsaw.

Ottesen Hansen, N. E. (1975). *Series Paper No. 7: Entrainment in Two-Layer Flows*, Institute of Hydrodynamics and Hydraulic Engineering, Technical University of Denmark.

Patterson, E. M. (1959). *Topology*, Oliver & Boyd, Edinburgh.

Peregrine, D. H. (1967). Long waves on a beach, *J. Fluid Mech.*, **27**, pp. 815–827.

Peregrine, D. H. (1972). Approximations for water waves and the approximations behind them, in *Waves on Beaches*, Academic Press, New York.

Peregrine, D. H. (1974). Discussion, *J. Hyd. Res.*, **1**, pp. 141–145.

Preissmann, A. (1961). Propagation des intumescences dans les canaux et rivieres, *1er Congres de l'Assoc.Française de Calcul*, Grenoble, pp. 433–442.

Preissmann, A. and Cunge, J. A. (1961). Calcul des intumescences sur machines électroniques, *Proceedings 9th Congress IAHR, Dubrovnik*, pp. 656–664.

Prenter, P. M. (1975). *Splines and Variational Methods*, Wiley-Interscience, New York.

Prigogine, I. (1955). *Thermodynamics of Irreversible Processes*, 3rd edn, Interscience, New York. (Reprinted 1967)

Rice, J. R. (1964). *Approximation of Functions*, Addison-Wesley, Reading, Massachusetts.

Riesz, F. and Nagy, B. Sz. (1960). *Functional Analysis* (trans. by L. F. Boron), Unger, New York.

Richtmyer, R. D. (1963). *NCAR Tech. Note 63–2*, National Center for Atmospherics Research, Boulder, Colorado.

Richtmyer, R. D. and Morton, K. W. (1967). *Difference Methods for Initial Value Problems*, 2nd edn., Interscience, New York.

Roache, P. J. (1972). *Computational Fluid Dynamics*, Hermosa, Albuquerque, New Mexico.

Rodi, W. (1978). Turbulence Models and their Applications in Hydraulics – A State of the Art Review, *Sonderforschungsbereich* **80**, Univ. Karlsruhe, Report SFB80/T/127.

Rosinger, E. E. (1982). *Nonlinear Equivalence, Reduction of PDEs to ODEs and Fast Convergent Numerical Methods*, Pitman, London.

Rouse, H. (1938). *Fluid Mechanics for Hydraulic Engineers*, McGraw-Hill, New York. (Reprinted: 1950)

Rouse, H. and Ince, S. (1957). *History of Hydraulics*, Dover, New York.

Rozhdestvensky, B. L. and Yanenko, N. N. (1968). *Quasilinear Equation Systems and their Applications to Gas Dynamics*, Nauka, Moscow. (In Russian)

Schwartz, L. (1957, 1959). *Théorie des Distributions*, Vols 1 and 2, Herman, Paris.

Shannon, C. E. and Weaver, W. (1949). *The Mathematical Theory of Communication*, University of Illinois Press, Urbana, Illinois.

Shokin, Yu. I., (1979). *Methods of Differential Approximation*, Nauka Publishing House, Siberian Division, Novosibirsk, (in Russian).

Skovgaard, O. and Bertelsen, J. Aa. (1974). Refraction computations for practical applications, *Proc. Int. Symp. Ocean Wave Measurement and Analysis*, ASCE, New Orleans, La, pp. 761–773.

Sobey, R. J. (1970). *Finite Difference Schemes Compared for Wave-Deformation Characteristics etc.*, Tech. Memor. No. 32, US Army Corps of Engineers, Coastal Engineering Research Center, Washington, DC.

Sobolev, S. (1936). Méthode nouvelle à résoudre de problème de Cauchy pour les équations linéaires hyperboliques normale, *Mat. Sbornik*, **1**, pp. 39–71.

Street, R. L., Chan, R. K. C. and Fromm, J. E. (1970). Two methods for the computation of the motion of long water waves – a review and applications, *8th Symp. Naval Hyd.* (ed. M. C. Plesset, T. Y. T. Wu and S. W. Doroff), ACR-179. Office of Naval Research, Department of the Navy, Arlington, Virginia.

Struik, D. J. (1948). *A Concise History of Mathematics*, Dover, New York.

Temple, G. and Bickley, W. G. (1933). *Rayleigh's Principle and its Applications in Engineering*, Oxford University Press, London. (Reprinted by Dover, New York: 1956)

Vasiliev, O. F., Gladyshev, M. T., Pritvits, N. A. and Sudobicher, V. G. (1965). Numerical methods for the calculation of shock wave formation in open channels, *Proceedings 11th Congress IAHR, Leningrad*, paper 3.44, 14 pp.

Vasiliev, O. F., Godunov, S. K. *et al.* (1963). Numerical method for computation of wave propagation in open channels; application to the problem of floods, *Dokl. Akad. Nauk SSSR*, **3**, p. 151. (In Russian)

Verwey, A. (1980). To appear in Cunge, J. A., F. M. Holley and A. Verwey, *Practical Aspects of Computational River Hydraulics*, Pitman, London.

Vliegenthart, A. C. (1968). *Berekeningen van Discontinuiteiten in ondiep water,* Math. Inst. der Techn. Hogeschool, Rapport NA-4, Delft. (In Dutch.) Published in English as *Dissipative Difference Schemes for Shallow Water Equations*, Report NA-5.

Weir, A. J. (1973). *Lebesgue Integration and Measure*, Cambridge University Press.

Welander, P. (1955). Studies on the general development of motion in a two-dimensional ideal fluid, *Tellus*, 7, No. 2, pp. 141–156.

Whitham, G. B. (1973). *Linear and Non-Linear Waves*, John Wiley, New York.

Wilson, A. H. (1957). *Thermodynamics and Statistical Mechanics*, Cambridge University Press.

Wittgenstein, L. (1956). *Lectures on the Foundations of Mathematics*, Harvester Press.

Wittgenstein, L. (1969). *Philosophische Gramatik*, Blackwell, Oxford.

Yanenko, N. N. (1968). *Méthode à Pas Fractionnaires* (trans. P. A. Nepomiastchy), Armand Colin, Paris.

Zemansky, M. W. (1957). *Heat and Thermodynamics*, 4th edn, McGraw-Hill, New York.

Index